Computer Simulations of Molecules and Condensed Matter

From Electronic Structures to Molecular Dynamics

Peking University–World Scientific Advanced Physics Series

ISSN: 2382-5960

Series Editors: Enge Wang *(Peking University, China)*
Jian-Bai Xia *(Chinese Academy of Sciences, China)*

Computer Simulations of Molecules and Condensed Matter

of Molecules and

Condensed Matter

From Electronic Structures
to Molecular Dynamics

Xin-Zheng Li
En-Ge Wang

W🌐 World Scientific

NEW JERSEY · LONDON · SINGAPORE · BEIJING · SHANGHAI · HONG KONG · TAIPEI · CHENNAI · TOKYO

Published by

World Scientific Publishing Co. Pte. Ltd.

5 Toh Tuck Link, Singapore 596224

USA office: 27 Warren Street, Suite 401-402, Hackensack, NJ 07601

UK office: 57 Shelton Street, Covent Garden, London WC2H 9HE

Library of Congress Cataloging-in-Publication Data

Names: Li, Xin-Zheng, 1978– author. | Wang, En-Ge, 1957– author.
Title: Computer simulations of molecules and condensed matter : from electronic structures to
    molecular dynamics / Xin-Zheng Li (Peking University, China),
    En-Ge Wang (Peking University, China).
Other titles: Peking University-World Scientific advance physics series ; v. 3.
Description: Singapore ; Hackensack, NJ : World Scientific, [2017] |
    Series: Peking University World Scientific advance physics series, ISSN 2382-5960 ; vol. 3 |
    Includes bibliographical references.
Identifiers: LCCN 2017032699| ISBN 9789813230446 (hardcover ; alk. paper) |
    ISBN 9813230444 (hardcover ; alk. paper)
Subjects: LCSH: Condensed matter--Computer simulation. |
    Molecular dynamics--Computer simulation.
Classification: LCC QC173.457.C64 L58 2017 | DDC 530.4/10113--dc23
LC record available at https://lccn.loc.gov/2017032699

British Library Cataloguing-in-Publication Data
A catalogue record for this book is available from the British Library.

Preface

State-of-the-art computer simulation of molecules and condensed matters, according to the Born–Oppenheimer (BO) approximation, requires an accurate *ab initio* treatment of both the electronic structures and the nuclei's motion on (and sometimes even beyond) the corresponding potential energy surfaces. As a student majoring in computational condensed matter physics, one of the authors (Xin-Zheng Li, XZL) had taken many years to understand this simple statement. During this time, his work in collaboration with Prof. Jianbai Xia and Prof. Enge Wang as well as other scholars in Europe (Prof. Angelos Michaelides, Prof. Dr. Matthias Scheffer, Dr. Ricardo Gómez-Abal, and Prof. Dr. Claudia Draxl, etc.) has luckily covered some topics in both regimes. Based on this limited, yet to a certain extent, unique experience, the authors want to share their understanding of this rapidly growing field with the readers, especially those Chinese graduate students majoring in computational condensed matter physics or chemistry. Special focus will be put on the basic principles underlying the present electronic structure calculations and the molecular dynamics simulations. A wide range of electronic structure theories will be introduced, including the traditional quantum chemistry method, density-functional theory, many-body perturbation theory, etc. Besides these electronic structures, motions of the nuclei will also be described using molecular dynamics, with extensions to enhanced sampling and free energy calculation techniques including umbrella sampling, metadynamics, integrated tempering sampling, etc., and the thermodynamic integration methods. As a further extension beyond the standard BO molecular dynamics, some simulation techniques for descriptions of the quantum nuclear effects will also be discussed, based on Feynman's path integral representation of the quantum mechanics. With such a choice of theories on both the electronic structures and molecular dynamics perspectives, hopefully, we can help those graduate students to find the proper method for tackling the physical/chemical problems they are interested in.

Contents

List of Figures

List of Tables

1
Introduction to Computer Simulations of Molecules and Condensed Matter

Since the discovery of electron as a particle in 1896–1897, the theory of electrons in matter has ranked among the great challenges in theoretical physics. The fundamental basis for understanding materials and phenomena ultimately rests upon understanding electronic structure [1].

It is without any hesitation that we assent to R. M. Martin's point of view and begin, with his quote, the introduction of this book on computer simulations of molecules and condensed matters. The electrons, being an interacting many-body entity, behave as a quantum glue which holds most of the matters together. Therefore, principles underlying the behavior of these electrons, a large extent, determine properties of the system (electronic, optical, magnetic, mechanical, etc.) we are going to investigate. As a consequence, an introduction to the computer simulation of molecules and condensed matters should naturally start with a discussion on theories of electronic structures.

One point implied in this statement is that the concept of electronic structures is polymorphous, in the sense that it covers all properties related to the electrons in matter. For example, it can refer to the total energy of the electrons, their density distribution, the energy needed for extracting (injecting) one electron out of (into) the system, their response to an external perturbation, etc. These properties are in principle measured by different experiments and described using different theoretical methods. Therefore, while discussing "electronic structures", one must point out the specific properties of the electrons referred to and the theories used. Among the various properties and theories used for describing the electronic system, we will focus on the ones concerning depiction of the total energies and spectroscopies of the system within the *ab initio* framework throughout this book. For electronic structure theories based on model Hamiltonian, e.g. the effective-mass envelope function and the tight-binding methods, which are equally important in molecular simulations but will not be discussed here, please refer to the seminal book of Martin in Ref. [1].

Besides the behavior of the electrons, the motion of the nuclei is another aspect one must accurately address in simulating the material properties, since a real material is composed of interacting electrons and nuclei. To describe such a correlated motion, some basic concepts underlying our daily research must be introduced, among which the Born–Oppenheimer (BO) approximation and the potential energy surface are the most crucial. Because of this, in the following, we start our discussions by introducing these two concepts. Using these concepts, we can categorize the majority of the tasks we want to fulfill in daily researches concerning simulations of material properties into two different regimes, i.e. those concerning mainly the electronic structures and those concerning mainly the nuclear motion. The whole book is then organized on the basis of such a categorization. In Chapters 2–4, we discuss different electronic structure theories and some technical details concerning their implementation. Chapters 5–7 focus on the molecular dynamics (MD) method and its various extensions in descriptions of the nuclear motion. With this choice of theories on both the electronic structures and the molecular dynamics levels, we hope that we can help the graduate students to find the proper method for tackling the physical/chemical problems they are interested in, in their practical researches.

1.1 Born–Oppenheimer Approximation and the Born–Oppenheimer Potential Energy Surface

Before we start, let us first present the key equation we are going to tackle in simulating properties of a real material. Any poly-atomic system can be viewed as an intermixture of two coupled subsystems, constituted by M nuclei and N electrons, respectively. In principle, the only prerequisite for the description of all the quantum mechanical properties of such a system, for simplicity in the non-relativistic regime, is the solution of the many-body Schrödinger equation

$$\hat{H}\Psi(\mathbf{r}_1, \mathbf{r}_2, \ldots, \mathbf{r}_N, \mathbf{R}_1, \mathbf{R}_2, \ldots, \mathbf{R}_M) = E\Psi(\mathbf{r}_1, \mathbf{r}_2, \ldots, \mathbf{r}_N, \mathbf{R}_1, \mathbf{R}_2, \ldots, \mathbf{R}_M),$$
$$(1.1)$$

where $\mathbf{r}_i$ stands for the Cartesian coordinate of the ith electron and $\mathbf{R}_i$ stands for that of the ith nucleus. The Hamiltonian operator is given by

$$\hat{H} = -\sum_{i=1}^{N} \frac{1}{2}\nabla_i^2 + \frac{1}{2}\sum_{i \neq i'} V(\mathbf{r}_i - \mathbf{r}_{i'}) - \sum_{j=1}^{M} \frac{1}{2M_j}\nabla_j^2$$

$$+ \frac{1}{2}\sum_{j \neq j'} V(\mathbf{R}_j - \mathbf{R}_{j'}) + \frac{1}{2}\sum_{i,j} V(\mathbf{r}_i - \mathbf{R}_j) \qquad (1.2)$$

in atomic units (a.u.) with Hartree as the unit of energy. The first two terms in Eq. (1.2) correspond to the kinetic energy and the Coulomb interaction potential, respectively, of the electrons. The third and fourth terms represent the same physical quantities, but for the nuclei. The fifth term is the Coulomb interaction potential between the electrons and the nuclei which couples the two subsystems.

We note that this atomic unit will be used in all equations throughout this book. The notation of the electron coordinate with $\mathbf{r}$ and that of the nucleus with $\mathbf{R}$ will be used when our discussions concern both the electrons and the nuclei. In Chapters 5 and 7, when propagations of purely the nuclei are discussed, $\mathbf{r}$ is also used to denote the position of the nucleus since in those cases, $\mathbf{R}$ is often used to denote other quantities in the literature.

The BO approximation, proposed by Born and Oppenheimer in 1927 [2] and also known as the "adiabatic approximation" in descriptions of electronic structures, makes use of the feature that the masses of the nuclei are several orders of magnitude larger than that of the electrons and their velocities are consequently much smaller with similar kinetic energy. Taking advantage of these extremely different dynamical regimes, this adiabatic approximation allows us to address the dynamics of the electronic subsystem separately from that of the nuclei by considering the latter as static. Pictorially, this is similar to the case when a slow-moving person accidently touches a hornets' nest and gets surrounded by the hornets (Fig. 1.1). Since the velocities of these hornets are much larger than that of the person, wherever he goes in order to escape, the hornets quickly adjust their positions so that this unlucky guy is always surrounded by the hornets and therefore has

Figure 1.1   Pictorial illustration of the BO approximation analogy. The electrons are much lighter than the nuclei and therefore with similar kinetic energy, they are much faster. When the nuclei move, the electrons quickly adjust their positions as if the nuclei are stationary. This is similar to the case when an unlucky slow-moving guy is surrounded by hornets. Wherever he goes, the hornets can easily adjust their positions as if this person is stationary. In other words, there is an "adiabatic" correlation between the movement of this person and that of the hornets, so that this person has to suffer the sting all the time.

to suffer being stung all the time. Similarly, in the electron–nuclei system, since the velocities of the electrons are some orders of magnitude larger than those of the nuclei, for a certain spatial configuration of the nuclei, we can always focus on describing the electronic system first and allow the electrons to relax to their eigenstates before the nuclei are ready to move to the next step. Using this concept, from the mathematical perspective, the poly-atomic quantum system including both electrons and nuclei as described by Eq. (1.1) can be simplified into a system that only includes the electrons as quantum particles. The Schrödinger equation for the electrons then reads

$$\hat{H}_e \Phi(\mathbf{r}_1, \mathbf{r}_2, \ldots, \mathbf{r}_N) = E_e \Phi(\mathbf{r}_1, \mathbf{r}_2, \ldots, \mathbf{r}_N), \tag{1.3}$$

where the Hamiltonian

$$\hat{H}_e = -\sum_{i=1}^{N} \frac{1}{2} \nabla_i^2 + \frac{1}{2} \sum_{i \neq i'} V(\mathbf{r}_i - \mathbf{r}_{i'}) + \frac{1}{2} \sum_{i,j} V(\mathbf{r}_i - \mathbf{R}_j) \tag{1.4}$$

depends only parametrically on the nuclear configuration. As simple as it is, the movements of the electrons and the nuclei become decoupled.

     The BO approximation, as introduced above, indicates that the total energy of the system at a certain spatial configuration of the nuclei equals

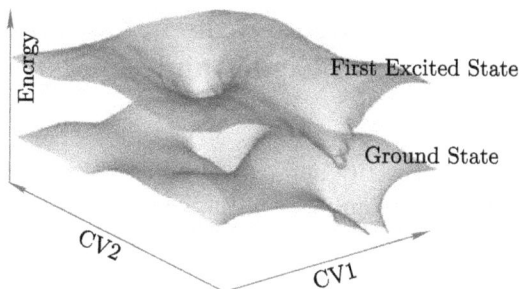

Figure 1.2 (Courtesy of Dr. Yexin Feng) PES is a plot of the system's total energy, defined as the total energy of the electronic system plus the classical Coulomb repulsion potential between the nuclei as a function of the nuclear geometric parameters. In the above case, two geometric parameters, labelled collective variables 1 and 2 (CV1 and CV2), are chosen to characterize the spatial configuration of the nuclei, and the PESs are a series of two-dimensional (2D) surfaces corresponding to the electronic ground and excited states. In principle, these surfaces representing different electronic states (ground electronic state, first excited electronic state, etc.) may intersect with each other.

the total energy of the electronic many-body system as described by Eq. (1.3) plus that of the classical internuclear Coulomb repulsion. The resulting quantity is a function of the nuclear geometry and it has a series of discrete eigenvalues corresponding to the electronic quantum mechanical ground and excited states. Consequently, the concept of the potential energy surface (PES) can be introduced as a plot of this total energy as a function of the geometry of the nuclei (Fig. 1.2). This concept of the PES is very helpful since it provides us a simple way to visualize the relationship between the geometry of the nuclei and the potential energy of the real poly-atomic system under investigation at a specific spatial configuration of the nuclei. Imagine that we can use two geometric parameters to describe this spatial configuration of the nuclei. At each spatial configuration, we then have a series of discrete total energies associated with the ground and excited states of the electrons as a function of these two geometric parameters. A gradual change in the nuclei's geometry induces a gradual change in these total energies. Through a continuous yet complete variation of these geometric parameters over the whole configurational space of the system under investigation, we can have a series of 2D PESs for this system. In fact, the nuclei are not stationary. At finite temperatures, the system fluctuates on these PESs and the purpose of material simulation simplifies into the reproducing of such fluctuations of the nuclei on these electronic ground- and excited-state PESs.

1.2   Categorization of the Tasks in Computer Simulations of Molecules and Condensed Matters

Using the concepts of the BO approximation and the PES as introduced above, we can separate the majority of the jobs we need to do in computer simulations of the molecules and condensed matter into two categories, namely those concerning mainly the electronic structures and those focusing on the motion or relaxation of the nuclei. Theories going beyond this BO approximation are beyond the scope of this book. We direct interested readers to see Refs. [3–5].

1.2.1   *Electronic Structure Calculations*

For a static spatial configuration of the nuclei, the equation we need to solve in simulating the material properties is Eq. (1.3), which depends only parametrically on the nuclear coordinates. However, solving such a many-body Schrödinger equation is still much harder said than done and one must rely on further approximations. Nowadays, there are different methods which can be used to fulfill this task, including the traditional quantum chemistry method, the density-functional theory (DFT), and the quantum Monte-Carlo (QMC) method, etc. within the *ab initio* framework and the effective-mass envelope function, tight-binding methods, etc. in the model Hamiltonian scheme. These methods have achieved great success within the last half–century as evidenced by such achievements as the Nobel Prize of Chemistry awarded to W. Kohn and J. Pople in 1998 etc. In this book, we will take some of them as examples to explain how the electronic structures are calculated in real poly-atomic systems.

Our discussion starts from the traditional quantum chemistry methods and DFT in Chapter 2. In practical implementations of these methods, however, a certain basis set must be chosen for the expansion of the electronic wave functions. These basis functions need to be both efficient and accurate in simultaneously depicting two extremely different regimes, i.e. the core state tightly bound to the nuclei, and the delocalized valence state, especially in the interstitial region. In Chapter 3, we will explain how they are designed in computer simulations in practice. In addition to this, the many-body perturbation theory, one of the standard methods used to depict the properties of the excited electronic states, will be described in Chapter 4. For the time being, in this introduction, let us just assume that we have the electronic structure of the poly-atomic system already properly described for each spatial configuration of the nuclei.

With this assumption, we can imagine that we have mapped out the BO PESs as introduced in Sec. 1.1 for the specific system we are interested in. The next problem therefore transforms into describing the motion of the nuclei on these PESs.

1.2.2   Geometry Optimization, Stationary Points on PES, Local Minimum, and Transition State

None of the molecules or condensed matter will be happy if they remain at an uncomfortable position on their PES. Taking the simplest case, a positively charged ion subjected to a 2D external potential in which the characteristic spatial scale for the variation of the potential is much larger than the dispersion of the electrons within this ion (in other words, the ion can be approximated as a point charge sitting on the 2D potential). As an example, the PES for the ground state of the entity composed by the electrons and the nucleus has a few local minima on the x–y plane. This is due to interactions between the electrons, the nucleus and in this case, most importantly, the external field (Fig. 1.3). There exist some local minima on this PES. At each local minimum, the force imposed on the nucleus is zero since $dE/dx = 0$. Away from them on the slope, the nucleus feels a force which leads it to the nearest local minimum (Fig. 1.3). In material simulations, if we imagine that we know nothing about the structure of the molecules or condensed matter, the first thing we need to do in order to understand something about this system is always to construct such a relaxation using chemical intuitions, starting from different geometries. Such a structural relaxation is called *geometry optimization*. It normally helps us to find the closest static local minimum on the PES of the system to be studied.

In real poly-atomic systems (molecules and condensed matter), since the configurational space is much more complicated (normally $3N$-$3D$ with N being the number of nuclei), the geometry optimization introduced above cannot guarantee that the system reaches a minimum after it is carried out. The only guaranteed property is that the resulting geometry satisfies $\partial E/\partial x = 0$ for all the spatial coordinates, denoted by x, of the system. The point on the PES corresponding to this geometry is called a stationary point. On a multi-dimensional PES, except for the local minima where the corresponding curvatures (namely, $\partial^2 E/\partial x_1 \partial x_2$) are all positive, stationary points also include those with negative curvatures. The most special

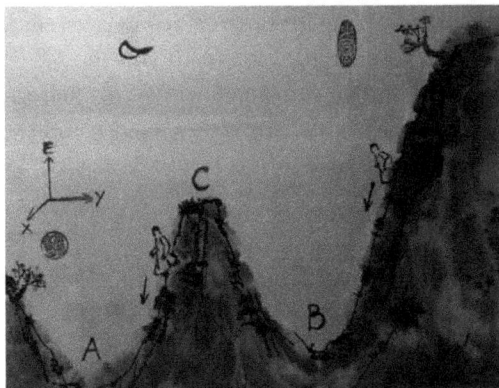

Figure 1.3   Illustration of the geometry optimization using a traditional Chinese painting, courtesy of Ms. Yehan Feng (XZL's wife). Imagine that one (x, y) coordinate gives the spatial configuration of the nuclei, and the landscape represents the PES. The system (denoted by the person) is unhappy when it is on the slope of the PES and consequently, forces exist which drag him to a local minima (e.g. points A and B).

ones among them, of chemical interest, are the so-called first-order saddle points, where the corresponding curvatures are all positive except for one. These first-order saddle points are interesting because in chemistry, two local minima on the PES can refer to the reactant and product states. The lowest first-order saddle point between them on the PES then corresponds to the transition state (TS) (see Fig. 1.4). Paths close to the TS allow chemical reactions to happen with the least energy cost compared to other paths. These terms will often be used in discussions on chemical reactions. At these stationary points (including the local minima and the first-order saddle point), the system does not feel the force for relaxation. Therefore, to distinguish between them, further calculations on vibrational properties are often needed. Close to stationary points other than those properly-defined local minima, there are imaginary-frequency phonons (or otherwise called soft phonons) associated with the negative curvature of the PES. Accordingly, vibrational spectrum calculation serves as an efficient way to distinguish a well-defined stable structure from those of other stationary points with negative curvatures.

1.2.3 *Metastable State and Transition State Searching*

So far, we have introduced the concepts of the local minima on the PES, the first-order saddle point (TS) on the PES, and geometry optimization.

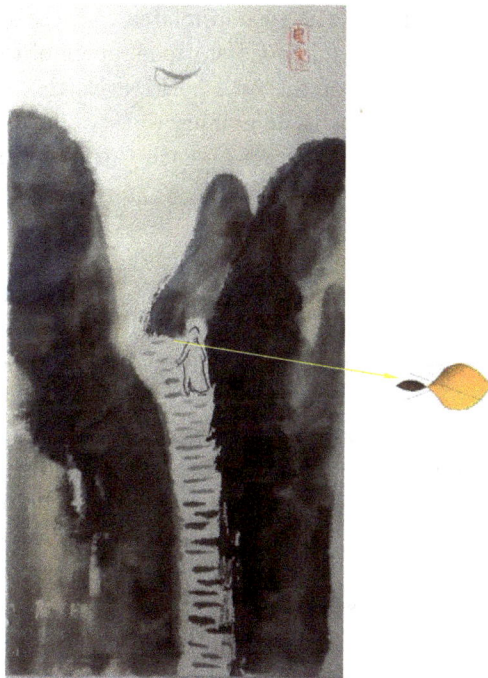

Figure 1.4   Similar to Fig. 1.3, this is a schematic representation of the concept of TS using a traditional Chinese painting. Courtesy of Ms. Yehan Feng. Again, one (x, y) coordinate is used to mean one spatial configuration of the nuclei, and the landscape represents the PES. The stationary points on the PES refer to those where the slope is zero (e.g. points A, B, and C in Fig. 1.3). However, the curvatures can be either positive or negative. Among all stationary points, the one with all curvatures are positive except for one which is called a first-order saddle point, or a TS, as indicated by the cartoon in yellow.

Structures of the molecules and condensed matter associated with the local minima and the first-order saddle points on the PES at the atomic level are of primary interest in simulations of material properties. Taking the local minima as an example, characterization of their structures can help us understand one of the most fundamental questions in material simulations, i.e. why is a material the way it is? Therefore, assuming that the electronic structures of the system at a specific spatial configuration of the nuclei can be satisfactorily solved, the primary task in material simulations associated with understanding the behavior of the system on its PES then simplifies into the characterization, on the atomic level, of the system's structures at these states-of-interest at the first step. The most rigorous way to identify geometries of the system related to these special points on the PES is to carry out a complete yet static mapping of the PES over the whole configurational space of the nuclei. However, this is not computationally doable in real materials. Imagining that we have N nuclei in the system and the N_G grid has been set on each Cartesian axis to represent the spatial coordinate of one nucleus, we need to calculate the total energy of the system at $[N_G^3]^N$ spatial configurations in order to map out this BO PES, which is an astronomical number in material simulations. Fortunately, we do not need to know all these details in order to understand the key properties of a material. Some simple, elegant, yet powerful computational schemes can be designed to identify the atomic structures of the materials at these states-of-interest in chemistry and physics, including the Crystal structure AnaLYsis by Particle Swarm Optimization (CALYPSO) method [7–12], the *ab initio* Random Structure Searching (AIRSS) method [13–15], and the Generic Algorithm [16–18], etc. Nowadays, this is a very active research topic in the field of computer simulations of material properties, especially for studies of the material property under different pressures. For people interested in research on this aspect, please see Refs. [7–18]. These methods are now very efficient in finding the stationary points on the PES. By resorting to the phonon calculations, one can further identify those properly defined local minima on the PES. And among them, the one with the lowest energy (or in the most rigorous manner free energy) is called the ground state. Other well-defined local minima are known as metastable states.

With the local minima on the PES identified, by comparing their structures and the electronic structures associated with them to the available experimental data, we can identify the ones consistent with experiments and carry out further investigations from these structures. Depending on what we want from our simulations, this includes TS searching, Monte-Carlo sampling, and MD simulations together with their various extensions. These concepts will be frequently used in our later discussions in this book. Among them, TS searching is often a concept at the static level, i.e. a concept in which we do not have to care about the thermal and quantum nuclear effects related to the real fluctuations of the nuclei, although the transition-state theory (TST) itself includes all these effects. In Monte-Carlo and MD simulations, these effects must be considered. To be clear, we start by explaining the simplest concept, i.e. TS searching, at the static limit. As a matter of fact, such searches for the TS are of primary interest in theoretical descriptions of the chemical reactions. What one needs to do is to find the lowest barrier between two local minima, which represent the reactant and product states, respectively. On a multi-dimensional PES, searches for this TS between these two local minima, however, are still computationally very difficult. Currently, there are various schemes on carrying out such searches, including the constrained optimization (CO) [19], nudged elastic band (NEB) [20–22], Dewar, Healy and Stewart (DHS) [23], Dimer [24–26], activation-relaxation technique (ART) [27, 28] and one-side growing string (OGS) [29] as well as their various combinations. The purpose of this section is to establish the concepts for material simulations related to discussions in the later chapters. Therefore, we will not go into details of such searches and instead, direct interested readers to the References section.

1.2.4 *Molecular Dynamics for the Thermal Effects*

These TS searching methods help us to identify the structures of the TS and the energy required for the corresponding chemical reactions to happen at the static level. In a more realistic description of these processes, however, statistical and dynamical effects, originating from the finite-temperature thermal and quantum effects of the nuclei, must also be taken into account. This leads us to the ultimate goal of material simulations as we have introduced above, i.e. an accurate description of a real system's propagation on its BO PESs. Nowadays, a very popular method to carry out such research is the so-called *ab initio* MD. In this method, prerequisite knowledge on the starting structures of the system is needed in order for a reasonable simulation to be carried out. It normally refers to a point on the PES close to the region relevant to the problem of interest. During the propagation of the nuclei, the electronic structures of the system are often described using a certain *ab initio* method within either the DFT or traditional quantum chemistry method framework. When the DFT is used, finite-temperature contributions to the electronic free energy can often be addressed with Mermin's finite-temperature version of the DFT [30–32]. With the electronic structures determined quantum mechanically, the forces on each nucleus can be calculated using the so-called Hellmann–Feynman theorem. The system then propagates on the PES subjected to the forces computed "on-the-fly" as the nuclei configuration evolves according to the Newton equation. We note, however, that empirical potentials can also be used to describe the interatomic interactions, which is outside the framework of the *ab initio* simulation method. We focus on the *ab initio* methods and consequently, the *ab initio* MD in this book. But it is worth noting that the principles underlying the propagation of the nuclei as introduced also apply to those when empirical potentials are used.

There are several schemes for the *ab initio* MD simulations to be carried out, depending on the ensemble used in the simulation. The simplest one is the so-called micro-canonical ensemble. In these simulations, the total number of particles in the system (N), the volume (V), and the total energy (E) of the system are kept constant. The system is completely isolated from the environment. Therefore, this ensemble is also called NVE ensemble in literature. Practically, however, experiments are often carried out at finite temperatures. Therefore, the temperature (T) of the system, instead of its total energy (E), should be kept constant in a more realistic simulation of such isothermal processes. This leads to the next class of the more often used ensembles in molecular simulations, i.e. the canonical ensemble (NVT). In this ensemble, the control of the temperature is often achieved by coupling the system to a heat bath (or thermostat) that imposes the desired temperature on the system. Depending on how this is carried out, we have different schemes for the ensemble to be simulated, e.g. the Andersen thermostat, the Nosé–Hoover thermostat, the Nosé–Hoover chain thermostat, or the Langevin thermostat, etc. All these different schemes help simulate an NVT process and will be discussed in detail in Chapter 5.

1.2.5 *Extensions of MD: Enhanced Sampling and Free-Energy Calculations*

The MD method as introduced above allows us to simulate the propagation of a system using a finite-time step so that the total energy of this system is conserved and the temperature is well in control. This means that the propagation needs to be done with a time interval on the order of femtoseconds, or even shorter. Processes of real chemical reactions, however, happen on a much longer time-scale, e.g. seconds. This indicates that $\sim 10^{15}$ electronic structures should be performed in order to see such a process happen. Plus, integrating the equation of motion, this means an astronomical computational cost in condensed matter physics or chemistry. The physical origin for this time-scale problem is from the morphology of PES as introduced in the earlier discussion. In complex chemical systems, notably those in condensed matter and in biology, the PESs typically have very complicated features, evident in the presence of many low-energy local minima and large barriers separating them. Transitions between these local minima are rare events and long-time simulations are required to evidence them. Meanwhile, complex morphology of the PES also indicates that there are shallow local minima separated by low barriers, and transitions between them happen much faster. Consequently, it is fair to say that one of the greatest challenges in MD simulations now is in numerically reproducing such a wide distribution of time scales in complex systems. More practical schemes, in which a more efficient sampling of the phase space can be guaranteed, are highly desired.

During the last few decades, great efforts have been made to accelerate the exploration of the multi-dimensional PES in complex systems. A number of methods, such as adaptive umbrella sampling [33], replica exchange [34–37], metadynamics [38, 39], multi-canonical simulations [40, 41], conformational flooding [42], conformational space annealing [43], and integrated tempering sampling [44–47] etc., have been proposed, each having its own strengths and weaknesses. For example, in the adaptive umbrella sampling, metadynamics, and configurational flooding methods, reaction coordinates need to be selected before the simulations can be carried out. Conversely, in replica exchange and multi-canonical simulations, different temperature simulations are required. This causes the trajectories obtained to lose their dynamical information. In Chapter 6, we will take the umbrella sampling, adaptive umbrella sampling, metadynamics, and integrated tempering sampling methods as examples to show how they work in practical simulations.

Another important extension of the MD simulation technique concerns the phase behavior of a given substance, in particular, transitions between two competing phases. The simplest example is the melting of a solid into liquid. In this example, the transition between the two competing phases is first-order and their transition curve can be calculated from the principle that at coexistence, the Gibbs free energies of the two phases are equal. In order to describe such an equality between two free energies, a method from which this free energy can be calculated should be available. In Chapter 6, we will also use the thermodynamic integration method as an example to show how this is done in practice [48–58].

1.2.6   *Path Integral Simulations for the Quantum Nuclear Effects*

In standard MD simulations, the nuclei are often treated as classical point-like particles and propagate according to the Newton equation. Statistically and dynamically, the quantum nature of the nuclei is completely neglected.

This classical treatment is normally a good approximation since the nuclear masses are several orders of magnitude larger than that of the electron. However, it needs to be pointed out that there are situations when the quantum nuclear effects (QNEs) must be accounted for, especially in systems where light elements such as hydrogen are involved. As a matter of fact, it has long been recognized that the statistical properties of hydrogen-bonded systems such as water heavily depend on which isotope of hydrogen is present. Taking the melting/boiling temperature of normal water (composed of H_2O) and heavy water (composed of D_2O) as an example, this value in heavy water is $3.8°C/1.4°C$ higher than that of normal water under the same ambient pressure. In classical statistical mechanics, it is well known that the statistical thermal effects do not depend on the isotope. Therefore, in descriptions of such effects, the quantum nature of the nuclei must also be taken into account.

One natural solution to describe such QNEs from the theoretical perspective is to construct a high-dimensional Schrödinger equation for the many-body entity of the nuclei. By solving the eigenstate wave functions and eigenvalues of this Schrödinger equation, one obtains both the statistical and dynamical properties of the system under finite temperature. However, it is worth noting that this many-body Schrödinger equation has a notorious scaling problem. Although we acknowledge that it is a rigorous method which has been extremely successful in descriptions of the gas phase reactions [59–62], due to the scaling problem associated with both mapping the *ab initio* high-dimensional PESs and solving the Schrödinger equation, its application is seriously limited to systems less than ∼6 atoms. When the system gets bigger, a practical method must be resorted to.

Thanks to the development of the path integral representation of quantum mechanics by Feynman [63–67], a framework where these QNEs can be described in a practical manner was systematically presented by Feynman and Hibbs in their seminal book in 1965 [67]. In Ref. [67, Chapter 10], when the statistical mechanics was discussed, the partition function of a quantum system was related to the partition function of a classical polymer using the method of path integral. This partition function of a classical polymer can be simulated using finite-temperature sampling methods such as MD or Monte-Carlo. Therefore, from the statistical perspective, the framework of such simulations, which we later called path integral molecular dynamics (PIMD) or path integral Monte-Carlo (PIMC), were already laid out. These PIMD and PIMC methods had experienced a golden time in the 1980s and 1990s, first purely from the statistical perspective [68–73], and then with extensions to the dynamical regime [74–84]. After 2000, a slight revision of the PIMD method, which was shown to be very successful in describing dynamical propagation and called ring-polymer molecular dynamics (RPMD), was also proposed [85–93]. In Chapter 7, we will give a detailed explanation of these methods.

1.3   Layout of the Book

Following the above introduction, we will use six chapters to introduce some important methods for the present computer simulations of molecules and condensed matter, covering the regimes of both electronic structures and molecular motion. Theories underlying the *ab initio* electronic structure calculations are used as the basis for these methods. Therefore, we will start with these and introduce some traditional quantum chemistry methods and the density-functional theory in Chapter 2. For real implementations of these methods, pseudopotentials are often used to save the computational cost, and a set of basis functions must be chosen to expand the electronic wave functions. Based on this consideration, the principles underlying the construction of a pseudopotential are explained and a frequently used basis set for all-electron calculations, i.e. the linearized augmented plane waves (APWS), is introduced in Chapter 3. Then, we go beyond the ground state electronic structures and introduce the many-body perturbation theory (or otherwise called Green's function method) as an example of the *ab initio* descriptions of the electronic excitations in Chapter 4, with a special focus on the popular *GW* approximation and its all-electron implementation.

From Chapter 5, we shift our attention to the nuclear motion. The MD method will be introduced in this chapter, with discussions covering its original micro-canonical form and extensions to simulations of other ensembles. One shortcoming of the standard MD method is that the multi-timescale problem in complex systems is beyond the scope of such simulations. Therefore, in Chapter 6, we will take the umbrella sampling [94–96], adaptive umbrella sampling [33], metadynamics [38, 39], and integrated tempering sampling [44–47] methods as examples to show how enhanced sampling works in practice. A popular method for the calculation of the free energy, i.e. the thermodynamic integration, will also be introduced.

The discussion in Chapters 5 and 6 assumes that the nuclei are classical point-like particles. In practice, they do have a quantum nature. To include the quantum features of the nuclei in the molecular simulations, the path integral and related methods are explained in Chapter 7. In addition to the original scheme of the path integral sampling, as given in the classical textbooks like Refs. [67, 97], a more or less complete explanation for the computational details in the statistical PIMD simulations, as well as its extensions to the zero-point energy calculations, dynamical calculations, and free energy calculations are also given. We hope that this limited, yet to a certain extent organized, introduction can provide the readers with an idea about what molecular simulations are and how we can carry out a molecular simulation in practice. We acknowledge that due to the limitation of our time and knowledge, lack of clarity and mistakes may exist at several places in the present manuscript. We sincerely welcome criticism and suggestions (please email xzli@pku.edu.cn), so that we can correct/improve in our next version.z

2
Quantum Chemistry Methods and Density-Functional Theory

The purpose of this chapter is to illustrate the basic principles underlying the current standard methods to accurately solve Eq. (1.3), which represents the main task for the *ab initio* calculations of the electronic structure in real poly-atomic systems. The different schemes to achieve this goal can be classified into two main categories: the wave function-based methods, traditionally known as quantum chemistry methods, and methods based on the density-functional theory (DFT). Both of these two methods have achieved great success within the last half-century, as evidenced by the Nobel Prize of Chemistry being awarded to W. Kohn and J. Pople in 1998. Hereinafter, we will address both of these schemes, with a special emphasis on DFT. Knowledge of their successes and limitations comprises the required point of departure for the discussions on the Green's function method afterwards.

2.1 Wave Function-Based Method

The variational principle, which can be viewed as another form of the many-body Schrödinger equation, states that any state vector of the electrons for which the average energy, defined as

$$E_e[\Phi] \equiv \frac{\langle \Phi | \hat{H}_e | \Phi \rangle}{\langle \Phi | \Phi \rangle}, \tag{2.1}$$

is stationary corresponds to an eigenvector of $\hat{H}_e$, with eigenvalue E_e. Furthermore, for any state of the system, the corresponding average energy satisfies

$$E_e[\Phi] \geqslant E_0, \tag{2.2}$$

where E_0 is the ground-state energy of the electronic system.

The essence of the wave function-based methods involves obtaining the stationary solutions of Eq. (2.1) within a trial-function space. The accuracy of the method is naturally determined by the choice of this trial-function space. By taking more sophisticated approximations for the trial many-body wave function of these electrons, the accuracy can be systematically improved. Unfortunately, at the same time its computational cost could increase. In this section, we will introduce some of these approximations in order of increasing complexity.

2.1.1   The Hartree and Hartree–Fock Approximations

Since the many-body wave function can always be written as a linear combination of products of one-particle wave functions, the simplest possible ansatz, first proposed by Hartree [98], for the many-body electronic wave function is to assume it as the product of the single-particle wave functions:

$$\Phi(\mathbf{r}_1, \mathbf{r}_2, \ldots, \mathbf{r}_N) = \varphi_1(\mathbf{r}_1)\varphi_2(\mathbf{r}_2)\ldots\varphi_N(\mathbf{r}_N). \tag{2.3}$$

Substituting this trial wave function into Eq. (2.1), making use of the variational principle, the many-body problem of this electronic system is mapped onto a set of single-particle, Schrödinger-like, equations

$$\hat{h}_i \varphi_i(\mathbf{r}_i) = \epsilon_i \varphi_i(\mathbf{r}_i), \tag{2.4}$$

with the Hamiltonian given by

$$\hat{h}_i = \left[-\frac{\nabla^2}{2} + V_{\text{ext}}(\mathbf{r}) + V^{\text{H}}(\mathbf{r}) + V_i^{\text{SIC}}(\mathbf{r}) \right]. \tag{2.5}$$

The solutions of Eq. (2.4) are coupled through the Hartree potential

$$V^{\text{H}}(\mathbf{r}) = \int \frac{n(\mathbf{r}')}{|\mathbf{r} - \mathbf{r}'|} d\mathbf{r}', \tag{2.6}$$

which depends on the electron density, defined as

$$n(\mathbf{r}) = \sum_{j=1}^{N} |\varphi_i(\mathbf{r})|^2. \tag{2.7}$$

The last term in Eq. (2.5) corrects for the interaction of the electron with itself, included in the Hartree potential.

$$V_i^{\text{SIC}}(\mathbf{r}) = -\int \frac{|\varphi_i(\mathbf{r}')|^2}{|\mathbf{r} - \mathbf{r}'|} d\mathbf{r}'. \tag{2.8}$$

Thus, the set of Eqs. (2.4)–(2.7) has to be solved self-consistently. Solving these equations, the ground-state energy of the electronic system can be calculated as

$$E_{\text{e}} = \sum_{i=1}^{N} \epsilon_i - \frac{1}{2} \sum_{i=1}^{N} \sum_{j\neq i}^{N} \iint \frac{|\varphi_i(\mathbf{r})|^2 |\varphi_j(\mathbf{r}')|^2}{|\mathbf{r} - \mathbf{r}'|} d\mathbf{r} d\mathbf{r}'. \tag{2.9}$$

With these, the principle underlying a calculation within the Hartree approximation for the many-body interactions is clear. In essence, it maps the many-particle problem into a set of independent particles moving in the mean Coulomb field of the other particles.

As simple and elegant as it looks, the Hartree approximation presents one of the first practical schemes for the calculation of a many-body electronic system's total energy. However, it needs to be pointed out that there is a major drawback in this method; namely, the many-particle wave function of the electrons does not obey the Pauli principle. The Pauli principle forbids two Fermi particles from occupying the same quantum mechanical state and therefore prevents electrons with the same spin from getting close to each other. It is a physical property of all many-body Fermi systems. Lacking this feature, the Hartree approximation generally underestimates the average distances between electrons. Consequently, the average Coulomb repulsion between them, as well as the total energy, are overestimated.

In order to fulfill the Pauli principle, the many-particle wave function has to be anti-symmetric among the exchange of two particles with the same spin. The simplest ansatz for a many-particle wave function, obeying Pauli's principle, is obtained by the anti-symmetrized product, known as the Slater determinant. This improved ansatz, proposed by Fock [99], is known as the Hartree–Fock approximation. For non-spin-polarized systems in which every electronic orbital is doubly occupied by two electrons with opposite spins (closed shell), this method can be introduced in a very simple form. For an N-electron system, the Slater determinant representing the many-particle wave function is written as

$$\Phi(\mathbf{r}_1,\sigma_1,\ldots,\mathbf{r}_N,\sigma_N) = \frac{1}{\sqrt{(N)!}} \begin{vmatrix} \varphi_1(\mathbf{r}_1,\sigma_1) & \varphi_1(\mathbf{r}_2,\sigma_2) & \cdots & \varphi_1(\mathbf{r}_N,\sigma_N) \\ \varphi_2(\mathbf{r}_1,\sigma_1) & \varphi_2(\mathbf{r}_2,\sigma_2) & \cdots & \varphi_2(\mathbf{r}_N,\sigma_N) \\ \vdots & \vdots & & \vdots \\ \varphi_N(\mathbf{r}_1,\sigma_1) & \varphi_N(\mathbf{r}_2,\sigma_2) & \cdots & \varphi_N(\mathbf{r}_N,\sigma_N) \end{vmatrix},$$

$$(2.10)$$

where σ_i represents the spin coordinate of the ith electron.

Substituting this equation into Eq. (2.1) and making use of the orthogonality of the space orbitals and spin states, one arrives at a set of equations for the single-particle orbitals:

$$\left[-\frac{\nabla^2}{2} + V_{\text{ext}}(\mathbf{r}) + V^{\text{H}}(\mathbf{r}) \right] \varphi_i(\mathbf{r},\sigma)$$

$$- \int d\mathbf{r}' \sum_j \frac{\varphi_j^*(\mathbf{r}',\sigma')\varphi_j(\mathbf{r},\sigma)}{|\mathbf{r}-\mathbf{r}'|} \delta_{\sigma,\sigma'} \varphi_i(\mathbf{r}',\sigma') = \epsilon_i \varphi_i(\mathbf{r},\sigma). \qquad (2.11)$$

The only difference from Eq. (2.5) is the last term, representing the exchange interaction between electrons, which is known as the Fock operator. It is non-local and affects only the dynamics of electrons with the same spin. Note that when $j = i$, the exchange term equals the corresponding term in the Hartree potential. This indicates that the Hartree–Fock approximation is self-interaction-free.

In terms of the single-particle orbitals, the total energy of the system can be written as

$$E_{\text{e}} = \langle \Phi | \hat{H}_{\text{e}} | \Phi \rangle = \sum_i^N H_{i,i} + \sum_i^N \sum_j^N \left(\frac{J_{i,j}}{2} - \frac{K_{i,j}}{4} \right), \qquad (2.12)$$

where

$$H_{i,i} = \int \varphi_i^*(\mathbf{r},\sigma) \left[-\frac{1}{2}\nabla^2 + V_{\text{ext}}(\mathbf{r}) \right] \varphi_i(\mathbf{r},\sigma) d\mathbf{r}d\sigma, \qquad (2.13)$$

$$J_{i,j} = \iint \varphi_i^*(\mathbf{r},\sigma)\varphi_i(\mathbf{r},\sigma)\frac{1}{|\mathbf{r}-\mathbf{r}'|}\varphi_j^*(\mathbf{r}',\sigma')\varphi_j(\mathbf{r}',\sigma')\mathrm{d}\mathbf{r}\mathrm{d}\sigma\mathrm{d}\mathbf{r}'\mathrm{d}\sigma',$$

(2.14)

$$K_{i,j} = \iint \varphi_i^*(\mathbf{r},\sigma)\varphi_j(\mathbf{r},\sigma)\frac{1}{|\mathbf{r}-\mathbf{r}'|}\varphi_j^*(\mathbf{r}',\sigma')\varphi_i(\mathbf{r}',\sigma')\mathrm{d}\mathbf{r}\mathrm{d}\sigma\mathrm{d}\mathbf{r}'\mathrm{d}\sigma'.$$

(2.15)

Here, $H_{i,i}$ is the non-interacting single-particle energy in the external field. $J_{i,j}$ is the "classical" Coulomb interaction between electrons in the states i and j and $K_{i,j}$ is the exchange interaction between them. The first two terms also appear in the Hartree method; the last term is introduced by imposing the Pauli principle on the many-body wave function and tends to reduce the total energy.

For spin-polarized systems, the method becomes more cumbersome. Since the number of electrons with each spin is not balanced anymore, the contribution from the Fock operator will be different for different spins. Consequently, the single-particle eigenvalues and eigenfunctions will also be different [100].

2.1.2   Beyond the Hartree–Fock Approximation

In the Hartree–Fock approximation, the electron–electron interaction is treated by means of a time-independent average potential. The fulfillment of Pauli's principle imposes a "static" correlation of the position of electrons with the same spin (exchange hole). However, it completely neglects dynamical effects due to the Coulomb interaction. Such dynamical effects due to Coulomb interaction arise from the fact that the movement of a given electron affects and is affected by, or in other words, "is correlated with", the movement of the other particles. Lacking this feature, many key physical processes representing the many-body interactions of the electronic system will be absent. Physically, this is the main limitation of the Hartree–Fock approximation. For atoms and small molecules, where these dynamical effects are the smallest and the Hartree–Fock approximation works best, this limitation leads to errors of around 0.5% in the total energy. For example, in a carbon atom, where the total energy is around $1000\,\mathrm{eV}$, this corresponds to $5\,\mathrm{eV}$, which already equals the order of magnitude of a chemical single-bond energy. Thus, to obtain a reliable description of chemical reactions, more sophisticated approximations are required.

Within the framework of the wave function-based methods, this can be achieved by making more elaborate approximations for the many-body wave function. Among the so-called post-Hartree–Fock methods, the configuration interaction (CI) method [101], the Møller–Plesset (MP) perturbation theory [102], and the coupled-cluster (CC) method [103, 104] have achieved great success in past years. As a general shortcoming of all these methods, their application is limited to atoms and small molecules, due to the scaling of computational costs with the system size.

2.2 Density-Functional Theory

As discussed in the previous section, the wave function-based post-Hartree–Fock methods can be very accurate in describing the properties of a many-body electronic system. However, the scaling behavior with respect to the system size is very poor. Taking the simplest post-Hartree–Fock method, i.e. the MP2 method, as an example, the usual scaling is already N^5, with N representing the number of electrons. Such scaling behavior has seriously limited its application to large poly-atomic systems. Starting from the 1980s, a method stemming from a completely different origin, i.e. the DFT, has achieved great success in the electronic structure calculations of molecules and condensed matter, largely due to a very good balance between computational cost and accuracy. This can be evidenced by the fact that despite being a physicist, W. Kohn was awarded the Nobel Prize of Chemistry in 1998. In the following, we will introduce the main principles underlying this theory and its main limitations.

The formalism of the DFT was introduced by Hohenberg and Kohn (HK) in 1964 [105]. In 1965, Kohn and Sham (KS) [106] presented a scheme to approximately treat the interacting electron system within this formalism. It is currently the most popular and successful method for studying ground-state electronic structures. Although far from a panacea for all physical problems in this domain, very accurate calculations can be performed with computational costs comparable to that for the Hartree method. In this chapter, we will present some major components of this theory. The discussion begins with its precursor: the Thomas–Fermi theory.

2.2.1 *Thomas–Fermi Theory*

The Thomas–Fermi theory was independently proposed by Thomas and Fermi in 1927 [107–109]. In its original version, the Hartree method was reformulated in a density-based expression for an electron gas with slowly varying density. The kinetic energy is locally approximated by that of a non-interacting homogeneous electron gas with the same density.

Later, Dirac introduced the exchange term into the model using the same local approximation [110]. The total energy of the electronic system, including the exchange term, is written as

$$E_e[n] = T + U^{\text{ext}} + U^{\text{H}} + U^{\text{X}}$$

$$= \frac{3}{10}(3\pi^2)^{2/3}\int d\mathbf{r}n(\mathbf{r})^{5/3} + \int d\mathbf{r}V_{\text{ext}}(\mathbf{r})n(\mathbf{r})$$

$$+ \frac{1}{2}\iint d\mathbf{r}d\mathbf{r}'\frac{n(\mathbf{r})n(\mathbf{r}')}{|\mathbf{r}-\mathbf{r}'|} - \frac{3}{4}\left(\frac{3}{\pi}\right)^{1/3}\int d\mathbf{r}n(\mathbf{r})^{4/3}.$$

(2.16)

This expression gives the exact energy for the homogeneous electron gas in the Hartree–Fock approximation. It is also a good approximation in systems with slowly varying electron densities.

Based on the formulism of the total energy, within this Thomas–Fermi theory, the density in a real system is obtained by minimizing $E_e[n]$ under the constraint of particle number conservation

$$\delta\left\{E_e - \mu\left[\int n(\mathbf{r})d\mathbf{r} - N\right]\right\} = 0.$$

(2.17)

Substituting Eq. (2.16) into Eq. (2.17), one obtains the Thomas–Fermi equation:

$$\frac{1}{2}(3\pi^2)^{2/3}n(\mathbf{r})^{2/3} + V_{\text{ext}}(\mathbf{r}) + V^{\text{H}}(\mathbf{r}) + V^{\text{x}}(\mathbf{r}) - \mu = 0,$$

(2.18)

where the Hartree potential, $V^{\text{H}}(\mathbf{r})$ is the same as defined in Eq. (2.6), and the exchange potential is given by

$$V^{\text{x}}(\mathbf{r}) = -\left[\frac{3}{\pi}n(\mathbf{r})\right]^{1/3}.$$

(2.19)

From Eq. (2.18), for a certain external potential $V^{\text{ext}}(\mathbf{r})$ and chemical potential μ, one can obtain the electron density of this system and consequently, the total energy (from Eq. (2.16)). However, in practice, as it is based on too-crude approximations, lacks the shell character of atoms and binding behavior, the Thomas–Fermi theory automatically fails in providing a proper description of real systems.

2.2.2 Density-Functional Theory

In 1964, HK formulated two theorems, which formally justified the use of density as the basic variable in determining the total energy of an inter-acting many-body system [105]. The first theorem proved the existence of a one-to-one correspondence between the external potential $V_{ext}(\mathbf{r})$, the ground-state many-body wave function Φ, and the ground-state density $n(\mathbf{r})$. Thus, the total energy of a system, which is a functional of the many-body wave function Φ, can also be reformulated as a functional of the density as

$$E_e[n(\mathbf{r})] = \int V_{ext}(\mathbf{r})n(\mathbf{r})d\mathbf{r} + F[n(\mathbf{r})]. \qquad (2.20)$$

$F[n(\mathbf{r})]$ contains the potential energy of the electronic interactions and the kinetic energy of the electrons. It is a universal functional independent of the external potential. Unfortunately, its exact form is unknown.

Since the expression for the Hartree energy as a functional of the density is known, the functional $F[n(\mathbf{r})]$ in Eq. (2.20) can be further decomposed as

$$F[n(\mathbf{r})] = U^H[n(\mathbf{r})] + G[n(\mathbf{r})], \qquad (2.21)$$

where the expression of the Hartree energy $U^H[n(\mathbf{r})]$ is already given by the third term on the right-hand side of Eq. (2.16). Like $F[n(\mathbf{r})]$, $G[n(\mathbf{r})]$ is an unknown universal functional of the density, independent of the external potential. The total energy can then be written as

$$E_e = \int V_{ext}(\mathbf{r})n(\mathbf{r})d\mathbf{r} + \frac{1}{2}\iint \frac{n(\mathbf{r})n(\mathbf{r}')}{|\mathbf{r}-\mathbf{r}'|}d\mathbf{r}d\mathbf{r}' + G[n(\mathbf{r})]. \qquad (2.22)$$

The second theorem proves that the exact ground-state energy of the electronic system corresponds to the global minimum of $E_e[n(\mathbf{r})]$, and the density $n(\mathbf{r})$ that minimizes this functional is the exact ground-state density $n_0(\mathbf{r})$.

These two theorems set up the foundation for the concept of "density functional". However, a practical scheme which can be used to calculate the density was still absent. This seminal contribution was given by KS. In 1965, they proposed a scheme to calculate the $G[n(\mathbf{r})]$ in Eq. (2.22) [106].

In this scheme, one can decompose this $G[n(\mathbf{r})]$ into two parts

$$G[n] = T^{s}[n] + E^{xc}[n]. \tag{2.23}$$

The first term is the kinetic energy of a non-interacting system with the same density. The second term is the exchange–correlation (XC) energy.

Minimizing the total energy in Eq. (2.20) under the constraint of particle number conservation (Eq. (2.17)), one gets

$$\int \delta n(\mathbf{r}) \left\{ V_{\text{eff}}(\mathbf{r}) + \frac{\delta T^{s}[n]}{\delta n(\mathbf{r})} - \mu \right\} d\mathbf{r} = 0, \tag{2.24}$$

where μ is the chemical potential and

$$V_{\text{eff}}(\mathbf{r}) = V_{\text{ext}}(\mathbf{r}) + \int \frac{n(\mathbf{r}')}{|\mathbf{r} - \mathbf{r}'|} d\mathbf{r}' + V^{xc}(\mathbf{r}). \tag{2.25}$$

$V^{xc}(\mathbf{r})$ is called the XC potential, given by

$$V^{xc}(\mathbf{r}) = \frac{\delta E^{xc}[n]}{\delta n(\mathbf{r})}. \tag{2.26}$$

Assuming a set of non-interacting particles with the same density,

$$n(\mathbf{r}) = \sum_{i=1}^{N} |\varphi_i(\mathbf{r})|^2, \tag{2.27}$$

Eq. (2.24) is equivalent to

$$\left\{ -\frac{1}{2}\nabla^2 + V_{\text{eff}}(\mathbf{r}) \right\} \varphi_i(\mathbf{r}) = \epsilon_i \varphi_i(\mathbf{r}). \tag{2.28}$$

Thus, the KS scheme maps the complex, interacting electronic systems into a set of fictitious independent particles moving in an effective, local potential. Since this effective potential depends on the density, Eqs. (2.25), (2.27), and (2.28) have to be solved self-consistently.

The total energy in the KS scheme is given by

$$E_e[n(\mathbf{r})] = \sum_{\text{occ}} \epsilon_i - \frac{1}{2} \iint \frac{n(\mathbf{r})n(\mathbf{r}')}{|\mathbf{r} - \mathbf{r}'|} d\mathbf{r}d\mathbf{r}' + E^{\text{xc}}[n(\mathbf{r})] - \int n(\mathbf{r})V^{\text{xc}}(\mathbf{r})d\mathbf{r}.$$

(2.29)

We note that in molecular simulations, another form of this functional is also used:

$$E_e[n(\mathbf{r})] = \sum_{\text{occ}} \int d\mathbf{r}\varphi_i^*(\mathbf{r}) \left(-\frac{1}{2}\nabla^2 \right) \varphi_i(\mathbf{r}) + \int n(\mathbf{r})V_{\text{ext}}(\mathbf{r})d\mathbf{r}$$

$$+ \frac{1}{2} \iint \frac{n(\mathbf{r})n(\mathbf{r}')}{|\mathbf{r} - \mathbf{r}'|} d\mathbf{r}d\mathbf{r}' + E^{\text{xc}}[n(\mathbf{r})].$$

(2.30)

These two equations are equivalent, as evidenced by inputting Eq. (2.28) to Eq. (2.29).

One limitation of the discussion above is that it is based on the original papers [105, 106] and is therefore restricted to the zero-temperature, non-degenerate, non-spin-polarized, and non-relativistic cases. Extensions to the finite-temperature calculations can be found in Refs. [30–32], to the *spin-polarized system* in Refs. [111–113], and to the *degenerate system* in Refs. [114–118]. For a comprehensive discussion on these extensions, please see Ref. [119, Chapter 3]. The inclusion of relativistic effect is addressed in Ref. [120]. Further extension of the theory to superconductors may be found in Refs. [121, 122]. Detailed discussion about the *v-representability* and related questions can be found in Refs. [114, 117]. An excellent review on the formal justification of the theory can be found in Ref. [119, Chapter 2].

2.2.3   *Exchange–Correlation Energy*

In the KS scheme, all the complexity of the many-body interaction is put into the XC energy $E^{\text{xc}}[n(\mathbf{r})]$. Unfortunately, the exact expression of this functional is unknown. In addition, different from the wave function-based post-Hartree–Fock method where the accuracy can be systematically improved by taking more complicated forms of the wave functions, a systematic series of approximations converging to the exact result is absent in the DFT. While semiempirical approaches allow one to obtain very precise results within the sample space fitting, their physical origin can be sometimes obscure and their precision outside that space unpredictable. In order to remain within the first-principles framework, the most universal and, to some extent, systematic scheme is the "constraint satisfaction" approach

[123]. These "constraints" consist of exact properties that the XC functional can be proven to fulfill. In this approach, the approximations to the XC functional are assigned to various rungs of the "so-called" *Jacob's ladder* [123] according to the number of ingredients they contain. The best non-empirical functional for a given rung is constructed to satisfy as many exact theoretical constraints as possible while providing satisfactory numerical predictions for real systems. Increasing the number of ingredients allows the satisfaction of more constraints, thus increasing, in principle, the accuracy.

The simplest approximation for $E^{xc}[n(\mathbf{r})]$ is the local-density approximation (LDA), proposed in the original paper of KS. It reads

$$E^{xc}[n(\mathbf{r})] = \int n(\mathbf{r})\epsilon_{xc}(n(\mathbf{r}))d\mathbf{r}, \qquad (2.31)$$

where $\epsilon_{xc}(n(\mathbf{r}))$ is the XC energy per particle of a homogeneous electron gas with the same density n [106]. The exchange contribution to ϵ_{xc} can be obtained analytically (Ref. [124]), giving

$$\epsilon_x(n) = -\frac{3}{4}\left(\frac{3n}{\pi}\right)^{1/3}. \qquad (2.32)$$

The correlation contribution has to be calculated numerically. In 1980, Ceperley and Alder performed a set of quantum Monte-Carlo (QMC) calculations for the homogeneous electron gas with different densities [125]. The correlation term of the LDA functionals used nowadays rely on different parametrizations of these results. One of the most used parametrizations is that proposed by Perdew and Zunger in 1981 [126]. The LDA is exact for the homogeneous electron gas and expected to be valid for inhomogeneous systems with slowly varying density. A large number of calculations, however, have also shown that it works remarkably well for several real systems with strongly inhomogeneous electron densities [113].

Naturally, are improvement to the LDA is the inclusion of the dependence on the gradient of the density. This approach gave rise to the generalized gradient approximations (GGAs) introduced in the late 1980s [127, 128]. The XC energy is written as

$$E^{xc}[n(\mathbf{r})] = \int n(\mathbf{r})\epsilon_{xc}(n(\mathbf{r}), \nabla n(\mathbf{r}))d\mathbf{r}. \qquad (2.33)$$

The so-called PBE functional [129] is the most commonly used non-empirical GGA functional nowadays. It is an improvement to the LDA for many properties, for example, the geometries and ground-state energy of molecules [129–132].

Further climbing Jacob's ladder, meta-GGA's functionals are found in the third rung. Its description, as well as prescriptions for the fourth and fifth rungs can be found in Ref. [133].

2.2.4   Interpretation of the Kohn–Sham Energies

The KS eigenvalues appear as formal Lagrange multipliers in Eq. (2.17) and correspond to the eigenstates of the fictitious, non-interacting KS particles. A crucial question is therefore whether they have any physical meaning.

Janak's theorem, together with Slater's transition-state theory (TST), provide a justification for the interpretation of the highest occupied state's eigenvalue as the ionization energy in extended systems [134, 135]. Later, this justification was extended to the finite systems [136]. For the other states, the KS eigenvalue, when calculated at half occupation, gives a good estimation of the corresponding total energy difference [137]. When the KS potential is continuous with respect to the electron density, these energy differences can even be approximated by the KS eigenvalues calculated with full occupation. However, since the exact form of $E^{xc}[n]$ is unknown, the comparison between the KS eigenvalues and experiments always relies on approximations of the XC potential.

For the LDA and GGAs, the result of such a comparison can be easily summarized: the work function and band structures in metals are found to be reasonably well described [138, 139], for semiconductors and insulators, universal underestimations of 50–100% for the fundamental band gaps are found. A well-known problem of the LDA is the self-interaction. By using the exact exchange optimized effective potential (OEPx), which is self-interaction-free, these band gaps are improved [140–144].

On the other hand, the fundamental band gap is determined from the ground-state energy of the $N-1$, N, and $N+1$-electron systems. The KS band gap is calculated as the difference between the lowest unoccupied and highest occupied KS eigenvalues in an N-electron system. It was proven that they differ by a term given by the discontinuity of the XC potential [145–147], i.e.

$$E_g = E_g^{KS} + \Delta_{xc},\qquad(2.34)$$

where

$$\Delta_{\mathrm{xc}} = \lim_{\delta \to 0^+} \left[V^{\mathrm{xc}}|_{N+\delta} - V^{\mathrm{xc}}|_{N-\delta} \right]. \tag{2.35}$$

In both LDA and GGA, this discontinuity is zero. In OEPx, it is not zero. Adding this term to the KS band gaps through the above equations, Grüning et al. have shown that fundamental band gaps similar to the Hartree-Fock method can be obtained [148]. The exact XC functional would allow the calculation of the band gap through Eq. (2.34). Nevertheless, for the description of the excited-state properties in general, a different theoretical approach is required. The standard treatment nowadays is the Green's function method we are going to address in Chapter 4.

3
Pseudopotentials, Full Potential, and Basis Sets

As shown in the previous chapter, in the Kohn–Sham (KS) scheme to the density-functional theory (DFT), the many-body electronic problem is reduced to an independent particle problem (Eq. (2.28)) under the action of an effective, density-dependent potential (Eqs. (2.25) and (2.26)). Any numerical implementation of this scheme to real poly-atomic systems has to deal with two extremely different regimes: the core states tightly bound to the nucleus, and the delocalized valence states, especially in the interstitial region. The former are represented by localized wave functions of atomic character, and their role in the bonding of the system is mainly concerned with the screening of the nuclear potential. The valence states play a determining role in the bonding, which reciprocally determines the characteristics of the wave functions of these states, going from localized states in ionic systems to fully itinerant ones in simple metals. However, itinerant the valence wave function may be, it also presents a fast oscillating behavior in the region close to the nuclei. In particular for periodic systems, where the reciprocal space representation is more efficient, plane waves are a natural basis set for the expansion of the single-particle wave functions. However, they are inefficient in representing both the strongly localized core states and the rapid oscillations of the valence wave functions in the nuclear region.

During the past decades, different strategies have been developed to address these coexistent regimes. They can be arranged in two big groups: all-electron and pseudopotential methods. The all-electron methods rely on the use of more sophisticated basis functions for the expansion of the wave functions, while keeping the potential genuine. These basis functions can address the oscillation of the wave functions in the nuclear region, as well as the core states, at reasonable computational cost. The pseudopotential methods, on the other hand, replace the strong nuclear potential and the core contribution to the Hartree and exchange–correlation (XC) potentials by an artificial effective ionic potential, namely the pseudopotential, which is designed to be much softer than the full potential in the atomic region. In this way, the valence wave functions behave smoothly in the nuclear region and can be efficiently expanded in plane waves. Only the valence electrons are treated explicitly.

In the following, we will give a brief introduction to the pseudopotential method using one of its classes, the norm-conserving pseudopotentials, as an example. Our focus is on the principles underlying the construction of a pseudopotential. Concerning the all-electron method, we will also introduce an important basis set, i.e. the full-potential-(linearized) augmented plane waves plus local orbitals (FP-(L)APW+lo). All-electron calculation employing this basis set is currently believed to be the most accurate form of DFT calculation in periodic systems.

3.1   Pseudopotential Method

The main idea behind the pseudopotential method is that, as long as the core electrons are tightly bound, they do not participate actively in the bonding process. Thus, the strong ionic potential, including contributions from the nucleus and the core electrons, can be replaced by an angular-dependent pseudopotential constructed from the free atom of the corresponding element [149]. In this way, we can only include the valence states explicitly in descriptions of the chemical bonding in the poly-atomic system, which significantly reduces the computational cost. Inside the core region, the pseudopotential is designed to be much softer than the ionic one. Outside the core region, it is required that the corresponding pseudowave function equals its all-electron counterpart in order to obtain the correct behavior over a wide range of chemical environments (transferability).

A pseudopotential fulfilling the above-mentioned prerequisites can be generated arbitrarily in many ways. The most used one is the "norm-conserving" scheme originally proposed by Hamann et al. [150] and later applied to elements from H to Pu by Bachelet et al. [151]. In this scheme, the integral of the pseudocharge-density inside the core region is required to agree with the all-electron one. This condition guarantees that the electrostatic potential produced outside the core radius is equal in both

cases. Furthermore, the energy dependence of the scattering properties of
the pseudopotential is of the second order and can be ignored without affect-
ing the transferability. Nevertheless, some cases, e.g. O $2p$ or Ni $3d$ orbitals,
have been found to be impossible to construct a pseudowave function much
softer than its all-electron counterpart for [152]. This is due to the fact that
in $2p$ and $3d$ orbitals, the radial function is already nodeless. Therefore,
when applying the constraint that the normal of the radial function does
not change, there is no way to make it softer.

Such a drawback indicates that for systems containing $2p$ and $3d$
valence states, within the framework of the norm-conserving pseudopotentials, a large cut-off radius for the plane wave basis set is still required to
describe behaviors of these states in the core region. As a matter of fact, in
the 1980s and 1990s, this drawback has seriously hindered the application of
the pseudopotential method-based DFT simulations to large systems, due
to the limited computing power available at that time. In the early 1990s,
a successful attempt to circumvent this limit was proposed by Vanderbilt
et al. in Refs. [152, 153]. The corresponding method is called ultra-soft pseudopotentials, in which the norm-conserving constraint is lifted. Nowadays,
due to the advances in computing power, the norm-conserving pseudopotentials are still one of the most frequently used pseudoptentials in practical
simulations. In this book, we restrict ourselves to this norm-conserving
scheme for a clear illustration of the principles underlying the construction
of a pseudopotential. Readers interested in ultra-soft pseudopotentials are
directed to Refs. [152, 153].

3.1.1 Generation of the Pseudopotential

The initial step for generating a norm-conserving pseudopotential is to perform an all-electron DFT calculation for the free atom. This corresponds to
obtaining a self-consistent solution of the radial, Schrödinger-like equation

$$\left[-\frac{1}{2}\frac{d^2}{dr^2} + \frac{l(l+1)}{2r^2} + V(r) - \epsilon_l \right] u_l(\epsilon_l, r) = 0, \qquad (3.1)$$

where $V(r)$ is equivalent to the effective potential V_{eff} in Eq. (2.25). It
should be obtained through a self-consistent solution of Eqs. (2.25)–(2.27),
with

$$\varphi_{l,m}(\mathbf{r}) = \frac{u_l(\epsilon_l, r)}{r} Y_{l,m}(\hat{r}).$$

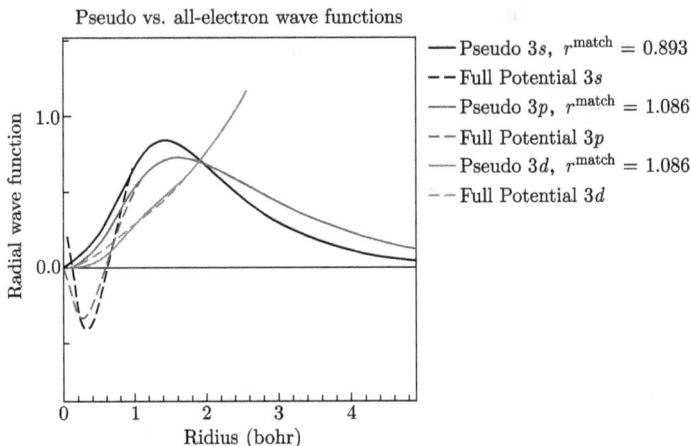

Figure 3.1 Pseudovalence wave functions in comparison with the all-electron ones. S atom is taken as the example. The $3s$, $3p$ and $3d$ all-electron and pseudopotential states are shown; the $3d$ state is an unbound state. The two wave functions agree with each other outside the matching radius, while the pseudo one is much softer inside this radius. Their norms are equal. The figure was generated using the fhi98PP pseudopotential program [154], which is open access and available at http://th.fhi-berlin.mpg.de/th/fhi98md/fhi98PP/.

After the self-consistent solution of these equations, the two full-potential functions $V(r)$ and $u_l(\epsilon_l, r)$ will be the key quantities from which the pseudopotential is generated.

For each angular momentum number "l" (from now on, we call this a "channel"), a cut-off radius (r_l^c) is chosen and the pseudovalence radial wave function $u_l^{ps}(\epsilon_l^{ps}, r)$ is derived from its all-electron counterpart $u_l(\epsilon_l, r)$ with the following minimal constraints:

(i) The pseudovalence state has the same eigenvalue as the all-electron one ($\epsilon_l^{ps} = \epsilon_l$).

(ii) $u_l^{ps}(\epsilon_l, r)$ equals $u_l(\epsilon_l, r)$ beyond the cut-off radius (designated by r^{match} in Fig. 3.1).

(iii) $u_l^{ps}(r)$ is nodeless. In order to obtain a continuous pseudopotential regular at the origin, it is also required to be twice differentiable and satisfy $\lim_{r \to 0} u_l^{ps} \propto r^{l+1}$.

(iv) The pseudovalence radial wave function is normalized (the norm-conserving constraint) which, together with (ii) implies

$$\int_0^{r'} |u_l^{ps}(\epsilon_l^{ps}; r)|^2 dr \equiv \int_0^{r'} |u_{n,l}(\epsilon_{n,l}; r)|^2 dr \quad \text{for } r' \geqslant r_l^c. \quad (3.2)$$

Once the pseudowave function is obtained, one can construct the screened[a] pseudopotential $V_l^{\text{ps,scr}}(r)$, which acts as the effective potential on the pseudovalence state, by inverting the radial Schrödinger equation, leading to

$$V_l^{\text{ps,scr}}(r) = \epsilon_l^{\text{ps}} - \frac{l\,(l+1)}{2r^2} + \frac{1}{2u_l^{\text{ps}}(r)}\frac{d^2}{dr^2}u_l^{\text{ps}}(\epsilon_l, r). \qquad (3.3)$$

In the last step, the pseudocharge-density $\tilde{n}_v^0$ is obtained by

$$\tilde{n}_v^0(r) = \sum_l^{\text{occ}} \left|\frac{u_l^{\text{ps}}(r)}{r}\right|^2, \qquad (3.4)$$

and the pseudopotential is *unscreened* by subtracting the Hartree and XC potential corresponding to this pseudocharge by

$$V_l^{\text{ps}}(r) = V_l^{\text{ps,scr}}(r) - V^{\text{H}}[\tilde{n}_v^0; r] - V^{\text{xc}}[\tilde{n}_v^0; r]. \qquad (3.5)$$

So defined, the ionic pseudopotential generated from Eq. (3.5) includes all the interactions of the valence electrons with the ion on the DFT level and is much softer in the core region than its all-electron counterpart (Fig. 3.2). The distinct procedures for generating a norm-conserving pseudopotential differ only in the way the pseudovalence radial wave functions (Fig. 3.1) are designed and the constraints they are required to fulfill.

In the Hamann scheme [150], an intermediate pseudopotential $V_l^{\text{ps,i}}(r)$ is constructed by cutting off the singularity of the full-potential $V(r)$ at the nucleus:

$$V_l^{\text{ps,i}}(r) = V(r)\left[1 - f\left(\frac{r}{r_{cl}}\right)\right] + c_l f\left(\frac{r}{r_{cl}}\right), \qquad (3.6)$$

where $f(x)$ is a cut-off function, which is unity at the origin, cuts off at $x \sim 1$ and decreases rapidly as $x \to \infty$. So defined, the intermediate pseudopotential $V_l^{\text{ps,i}}(r)$ approaches its all-electron counterpart $V(r)$ rapidly and continuously as x increases outside the cut-off radius r_{cl}. The free parameter c_l is then adjusted so that the nodeless solution w_l of the radial

---

[a]In this context, the term "screened" is used in the sense that $V_l^{\text{ps,scr}}(r)$ also contains the interaction between valence states.

The ionic pseudopotential

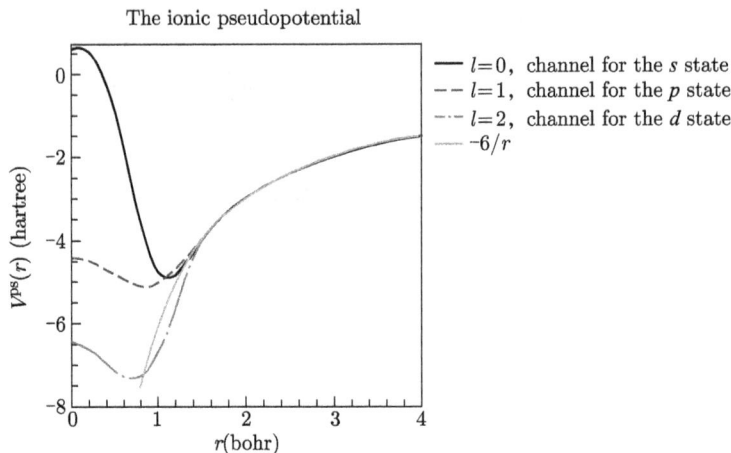

Figure 3.2   Ionic pseudopotentials for $l = 0, 1, 2$ in S atom. The pseudopotential is much softer in the region close to the nucleus compared with the behavior of $-Z/r$, where Z is the ionic charge. Similar to Fig. 3.1, the fhi98PP program is used [154].

equation

$$\left[-\frac{1}{2}\frac{d^2}{dr^2} + \frac{l(l+1)}{2r^2} + V_l^{ps,i}(r) - \epsilon_l^{(i)} \right] w_l(\epsilon_l, r) = 0 \qquad (3.7)$$

has the same eigenvalue as its all-electron counterpart $(\epsilon_l^{(i)} = \epsilon_l)$.

Since both w_l and u_l are solutions of the same potential outside the core region, one can write

$$\gamma_l w_l(r) \rightarrow u_l(r), \quad r > r_{cl}. \qquad (3.8)$$

The pseudowave function is now modified by adding a correction in the core region

$$u_l^{ps}(r) = \gamma_l [w_l(r) + \delta_l g_l(r)], \qquad (3.9)$$

where $g_l(r)$ must vanish as r^{l+1} for small r to give a regular pseudopotential at the origin, and it must vanish rapidly for $r > r_{cl}$ since $\gamma_l w_l(r)$ is already the desired solution in that region. At this point, the normalization condition is used to set the value of δ_l,

$$\gamma_l^2 \int |w_l(r) + \delta_l g_l(r)|^2 dr = 1. \qquad (3.10)$$

The pseudowave functions generated by this procedure fulfill conditions (i), (iii), and (iv). Condition (ii) is reached exponentially beyond the cut-off radius. Even within this procedure, the pseudopotential is not unique. It depends on the choice of states included in the core, the selection of the cut-off functions and the core radii (see Ref. [151] for details). Later on, Hamann [155] also extended this procedure to generate pseudopotentials for the atomic unbound states.

In the Troullier–Martins scheme [156], the pseudovalence radial wave function exactly equals the all-electron one outside the cut-off radius. Inside the cut-off radius, it is assumed to have the following analytic form:

$$u_l^{ps}(r) = r^{(l+1)} e^{p(r)}, \tag{3.11}$$

where $p(r)$ is a polynomial of sixth order in r^2. The coefficients are determined from conditions (ii)–(iv), plus the additional constraints of continuity of the first four derivatives at r_l^c and zero curvature of the screened pseudopotential at the origin ($\frac{d^2}{dr^2} V_l^{ps,scr}(r)|_{r=0} = 0$). Condition (i) is accomplished directly by solving Eq. (3.3). As a consequence of the additional requirements, the Troullier–Martins pseudopotentials are softer than the Hamann ones.

The different radial dependence of the above-defined pseudopotentials for each channel results in the total pseudopotential being semilocal (i.e. non-local in the angular coordinates, but local in the radial one)

$$V^{ps}(\mathbf{r}, \mathbf{r}') = V^{loc}(r)\delta(\mathbf{r} - \mathbf{r}') + \sum_{l=0}^{l_{max}} \sum_{m=-l}^{m=l} Y_{l,m}^*(\hat{r})\delta V_l^{ps}(r)\frac{\delta(r-r')}{r^2} Y_{l,m}(\hat{r}'), \tag{3.12}$$

where $\delta V_l^{ps}(r) = V_l^{ps}(r) - V^{loc}(r)$. The l-independent term ($V^{loc}(r)$) is chosen such that the semilocal terms (δV_l^{ps}) are confined to the core region and eventually vanish beyond some l_{max}.

Kleinman and Bylander (KB) [157] proposed a transformation of the semilocal terms into a fully non-local form defining

$$\delta V^{KB}(\mathbf{r}, \mathbf{r}') = \sum_{l,m} \frac{\delta V_l^{ps}(r)\tilde{\varphi}_{l,m}(\mathbf{r})\tilde{\varphi}_{l,m}^*(\mathbf{r}')\delta V_l^{ps}(r')}{\langle \tilde{\varphi}_{l,m}|\delta V_l^{ps}|\tilde{\varphi}_{l,m}\rangle}, \tag{3.13}$$

where $\tilde{\varphi}_{lm}(\mathbf{r}) = \frac{u_l^{ps}(\epsilon_l,r)}{r} Y_{lm}(\hat{r})$. It can be easily verified that

$$\int \delta V^{KB}(\mathbf{r}, \mathbf{r}') \, \tilde{\varphi}_{l,m}^*(\mathbf{r}') d^3\mathbf{r}' = \delta V_l^{ps}(r)\tilde{\varphi}_{l,m}(\mathbf{r}), \tag{3.14}$$

that is, the KB form is equivalent to the semilocal one in the sense that it produces the same atomic pseudoorbitals. At the expense of a more complicated expression in real space, the KB form is fully separable, strongly reducing the number of integrations necessary, e.g. in a plane wave basis set, to calculate the Hamiltonian matrix elements. It is used in most of the electronic structure codes nowadays.

3.1.2 *Implicit Approximations*

3.1.2.1 *Frozen Core*

The main assumption in the pseudopotential method is that the core states, strongly bound to the nucleus and localized, are insensitive to the environment surrounding the atom. Therefore, they can be excluded from the self-consistent calculation in the poly-atomic system. This is the "frozen core" approximation.

The pseudopotential is defined by the requirements that the wave functions and eigenvalues are accurately reproduced, however, no conditions on total energies are imposed. In 1980, von Barth and Gelatt [158] demonstrated that the error in the total energy is of second order in the difference between frozen and true-core densities. Their calculations for Mo further confirmed this conclusion, thus validating the application of the pseudopotential method in total energy calculations from this perspective.

3.1.2.2 *Core–Valence Linearization*

The definition of the pseudopotential in Eq. (3.5) implies that the self-consistent total XC potential in a condensed matter system is written as

$$V^{\mathrm{xc}}[n(\mathbf{r})] = \{V^{\mathrm{xc}}[n^0(\mathbf{r})] - V^{\mathrm{xc}}[\tilde{n}_v^0(\mathbf{r})]\} + V^{\mathrm{xc}}[\tilde{n}_v(\mathbf{r})], \qquad (3.15)$$

where the terms in curly brackets are included in the pseudopotential. Equation (3.15) would be exact, within the frozen-core approximation, if the XC potential were a linear functional of the density.[b] As it is clearly not the case, the assumption of validity of Eq. (3.15) is known as *core–valence linearization*.

---

[b]Note that Eq. (3.15) is also exact in a non-self-consistent calculation, since in that case, $n(\mathbf{r}) = n^0(\mathbf{r})$ and $\tilde{n}_v(\mathbf{r}) = \tilde{n}_v^0(\mathbf{r})$.

However, the errors due to this approximation are small in most cases, as long as the overlap between the core and valence densities are not significant. Louie *et al.* [159] developed a method for the generation and usage of pseudopotentials that explicitly treats the nonlinear core–valence XC interaction. The method consists in modifying Eq. (3.5) to

$$V_l^{\mathrm{ps}}(r) = V_l^{\mathrm{ps,scr}}(r) - V^{\mathrm{H}}[\tilde{n}_v^0; r] - V^{\mathrm{xc}}[\tilde{n}_v^0 + \tilde{n}_c; r], \qquad (3.16)$$

where $\tilde{n}_c$ is a partial core density. It reproduces the full core density in the region where it overlaps with the valence density, outside a chosen cut-off radius r^{nlc}. Inside this radius, it is chosen to match the true density at r^{nlc}, minimize the integrated density and be easily Fourier transformed in order to optimize its use within the plane wave basis set. This density has to be added to the pseudovalence density in the self-consistent calculation whenever V^{xc} or E^{xc} are computed.

3.1.2.3 *Pseudoization*

By *pseudoization*, we refer to the fact that the wave functions of the valence states in the pseudopotential method are, by construction, nodeless and much smoother than their all-electron analogue. It is only observable in the core region, which constitutes a small portion of space. As long as the full potential is local, the errors in the energies, within this region, are taken care of in the pseudopotential by construction. Furthermore, the norm-conserving constraint ensures that the Hartree potential generated by the pseudocharge outside the core region is the same as in the all-electron treatment. Nevertheless, whether it is also negligible in the calculation of non-local operators is unclear, as mentioned in Ref. [160]. The fact that *pseudoization* can lead to qualitative differences between PP and AE calculations has been pointed out in Ref. [161], where significant discrepancies in the electron–hole distribution function of LiF were observed. Because of these, it is worth noting that this *pseudoization* is also a possible source of error in practical *ab initio* electronic structure calculations, especially when non-local operators are concerned [162, 163].

3.2   FP-(L)APW+lo Method

The FP-(L)APW+lo method is a development of the augmented plane wave (APW) method originally proposed by Slater [164]. Thus, we start our discussions here with a short introduction of the APW method. The essential

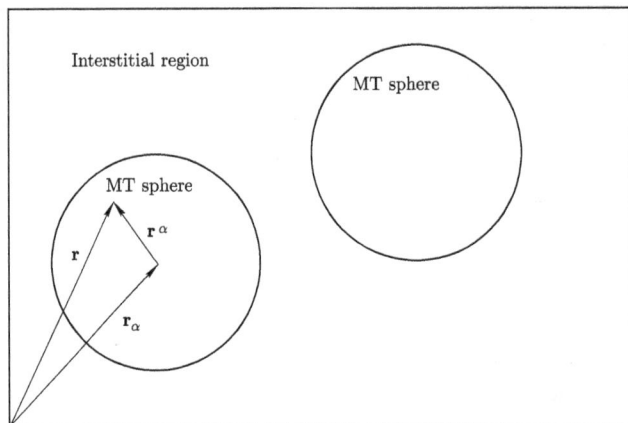

Figure 3.3   Schematic view of the space partition in the APW method. The space is divided into the interstitial region and a series of non-overlaping muffin-tin (MT) regions. The potential in the MT region is atomic-like, while that in the interstitial region is much softer.

idea underlying this method is that in the region close to nuclei, the potentials and wave functions are similar to those in the free atoms, strongly varying but nearly spherical. In the space between the atoms, both of them are smooth. In his seminal work, Slater proposed a division of the space in the unit cell into a set of non-overlapping spheres centered at each atom and the interstitial region between them (Fig. 3.3). The potential was taken as spherically symmetric inside the spheres and constant outside (later on known as the muffin-tin approximation, for obvious reasons). Accordingly, the eigenfunctions of the Hamiltonian corresponding to each of the regions are taken as basis functions, namely, plane waves in the interstitial and atomic orbitals in the "muffin-tin" (MT) spheres. Adding the continuity condition at the sphere boundary, the APWs were born

$$\phi_{\mathbf{G}}^{\mathbf{k}}(\mathbf{r}) = \begin{cases} \dfrac{1}{\Omega} e^{i(\mathbf{G}+\mathbf{k})\cdot\mathbf{r}}, & r \in \text{interstitial}, \\ \displaystyle\sum_{l,m} A_{l,m}(\mathbf{k}+\mathbf{G}) \, u_l(r^\alpha, \epsilon) \, Y_{l,m}(\mathbf{r}^\alpha), & r \in \text{MT}. \end{cases}$$

$$(3.17)$$

Inside each MT sphere, the radial wave function $u_l(r^\alpha, \epsilon)$ at the reference energy ϵ_l, is obtained from

$$\left\{ -\frac{d^2}{dr^2} + \frac{l(l+1)}{r^2} + V(r) - \epsilon_l \right\} r u_l(r, \epsilon_l) = 0. \qquad (3.18)$$

The augmentation coefficients ensuring the continuity of the wave function at the sphere boundary are given by

$$A_{l,m}(\mathbf{k}+\mathbf{G}) = \frac{4\pi i^l}{\Omega^{1/2}u_l(R^{\alpha}_{\mathrm{MT}},\epsilon_l)}j_l(|\mathbf{k}+\mathbf{G}|R^{\alpha}_{\mathrm{MT}})Y^*_{l,m}(\mathbf{k}+\mathbf{G}). \qquad (3.19)$$

This optimized choice of basis functions in different regions is the essence of the augmented methods and all its descendants. The wave function is expanded in terms of these APWs as

$$\varphi_{n,\mathbf{k}}(\mathbf{r}) = \sum_{\mathbf{G}} C^n_{\mathbf{G}}\phi^{\mathbf{k}}_{\mathbf{G}}(\mathbf{r}). \qquad (3.20)$$

The coefficients $C^n_{\mathbf{G}}$ should be obtained by solving the eigenvalue equation

$$\sum_{\mathbf{G'}}(H_{\mathbf{G},\mathbf{G'}} - \epsilon_n S_{\mathbf{G},\mathbf{G'}})C^n_{\mathbf{G}} = 0 \qquad (3.21)$$

for each $\mathbf{k}$ where $H_{\mathbf{G},\mathbf{G'}}(S_{\mathbf{G},\mathbf{G'}})$ are the Hamiltonian (overlap) matrix elements in the APW basis.

The major drawback of this method is that, inside the MT sphere, the APWs are solutions of the Schrödinger equation only at the reference energy (i.e. $\epsilon_l = \epsilon_n$). Thus, the eigenvalue equation (3.21) becomes nonlinear and its solution much more computationally demanding for each $\mathbf{k}$-point. Furthermore, it is hard, though not impossible (Refs. [165, 166]) to extend the method to the full potential case. When the potential inside a MT sphere is not spherical, the exact solution of the particle's wave function inside this MT sphere does not correspond to the solution of the radial Schrödinger equation with the same eigenvalue.

Another shortcoming of the APW method, known as the asymptote problem, is related to the indetermination of the augmentation coefficients when the radial function has a node at the MT radius ($u_l(R_{\mathrm{MT}})$ in the denominator of Eq. (3.19)). In the vicinity of this region, the relation between $A_{l,m}$ and $C_{\mathbf{G}}$ becomes numerically unstable.

With the aim of overcoming these limitations, Andersen [167] proposed a modification of the APW method in which the wave functions and their derivatives are made continuous at the MT radius by matching the interstitial plane waves to the linear combination of a radial function, and its energy derivative, calculated at a fixed reference energy. The method, known as the linearized augmented plane waves (LAPWs) method, rapidly demonstrated its power and accuracy, becoming, during decades, the benchmark for electronic structure calculations within the KS scheme. Recently, Sjösted *et al.* [168] proposed an alternative method in which the APW wave functions are recovered, but with the radial functions calculated at a fixed energy. The flexibility of the basis is achieved by adding a set of local orbitals constructed as linear combinations of the same radial functions and their energy derivatives, with the condition that the function cancels at the sphere radius. This method, called APW plus local orbitals (APW+lo), requires fewer plane waves for an accurate description of the electronic structure properties, thus increasing the computational efficiency. However, this improvement is limited by the large number of local orbitals required for large l's. Nowadays, the state-of-the-art method involves a combination of both, using APW+lo's for small l and LAPWs for the large ones [169], known as (L)APW+lo method. The rest of the chapter is devoted to an overview of these methods.

3.2.1   *LAPW Basis Functions*

The LAPW basis set is defined by

$$\phi_{\mathbf{k+G}}\left(\mathbf{r}\right)$$

$$= \begin{cases} \dfrac{1}{\Omega}e^{i(\mathbf{G+k})\cdot\mathbf{r}}, & r \in \text{interstitial}, \\[2mm] \displaystyle\sum_{l,m}[A_{l,m}(\mathbf{k}+\mathbf{G})u_l(r^\alpha,\epsilon_l) \\[1mm] \qquad +B_{l,m}\left(\mathbf{k}+\mathbf{G}\right)\dot{u}_l(r^\alpha,\epsilon_l)]Y_{l,m}\left(\mathbf{r}^\alpha\right), & r \in \text{MT}, \end{cases} \tag{3.22}$$

where $(\dot{u}_l(r,\epsilon_l) = \partial u_l(r,\epsilon)/\partial\epsilon|_{\epsilon=\epsilon_l})$. The augmentation coefficients $A_{l,m}$ and $B_{l,m}$ are obtained by requiring both the value and the slope of the basis function to be continuous on the MT sphere boundary.

Making a Taylor expansion of the radial wave function around the reference energy ϵ_l, one has

$$u_l(r,\epsilon) = u_l(r,\epsilon_l) + (\epsilon - \epsilon_l)\dot{u}(r,\epsilon_l) + O((\epsilon - \epsilon_l)^2), \tag{3.23}$$

which means that in the linearized treatment, the error in the wave function is of the second order in $\epsilon - \epsilon_l$. Taking into account the variational principle, this leads to an error of fourth order, $(\epsilon - \epsilon_l)^4$, in the band energy. In other words, the LAPWs form a good basis over a relatively large energy region, typically allowing the calculation of all the valence bands with a single set of reference energies, i.e. by a single diagonalization of Eq. (3.21).

However, there are situations in which the use of a single set of reference energies is inadequate for all the bands of interest. Such a situation arises, for example, when two (or more, but rarely) states with the same l participate in the chemical bonding (semicore states), or when bands over an unusually large energy region are required, like for high-lying excited states. To address such cases, the local orbitals were introduced by Singh in 1991 [170]:

$$
\phi_{\text{LAPW}}^{\text{LO}}(\mathbf{r})
$$
$$
= \begin{cases}
0, & r \in \text{interstitial}, \\
[A_{l,m}^{\alpha} u_l(r^{\alpha}, \epsilon_l) + B_{l,m}^{\alpha} \dot{u}_l(r^{\alpha}, \epsilon_l) \\
\quad + C_{l,m}^{\alpha} u_l(r^{\alpha}, \epsilon_l^{(2)})] Y_{l,m}(\mathbf{r}^{\alpha}), & r \in \text{MT}.
\end{cases}
\tag{3.24}
$$

In this way, a second set of energy parameters $\epsilon_l^{(2)}$ is introduced to provide the additional variational freedom required for an accurate representation of the different states with the same l. The coefficients $A_{l,m}$s, $B_{l,m}$s and $C_{l,m}$s are determined by requiring the local orbital and its radial derivative to be zero at the MT sphere boundary and normalized.

3.2.2   APW+lo Basis Functions

The LAPW basis set is designed to be flexible in describing the wave functions in the vicinity of the reference energy. However, the requirement of continuous derivatives at the MT radius increases the number of plane waves needed to achieve a given level of convergence with respect to the APW method.

Recently, Sjösted et al. [168] proposed an alternative way to linearize the APW method in which the continuous derivative condition is released. In this method, the eigenvalue problem of the original APW method is linearized by choosing fixed linearization energies (ϵ_l) for the APW basis functions in Eq. (3.17). Then, the flexibility of the basis set with respect to the reference energy is obtained by adding a set of local orbitals (lo):

$$
\phi_{\text{APW}}^{\text{lo}}(\mathbf{r}) = \begin{cases}
0, & r \in \text{interstitial}, \\
[A_{l,m} u_l(r^{\alpha}, \epsilon_l) + B_{l,m} \dot{u}_l(r^{\alpha}, \epsilon_l)] Y_{l,m}(\mathbf{r}^{\alpha}), & r \in \text{MT},
\end{cases}
\tag{3.25}
$$

using the same linearization energies. The coefficients are obtained by requiring the function to be zero at the sphere boundary and normalized.

The APW basis functions keep the convergence behavior of the original APW method while the local orbitals (lo) make it flexible with respect to choice of the reference energy. The complete APW+lo basis set therefore consists of two different types of basis functions, the APWs (Eq. (3.17) at fixed linearization energies) and the lo's (Eq. (3.25)). As in the LAPW method, when different states with the same l (semicore states) have to be treated, a second set of local orbitals of the form:

$$\phi_{APW}^{LO}(\mathbf{r}) = \begin{cases} 0, & r \in \text{interstitial}, \\ [A_{l,m}u_l(r^\alpha, \epsilon_l) \\ \quad + C_{l,m}u_l^{(2)}(r^\alpha, \epsilon_l^{(2)})]Y_{l,m}(\mathbf{r}^\alpha), & r \in \text{MT} \end{cases} \qquad (3.26)$$

can be added. The coefficients are determined by matching the function to zero at the muffin-tin radius, with no condition on the slope.

3.2.3   Core States

As already mentioned, the (L)APW+lo is an all-electron method. However, it does not mean that core and valence states are treated in the same way. While the latter are expanded in the previously described basis set using the crystal potential, the former are calculated numerically by solving the relativistic radial Schrödinger equation for the atom. The influence of the core states on the valence is carried out by the inclusion of the core density in the Hartree and XC potentials. Reciprocally, the core states are calculated using the spherical average of the crystal potential in the muffin-tin sphere. Thus, both core and valence states are calculated self-consistently.

In the Wien2k code, the wave function of each core state is represented as

$$\tilde{\varphi}_{a,n,j,m_j}^{core}(\mathbf{r}) = u_{a,n,\kappa}(r^a)|jm_j\rangle_l, \qquad (3.27)$$

where

$$|jm_j\rangle_l \equiv \sum_{\sigma=-\frac{1}{2}}^{\frac{1}{2}} \left(l \, \frac{1}{2} \, m_l \, \sigma | j \, m_j\right) Y_{lm_l}(\hat{r}^a) |\sigma\rangle \delta_{m+\sigma,m_j}, \qquad (3.28)$$

and $(l \, \frac{1}{2} \, m_l \, \sigma | j \, m_j)$ is the corresponding Clebsch–Gordon coefficient (Ref. [171]).

Table 3.1   Relativistic quantum numbers.

l		$j = l + \dfrac{s}{2}$		κ		Max. occupation	
		$s = -1$	$s = +1$	$s = -1$	$s = +1$	$s = -1$	$s = +1$
s	0		1/2		−1		2
p	1	1/2	3/2	1	−2	2	4
d	2	3/2	5/2	2	−3	4	6
f	3	5/2	7/2	3	−4	6	8

The radial wave function is defined by the relativistic quantum number $\kappa = -s(j + \frac{1}{2})$ as shown in Table 3.1.

3.2.4   *Potential and Density*

The representation of the density and the potential has to confront the same difficulties as the representation of the wave functions, namely, rapid variations in the muffin-tin spheres and soft oscillations in the interstitial. The use of a dual representation as for the wave functions, which is the basis of the (L)APW+lo efficiency, seems the natural choice. However, an expansion in spherical harmonics inside the spheres and plane waves in the interstitial is clearly inefficient. The complete representation of the density requires a basis set at least eight times larger than the basis required for the wave functions. Since the number of augmentation functions in the MT sphere also increases four times, the number of augmentation coefficients is 2^5 times larger.

This can be reduced by exploiting the symmetries of the density (potential), namely:

(i) Inside the muffin-tin sphere, they respect the symmetry of the corresponding nuclear site.
(ii) In the interstitial region, they have the symmetry of the corresponding space group.
(iii) Both are real quantities.

Inside the muffin-tin spheres, properties (i) and (iii) allow the representation of the density in a lattice harmonic expansion [172]. For the interstitial region, the use of stars ensures both properties (ii) and (iii) to be fulfilled with a minimum number of coefficients. More details can be found in Ref. [149].

4
Many-Body Green's Function Theory and the *GW* Approximation

In Chapter 2, we pointed out that although the Kohn–Sham (KS) eigen-values provide a good zeroth-order approximation for the single-particle excitation energies, LDA/GGA fails to give a satisfactory description of the fundamental band gaps in semiconductors and insulations. The many-body Green function theory (or otherwise called many-body perturbation theory), on the other hand, provides the formal basis for evaluating the experimentally observed quasi-particle band structures. Nowadays, it has become routine practice to resort to this many-body perturbation theory (especially within its so-called "*GW* approximation") for the descriptions of the single-particle excitation in real poly-atomic systems. Because of this, we will present an overview of the Green function method in the many-body electronic systems, with a special emphasis on the *GW* approximation in this chapter.

In order to provide a simple introduction, we begin with a general description of the main ingredients in this theory in Sec. 4.1. This description starts from the definition of the single-particle Green function, the central physical quantity in this method. In Sec. 4.1.1, we explain the correspondence between the poles of this single-particle Green function on the frequency axis and the single-particle excitation energies of a many-body system. A numerical description of this single-particle Green function in a many-body system, however, is much easier said than done. In practical simulations, one needs to resort to the single-particle Green function of a non-interacting system. The equation, which relates the single-particle Green function of the many-body system and that of the non-interacting system, is called the Dyson equation. In Sec. 4.1.2, we show how this equation is deduced. In the Dyson equation, all the complexity of the many-body interaction is put into a term which is called the "self-energy". Then in Sec. 4.1.3, we devote our discussions to the concept of this self-energy and its expansion in terms of the dynamically screened Coulomb potential, i.e. the Hedin equations. The concept of the quasi-particle is introduced in Sec. 4.1.4.

Then we go to the practical schemes for the numerical treatment of the Hedin equations in real poly-atomic systems. The simplest approximation to the self-energy, which includes the dynamical screening effects, is the so-called GW approximation. Here, G means the Green function and W means the screening Coulomb potential. Their product presents the simplest approximation for the self-energy, which, as was said, includes the important dynamical screening feature of the many-body interactions. The principles underlying this GW approximation will be introduced in Sec. 4.2.

This GW approximation, in its most rigorous form, means that the Dyson equation needs to be solved self-consistently. In practical simulations of real poly-atomic systems, however, a further approximation beyond it is often used. This further approximation treats the self-energy, obtained in a GW manner, only as a first-order correction to the exchange–correlation (XC) potential of the fictitious non-interacting particles, which can correspond to either the KS or the Hartree–Fock orbitals. It is nowadays often called the G_0W_0 approximation in the literature. Over the last 30 years, it has achieved great success in describing the single-particle excitations in a wide range of poly-atomic systems, ranging from molecules to periodic semiconductors, to even metals. This G_0W_0 approach will be described in Sec. 4.3.

With these, we hope that we have relayed the key messages of the principles underlying this single-particle excitation theory in a relatively clear manner. As an example of its implementation in computer programming, we use an all-electron GW code, i.e. the FHI-gap,[a] to show how some technical details are treated in Sec. 4.4. We note that due to the heavy computational load associated with the LAPW basis, this code might not be the best choice for the GW calculation of large poly-atomic systems. When large simulation cells must be used, other computational codes such as BerkeleyGW[b] [173], which employs the pseudopotentials and plane wave

---

[a]For details, please see http://www.chem.pku.edu.cn/jianghgroup/codes/fhi-gap.html.
[b]For details, please see http://www.berkeleygw.org/.

basis set, should be a much better choice. Nevertheless, because of the accuracy of this LAPW basis set in descriptions of the electronic states in a full-potential (FP), all-electron manner, we would like to recommend the FHI-gap code as the benchmark for excited state single-particle electronic structure calculations of the periodic systems. It is important to note that most of this FHI-gap code was written by Dr. Ricardo Gómez-Abal starting from 2002, during his postdoc years in the group of Prof. Matthias Scheffler in the Fritz Haber Institute of the Max Planck Society, Berlin. From 2004 to 2008, one of the authors of this book (XZL) did his Ph.D. in this group. His thesis focused on the development of this all-electron *GW* code and its applications to conventional semiconductors. Dr. Ricardo Gómez-Abal supervised him directly during this time. Starting from 2006, Hong Jiang (who is currently a Professor in the College of Chemistry and Molecular Engineering, Peking University, China) joined this project and significantly improved its completeness and efficiency. After 2008, this code was maintained and further developed mainly by Prof. Hong Jiang.

4.1   Green Function Method

4.1.1   *The Green Function*

The single-particle Green function is defined as

$$G\left(\mathbf{r},t;\mathbf{r}',t'\right) = -i\langle N|\hat{T}\{\hat{\psi}\left(\mathbf{r},t\right)\hat{\psi}^{\dagger}\left(\mathbf{r}',t'\right)\}|N\rangle, \qquad (4.1)$$

where $\hat{\psi}(\mathbf{r},t)$ ($\hat{\psi}^{\dagger}(\mathbf{r}',t')$) is the quantum field operator describing the annihilation (creation) of one electron at position $\mathbf{r}$ and time t (position $\mathbf{r}'$ and time t'). The operator $\hat{T}$ is the time-ordering operator, which reorders the field operators in ascending time order from right to left. $|N\rangle$ is the ground-state eigenfunction of the N-electron system. Making use of the Heaviside function (Appendix A), and the commutation relations for fermionic operators, Eq. (4.1) can be rewritten as

$$G\left(\mathbf{r},t;\mathbf{r}',t'\right) = -i\langle N|\hat{\psi}\left(\mathbf{r},t\right)\hat{\psi}^{\dagger}\left(\mathbf{r}',t'\right)|N\rangle\Theta\left(t-t'\right)$$
$$+ i\langle N|\hat{\psi}^{\dagger}\left(\mathbf{r}',t'\right)\hat{\psi}\left(\mathbf{r},t\right)|N\rangle\Theta\left(t'-t\right), \qquad (4.2)$$

making evident that for $t > t'$ ($t < t'$), the Green function describes the propagation of an added electron (hole) in the system.

In the Heisenberg representation, the field operator is written as

$$\hat{\psi}(\mathbf{r}, t) = e^{i\hat{H}t}\hat{\psi}(\mathbf{r})e^{-i\hat{H}t}, \tag{4.3}$$

where $\hat{H}$ is the Hamiltonian operator and $\hat{\psi}(\mathbf{r})$ is the field operator in the Schrödinger representation.

Inserting Eq. (4.3) into Eq. (4.2) and making use of the completeness relation in the Fock space:

$$1 = \sum_{n=0}^{\infty}\sum_{s}|n, s\rangle\langle n, s|, \tag{4.4}$$

where $|n, s\rangle$ corresponds to the sth eigenstate of the the n-electron system, we can transform Eq. (4.2) into

$$G(\mathbf{r}, t; \mathbf{r}', t') = -i\sum_{s}\langle N|\hat{\psi}(\mathbf{r})|N+1, s\rangle e^{-i\left(E_{N+1}^s - E_N\right)(t-t')}$$

$$\cdot\langle N+1, s|\hat{\psi}^{\dagger}(\mathbf{r}')|N\rangle\Theta(t-t') + i\sum_{s}\langle N|\hat{\psi}^{\dagger}(\mathbf{r}')|N-1, s\rangle$$

$$\cdot e^{-i\left(E_{N-1}^s - E_N\right)(t'-t)}\langle N-1, s|\hat{\psi}(\mathbf{r})|N\rangle\Theta(t'-t). \tag{4.5}$$

Here, E_N stands for the ground-state energy of the N-electron system, and $E_{N\pm1}^s$ for the sth excited-state energy of the $N\pm1$ electronic system.

Using the excitation energy ϵ_s and amplitude $\psi_s(\mathbf{r})$ defined by

$$\begin{aligned}\epsilon_s &= E_{N+1}^s - E_N, \quad \psi_s(\mathbf{r}) = \langle N|\hat{\psi}(\mathbf{r})|N+1, s\rangle, \quad \text{for } \epsilon_s \geqslant \mu,\\ \epsilon_s &= E_N - E_{N-1}^s, \quad \psi_s(\mathbf{r}) = \langle N-1, s|\hat{\psi}(\mathbf{r})|N\rangle, \quad \text{for } \epsilon_s < \mu,\end{aligned} \tag{4.6}$$

where μ is the chemical potential of the N-electron system ($\mu = E_{N+1} - E_N$),[c] we can further simplify Eq. (4.5) into the form

$$G(\mathbf{r}, \mathbf{r}'; t-t') = -i\sum_{s}\psi_s(\mathbf{r})\psi_s^*(\mathbf{r}')e^{-i\epsilon_s(t-t')}$$

$$[\Theta(t-t')\Theta(\epsilon_s - \mu) - \Theta(t'-t)\Theta(\mu - \epsilon_s)]. \tag{4.7}$$

---

[c]In Fig. 4.1, this corresponds to the energy of the lowest unoccupied state when $T \to 0$ K; for its derivation, see Ref. [174].

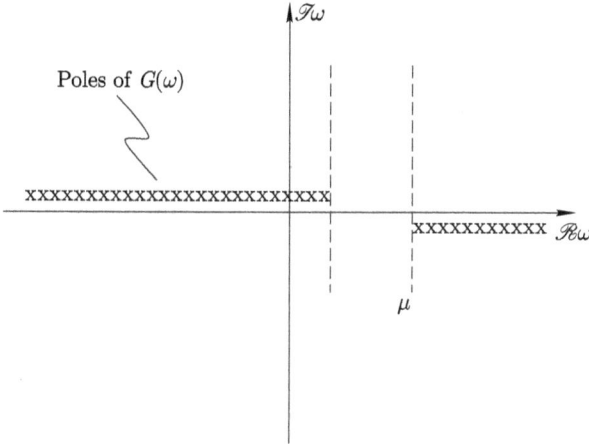

Figure 4.1   Position of the poles of the Green function (Eq. (4.8)) in the complex frequency plane. Those corresponding to the unoccupied states are slightly below the real frequency axis while those corresponding to the occupied states are slightly above it.

Performing a Fourier transform to the frequency axis, we obtain the spectral, or Lehmann [175], representation

$$G\left(\mathbf{r},\mathbf{r}',\omega\right) = \lim_{\eta\to0^+} \sum_s \psi_s\left(\mathbf{r}\right)\psi_s^*\left(\mathbf{r}'\right)\left[\frac{\Theta\left(\epsilon_s-\mu\right)}{\omega-\left(\epsilon_s-i\eta\right)} + \frac{\Theta\left(\mu-\epsilon_s\right)}{\omega-\left(\epsilon_s+i\eta\right)}\right].$$
(4.8)

The key feature of Eq. (4.8) is that the Green function has single poles corresponding to the exact excitation energies of the many-body system. For excitation energies larger (smaller) than the chemical potential, these singularities lie slightly below (above) the real axis in the complex frequency plane (Fig. 4.1).

It can be easily shown that in the non-interacting case, Eq. (4.8) reduces to

$$G_0\left(\mathbf{r},\mathbf{r}',\omega\right) = \lim_{\eta\to0^+} \sum_n \varphi_n\left(\mathbf{r}\right)\varphi_n^*\left(\mathbf{r}'\right)\left[\frac{\Theta\left(\epsilon_n-\epsilon_F\right)}{\omega-\left(\epsilon_n-i\eta\right)} + \frac{\Theta\left(\epsilon_F-\epsilon_n\right)}{\omega-\left(\epsilon_n+i\eta\right)}\right],$$
(4.9)

where $\epsilon_n(\varphi_n)$ is the eigenvalue (eigenfunction) of the single-particle Hamiltonian and ϵ_F is the Fermi energy.

4.1.2   The Dyson Equation

The time evolution of the field operator in the Heisenberg representation is given by the equation of motion:

$$i\frac{\partial}{\partial t}\hat{\psi}\left(\mathbf{r},t\right)=\left[\hat{\psi}\left(\mathbf{r},t\right),\hat{H}\right]\qquad(4.10)$$

with the Hamiltonian operator given by

$$\hat{H}=\int\mathrm{d}\mathbf{r}\mathrm{d}t\hat{\psi}^{\dagger}\left(\mathbf{r},t\right)\left[-\frac{1}{2}\nabla^{2}+V_{\mathrm{ext}}\left(\mathbf{r}\right)\right]\hat{\psi}\left(\mathbf{r},t\right)$$

$$+\frac{1}{2}\iint\mathrm{d}\mathbf{r}\mathrm{d}t\mathrm{d}\mathbf{r}'\mathrm{d}t'\hat{\psi}^{\dagger}\left(\mathbf{r},t\right)\hat{\psi}^{\dagger}\left(\mathbf{r}',t'\right)v\left(\mathbf{r},t;\mathbf{r}',t'\right)\hat{\psi}\left(\mathbf{r}',t'\right)\hat{\psi}\left(\mathbf{r},t\right),$$

$$(4.11)$$

where $v(\mathbf{r},t;\mathbf{r}',t') = \delta(t-t')/|\mathbf{r}-\mathbf{r}'|$ is the Coulomb interaction. By evaluating the commutator in Eq. (4.10), the equation of motion for the single-particle Green function can be obtained:

$$\left[i\frac{\partial}{\partial t}+\frac{1}{2}\nabla^{2}-V_{\mathrm{ext}}\left(\mathbf{r}\right)\right]G\left(\mathbf{r},t;\mathbf{r}',t'\right)$$

$$+i\int\mathrm{d}\mathbf{r}_{1}\frac{1}{|\mathbf{r}-\mathbf{r}_{1}|}\langle N|T[\hat{\psi}^{\dagger}\left(\mathbf{r}_{1},t\right)\hat{\psi}\left(\mathbf{r}_{1},t\right)\hat{\psi}\left(\mathbf{r},t\right)\hat{\psi}^{\dagger}\left(\mathbf{r}',t'\right)]|N\rangle$$

$$=\delta\left(\mathbf{r}-\mathbf{r}'\right)\delta\left(t-t'\right).\qquad(4.12)$$

The quantity in the integrand of the second term is the two-particle Green function. Following the same procedure to obtain the equation of motion for the two-particle Green function will give a term depending on the three-particle Green function, and so on.

To break this hierarchy, a mass operator can be introduced, defined by

$$\int\mathrm{d}\mathbf{r}_{1}\mathrm{d}t_{1}M\left(\mathbf{r},t;\mathbf{r}_{1},t_{1}\right)G\left(\mathbf{r}_{1},t_{1};\mathbf{r}',t'\right)$$

$$=-i\int\mathrm{d}\mathbf{r}_{1}v\left(\mathbf{r}-\mathbf{r}_{1}\right)\langle N|T[\hat{\psi}^{\dagger}\left(\mathbf{r}_{1},t\right)\hat{\psi}\left(\mathbf{r}_{1},t\right)\hat{\psi}\left(\mathbf{r},t\right)\hat{\psi}^{\dagger}\left(\mathbf{r}',t'\right)]|N\rangle.$$

$$(4.13)$$

With this operator, Eq. (4.12) can be rewritten as

$$\left[i\frac{\partial}{\partial t} + \frac{1}{2}\nabla^2 - V_{\text{ext}}(\mathbf{r})\right] G(\mathbf{r}, t; \mathbf{r}', t')$$

$$- \int d\mathbf{r}_1 dt_1 M(\mathbf{r}, t; \mathbf{r}_1, t_1) G(\mathbf{r}_1, t_1; \mathbf{r}', t') = \delta(\mathbf{r} - \mathbf{r}')\delta(t - t').$$

$$(4.14)$$

Since the Hartree interaction is a one-particle operator, it is usually separated from the mass operator M to define the self-energy, $\Sigma = M - V_{\text{H}}$. Replacing the mass operator in Eq. (4.14), we arrive at

$$\left[i\frac{\partial}{\partial t} - H_0(\mathbf{r})\right] G(\mathbf{r}, t; \mathbf{r}', t')$$

$$- \int d\mathbf{r}_1 dt_1 \Sigma(\mathbf{r}, t; \mathbf{r}_1, t_1) G(\mathbf{r}_1, t_1; \mathbf{r}', t') = \delta(\mathbf{r} - \mathbf{r}')\delta(t - t'),$$

$$(4.15)$$

where

$$H_0(\mathbf{r}) = -\frac{1}{2}\nabla^2 + V_{\text{ext}}(\mathbf{r}) + V_{\text{H}}(\mathbf{r}). \qquad (4.16)$$

In the Hartree approximation, Eq. (4.15) becomes

$$\left[i\frac{\partial}{\partial t} - H_0(\mathbf{r})\right] G_0(\mathbf{r}, t; \mathbf{r}', t') = \delta(\mathbf{r} - \mathbf{r}')\delta(t - t'). \qquad (4.17)$$

Multiplying Eq. (4.15) by G_0 on the left, using the hermiticity of the single-particle operator together with Eq. (4.17) and then integrating yields the well-known Dyson equation:

$$G(\mathbf{r}, t; \mathbf{r}', t') = G_0(\mathbf{r}, t; \mathbf{r}', t')$$

$$+ \iint d\mathbf{r}_1 dt_1 d\mathbf{r}_2 dt_2 \, G_0(\mathbf{r}, t; \mathbf{r}_2, t_2) \Sigma(\mathbf{r}_2, t_2; \mathbf{r}_1, t_1)$$

$$\times G(\mathbf{r}_1, t_1; \mathbf{r}', t'). \qquad (4.18)$$

Recurrently replacing G on the right-hand side by $G_0 + G_0 \Sigma G^d$ leads to the series expansion:

$$G = G_0 + G_0 \Sigma G_0 + G_0 \Sigma G_0 \Sigma G_0 + \cdots, \qquad (4.19)$$

---

[d]In this symbolic notation, products imply an integration, as a product of matrices with continuous indices, i.e. $AB = \int A(1,3)B(3,2)d3$.

which shows that the single-particle propagator $G(\mathbf{r}, t; \mathbf{r}', t')$ is equal to the "free" particle propagator $G_0(\mathbf{r}, t; \mathbf{r}', t')$ plus the sum of the probability amplitudes propagating from $(\mathbf{r}, t)$ to $(\mathbf{r}', t')$ after single, double, etc., scattering processes, with Σ playing the role of the scattering potential. Diagrammatically, this relation is shown as

where the double plain arrow represents the interacting Green function and the plain arrow represents the non-interacting one.

4.1.3   Self-Energy: Hedin Equations

For an electron propagating in a solid or condensed phase, the origin of the scattering processes lies in the Coulomb interaction with the Fermi sea. Thus, it is natural to expand the self-energy in terms of the bare Coulomb interaction. In the diagrams below, we show examples of some simple (low-order) scattering processes. Diagram (a) is a first-order scattering process that describes the propagating electron instantaneously exchanging its position with one electron from the Fermi sea via the Coulomb interaction. It corresponds to the exchange interaction. Solving the Dyson equation (4.15) including only this term in the self-energy and updating the Green function self-consistently yields the Hartree–Fock approximation. In Diagram (b), the interaction of the probe electron with the Fermi sea excites an electron out of it, generating an electron–hole pair, which annihilate each other at a later time, interacting again with the probe electron. This second-order scattering process, depicted in a "bubble" diagram, represents an electron repelling another from its neighborhood, thus generating a positive charge cloud around it. It is one of the simplest dynamical screening processes. In Diagram (c), the excited electron in the electron–hole pair of Diag. (b) further excites another electron–hole pair from the Fermi sea, changing the positive charge cloud around the probe electron again. Nevertheless, the long range of the bare Coulomb interaction results in a poor convergence of this expansion for the self-energy; in fact, it diverges for metals.

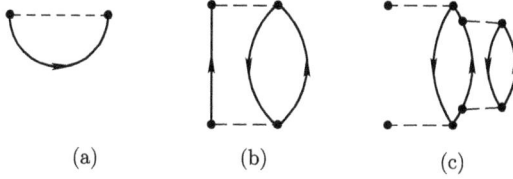

(a)   (b)   (c)

To solve this convergence problem related to the expansion of the self-energy in terms of the bare Coulomb interaction, in 1965, Hedin [176] proposed a different approach for obtaining this self-energy. In this approach, the self-energy is obtained by expanding it in terms of a dynamically screened Coulomb potential instead of the bare one. The derivation using the functional derivative technique can be found in Refs. [176–178]. Here, we just present the resulting set of equations:

$$\Gamma\left(1,2,3\right) = \delta\left(1,2\right)\delta\left(2,3\right) + \int d\left(4,5,6,7\right) \frac{\delta\Sigma\left(1,2\right)}{\delta G\left(4,5\right)} G\left(4,6\right)$$

$$\times G\left(7,5\right)\Gamma\left(6,7,3\right), \tag{4.20a}$$

$$P\left(1,2\right) = -i \int G\left(2,3\right)G\left(4,2\right)\Gamma\left(3,4,1\right)d\left(3,4\right), \tag{4.20b}$$

$$W\left(1,2\right) = v\left(1,2\right) + \int W\left(1,3\right)P\left(3,4\right)v\left(4,2\right)d\left(3,4\right), \tag{4.20c}$$

$$\Sigma\left(1,2\right) = i \int d\left(3,4\right)G\left(1,3^{+}\right)W\left(1,4\right)\Gamma\left(3,2,4\right). \tag{4.20d}$$

In these equations, we have used the Indo-Arabic numbers to denote the events and simplify the notation, e.g. $1 = \left(\mathbf{r}_1, t_1\right)$. Γ is a vertex function, P is the polarizability and W is the dynamically screened Coulomb potential. In Eq. (4.20a), the vertex function is written in terms of a four-point kernel (given by the functional derivative of the self-energy). Replacing the self-energy by the expression in Eq. (4.20d) would allow the expansion of the vertex function in terms of the screened Coulomb potential. For the purpose of this book, it will nevertheless be sufficient to represent it by a filled triangle:

Equations (4.20b)–(4.20d) can then be represented diagrammatically as

and

where the double wiggly line represents the screened Coulomb potential.

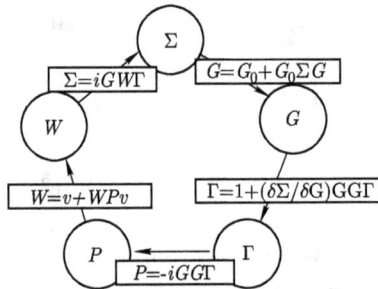

Figure 4.2   Schematic representation of the self-consistent solution of the Hedin equations in conjunction with the Dyson equation for the determination of the Green function (G) and the self-energy (Σ). Entries in boxes symbolize the mathematical relations that link Σ, G, Γ, P and W.

The set of Eq. (4.20), together with the Dyson Equation (4.18), constitute the definitive solution of the quantum mechanical many-body problem. One just needs to solve them self-consistently to obtain the single-particle Green function of the interacting system (see Fig. 4.2). However, one needs to note that a direct numerical solution is prevented by the functional derivative in Eq. (4.20a), and, as usual, one has to rely on approximations. This will be the subject of the second part of this chapter.

4.1.4   *The Quasi-Particle Concept*

Defining the excitation energies ϵ_s and amplitudes $\psi_s(\mathbf{r})$ (Eq. (4.6)) allowed us to write the Green function of the interacting system in the spectral representation (Eq. (4.8)). The expression obtained has the same form as the Green function of the non-interacting system (Eq. (4.9)), with the excitation energies (amplitudes) playing the role of the single-particle eigenvalues (eigenfunctions). We may ask, under which condition can the object defined by $\psi_s(\mathbf{r})$ and ϵ_s be interpreted as a "particle" that can be measured experimentally.

The experimentally obtained quantity in photoemission experiments is the spectral function,[e] i.e. the density of the excited states that contribute to the spectrum. For a finite system, it is defined by (Fig. 4.3(a))

$$A(\mathbf{r},\mathbf{r}';\omega) = \sum_s \psi_s(\mathbf{r})\,\psi_s^*(\mathbf{r})\,\delta(\omega - \epsilon_s), \tag{4.21}$$

and the Green function can be rewritten as

$$G(\mathbf{r},\mathbf{r}',\omega) = \lim_{\eta\to 0^+} \int d\omega' \frac{A(\mathbf{r},\mathbf{r}';\omega')}{\omega - \omega' + i\,\mathrm{sgn}(\omega' - \mu)\eta}. \tag{4.22}$$

Therefore, the interpretation of the excitation as a "particle" presents no difficulty in a finite system. Furthermore, inserting the expression for the Green function in terms of ϵ_s and $\psi_s(\mathbf{r})$ (Eq. (4.7)) in Eq. (4.15), it can be shown that they are solutions of the quasi-particle equation[f]

$$\left[-\frac{1}{2}\nabla^2 + V_{\text{ext}}(\mathbf{r}) + V_{\text{H}}(\mathbf{r})\right]\psi_s(\mathbf{r}) + \int d\mathbf{r}'\Sigma(\mathbf{r},\mathbf{r}';\epsilon_s)\,\psi_s(\mathbf{r}') = \epsilon_s\psi_s(\mathbf{r}). \tag{4.23}$$

In an extended system, the delta functions in Eq. (4.21) form a continuous spectrum (Fig. 4.3(b)). However, if in a given energy window, the spectrum can be described by a series of Lorenzian peaks with finite widths,

---

[e] Assuming the cross-section of the perturbation to be independent of the energy and neglecting experimental errors.
[f] This equation was first derived by Schwinger in Ref. [179]. It was applied to the many-body electronic system by Pratt in Refs. [180, 181] and later systematically by Hedin in Refs. [176, 182].

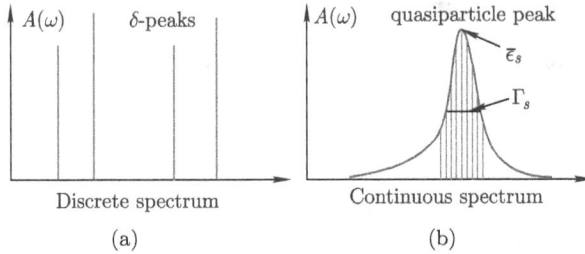

Figure 4.3   Spectral function for a discrete (a) and a continuous (b) spectrum.

so that the spectral *density* function can be written as

$$A(\mathbf{r}, \mathbf{r}'; \omega) = \sum_s \psi_s(\mathbf{r}) \psi_s^*(\mathbf{r}') \frac{\Gamma_s}{(\omega - \bar{\epsilon}_s)^2 + \Gamma_s^2}, \qquad (4.24)$$

where $\bar{\epsilon}_s$ is the center of the peak and Γ_s the width, Eq. (4.22) can be integrated analytically. Thus, the results of Eqs. (4.7) and (4.8) are recovered, provided one redefines $\epsilon_s = \bar{\epsilon}_s + i\Gamma_s$. In this case, the object defined by $\psi_s(\mathbf{r})$ and the complex ϵ_s is called a "quasi-particle". It describes the group behavior of a set of excitations with continuous excitation energies. The real part of ϵ_s corresponds to the average energy of these related excitations. The imaginary part leads to a decaying factor $e^{-\Gamma_s t}$, i.e. the excitation has a finite lifetime given by $\tau = \Gamma_s^{-1}$. That the quasi-particle "disappears" can be physically understood, taking into account that one is dealing with an infinite system. In other words, the quasi-particle can decay to the "infinite" reservoir. The quasi-particle equation (4.23) remains valid, provided one performs an analytic continuation of the self-energy to the complex frequency plane. A more detailed discussion on the subject can be found in Ref. [183].

4.2 *GW* Approximation

In the above discussions, we have shown that in descriptions of the single-particle excitation of a many-body system, the key quantity is the self-energy and the key equation is the Dyson equation. The self-energy contains all the complexity of the many-body interactions and in practical simulations, one must resort to approximations of this quantity. The simplest approximation of this quantity, which contains the dynamic feature of the screened Coulomb interaction, is the so-called *GW* approximation. It was first proposed by Hedin in 1965 [176]. Mathematically, it amounts to taking the zeroth-order expansion of the vertex function in terms of W. Thus, we are left with

$$\Gamma\left(1,2,3\right) = \delta\left(1,2\right)\delta\left(2,3\right), \tag{4.25a}$$

$$P\left(1,2\right) = -iG\left(1,2\right)G\left(2,1\right), \tag{4.25b}$$

$$W\left(1,2\right) = v\left(1,2\right) + \int d\left(3,4\right)W\left(1,3\right)P\left(3,4\right)v\left(4,2\right), \tag{4.25c}$$

$$\Sigma\left(1,2\right) = iG\left(1,2\right)W\left(1^{+},2\right). \tag{4.25d}$$

Diagrammatically, the three-point Γ function is collapsed into a point. The elementary unit in this set of equations is the bubble diagram of the polarizability operator:

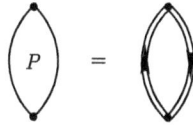

This approximation for the polarizability is known as the random phase approximation (RPA). Physically, it represents the polarization generated by the creation and annihilation of a dressed electron–hole pair, while the interaction between the (dressed) electron and hole is neglected. In other words, scattering processes where the electron or the hole in the electron–hole pair interacts with the medium are taken into account. For example, the process represented by the following diagram:

can be included. However, processes like

where the electron and the hole interact with each other are neglected.

The screened Coulomb interaction resulting from Eq. (4.25c) is the same as in Eq. (4.20c):

except that now the polarizability is represented in the RPA.

In Eq. (4.25d), the self-energy is written as a product of the Green function and the screened Coulomb interaction diagrammatically as

The shape of this diagram is similar to the Hartree–Fock approximation, with the instantaneous bare Coulomb potential replaced by the dynamically screened Coulomb one. This approximation to the self-energy includes processes represented, for example, by the diagrams:

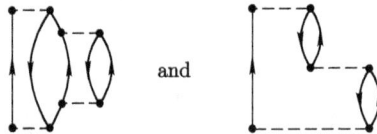

and

through the screened Coulomb potential. Also, processes like

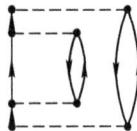

are included through the interacting Green function. However, diagrams like

where the added electron interacts with that of the electron–hole pair are neglected.

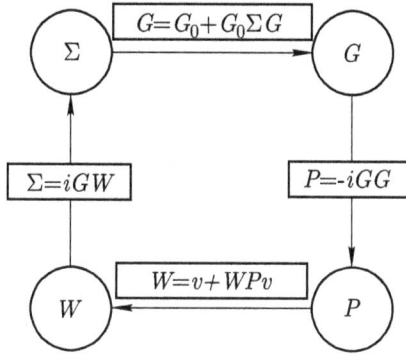

Figure 4.4   Schematic representation of the self-consistent solution of Hedin equations in *GW* approximation. Entries in boxes symbolize the mathematical relations that link Σ, G, P and W.

In Fig. 4.4, we show a sketch of the self-consistent procedure required to solve the *GW* equations (4.25) together with the Dyson equation. However, this procedure is still extremely computationally demanding and rarely carried out.

4.3 $G_0 W_0$ Approximation

A popular method in avoiding the self-consistent solution of the Dyson equation, which contains the essence of the many-body interactions at the *GW* level in descriptions of the self-energy, is the so-called $G_0 W_0$ approximation. The key point of this approximation is the assumption that one can count on an effective single-particle potential $V^{\text{xc}}(\mathbf{r})$, which contains some of the XC effects in a many-body system and approximates the self-energy reasonably well. In other words, the solutions of the single-particle equation

$$\hat{H}_{\text{eff}}(\mathbf{r})\varphi_i(\mathbf{r}) = \epsilon_i\varphi_i(\mathbf{r}) \tag{4.26}$$

with

$$\hat{H}_{\text{eff}}(\mathbf{r}) = -\frac{1}{2}\nabla^2 + V_{\text{ext}}(\mathbf{r}) + V^{\text{H}}(\mathbf{r}) + V^{\text{xc}}(\mathbf{r}) \tag{4.27}$$

are such that $\varphi_i(\mathbf{r}) \approx \psi_s(\mathbf{r})$, $\epsilon_i \approx \Re(\epsilon_s)$ ($\psi_s(\mathbf{r})$ and ϵ_s are the solutions of Eq. (4.23)).

For a convenient understanding of the perturbative treatment underlying this approximation, we can rewrite the quasi-particle equation (4.23) as

$$\left[-\frac{1}{2}\nabla^2 + V_{\text{ext}}\left(\mathbf{r}\right) + V^{\text{H}}\left(\mathbf{r}\right) + V^{\text{xc}}\left(\mathbf{r}\right) \right] \psi_s\left(\mathbf{r}\right)$$

$$+ \int \mathrm{d}\mathbf{r}' \Delta\Sigma\left(\mathbf{r}, \mathbf{r}'; \epsilon_s\right) \psi_s\left(\mathbf{r}'\right) = \epsilon_s \psi_s\left(\mathbf{r}\right), \qquad (4.28)$$

where

$$\Delta\Sigma\left(\mathbf{r}, \mathbf{r}'; \epsilon_s\right) = \Sigma\left(\mathbf{r}, \mathbf{r}'; \epsilon_s\right) - V^{\text{xc}}\left(\mathbf{r}'\right)\delta(\mathbf{r} - \mathbf{r}'). \qquad (4.29)$$

Since, according to our assumption, the correction due to $\Delta\Sigma$ is small, one can obtain the quasi-particle energies by applying the first-order perturbation theory through

$$\psi_i(\mathbf{r}) = \varphi_i(\mathbf{r}),$$

$$\epsilon_i^{\text{qp}} = \epsilon_i + \langle \varphi_i\left(\mathbf{r}_1\right) | \Re\left[\Delta\Sigma\left(\mathbf{r}_1, \mathbf{r}_2; \epsilon_i^{\text{qp}}\right)\right] | \varphi_i\left(\mathbf{r}_2\right) \rangle. \qquad (4.30)$$

Taking the self-energy in the GW approximation, and further assuming that the non-interacting Green function G_0 corresponding to $\hat{H}_{\text{eff}}$ is a good approximation to the interacting one, the self-energy can be calculated through

$$P_0\left(1, 2\right) = -iG_0\left(1, 2\right)G_0\left(2, 1\right), \qquad (4.31a)$$

$$W_0\left(1, 2\right) = v\left(1, 2\right) + \int d\left(3, 4\right) W_0\left(1, 3\right) P\left(3, 4\right) v\left(4, 2\right), \qquad (4.31b)$$

$$\Sigma\left(1, 2\right) = iG_0\left(1, 2\right) W_0\left(1^+, 2\right). \qquad (4.31c)$$

With these, one can easily calculate the quasi-particle corrections to the single-particle excitation energy in Eq. (4.30).

We note that this treatment of the quasi-particle excitation energies is obviously not done in a self-consistent manner, as illustrated pictorially by the diagram in Fig. 4.4. Due to the fact that the self-energy is represented by the product of G_0 and W_0, this further approximation beyond the self-consistent treatment of the GW approximation is known as the G_0W_0 approximation in the literature.

With this framework of the G_0W_0 approximation set up, the next thing we need to do is to find a good zeroth-order approximation for the quasi-particle energies and wave functions. The KS scheme to the density-functional theory naturally provides such a good zeroth-order approximation since its single-particle effective potential includes quite some XC effects. Thus, one usually starts from the KS Green function:

$$G_0(1,2) = -i \sum_j \varphi_j^{KS}(\mathbf{r}_1) \, \varphi_j^{KS*}(\mathbf{r}_2) \, e^{-i\epsilon_j^{KS}(t_1-t_2)}$$

$$\cdot \left[\Theta(t_1 - t_2) \, \Theta\left(\epsilon_j^{KS} - \mu\right) - \Theta(t_2 - t_1) \, \Theta\left(\mu - \epsilon_j^{KS}\right) \right]. \qquad (4.32)$$

We note that other fictitious single-particle states can also be used as the starting guess of the quasi-particle states. The results depend on this starting guess and normally correct them in the right direction. For example, the LDA-based KS orbitals give smaller band gaps in molecules and semi-conductors (in the case of a molecule, it is the difference between the lowest unoccupied molecular state energy and the highest occupied molecular state energy). After the G_0W_0 correction, these band gaps increase. On the other hand, when the Hartree–Fock band gaps are used, the starting point has larger band gaps than those in experiments. Then when the G_0W_0 correction is used, these values decrease toward the experimental results. In spite of this dependence, the KS orbitals are still the most popular choice in calculations of the G_0W_0 corrections. As an example, in Fig. 4.5, we show how the LDA-based KS band diagram is compared with the G_0W_0 band diagram in Si. It is clear that after the G_0W_0 correction is added, the agreement between the theoretically calculated band diagram (in an *ab initio* manner) and experiment significantly improves.

For the completeness of this introduction to the *GW* method, in the following, we use three more paragraphs to summarize the history behind the development of this method and its present status.

The first application of the G_0W_0 method to real materials was carried out by Hybertsen and Louie in 1985 [185]. Since then, it has achieved an impressive success in reproducing the experimental single-particle excitation spectra for a wide range of systems from simple metals [186, 187] to weakly correlated semiconductors and insulators [188–190]. Nowadays, it has become the standard method used to calculate the single-electron excitations of condensed matter systems. For these early implementations, one notes that the pseudopotential method and its associated plane waves for the expansion of the wave functions were often resorted to due to their simplicity of implementation and computational efficiency. In the last 10 years, all-electron method-based implementations of the

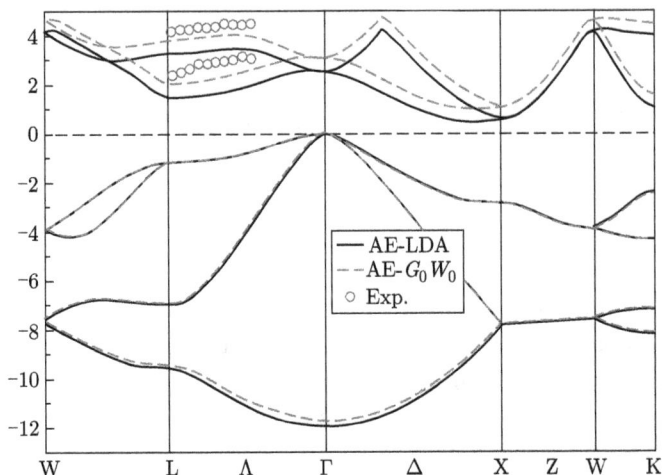

Figure 4.5   Band diagram of Si (in eV, referenced to the valence band maximum) obtained from the all-electron LDA (labels as AE-LDA in solid lines) and G_0W_0 (labelled as AE-G_0W_0 in dash lines) calculations. The experimental results (circles) are taken from Ref. [184].

GW approximation also appeared [160–163, 178, 191–193]. The earliest results for simple semiconductors such as Si and GaAs show that when the all-electron method is used, the G_0W_0 band gaps of these semiconductors become smaller compared with the pseudopotential-based results [160, 191] (see Table 4.1), and their agreement with the experimental values gets worse.

For a better understanding of this discrepancy, we take Si as the guiding example and explain how its fundamental G_0W_0 band gap evolves with time in the last 30 years. Early pseudopotential-based G_0W_0 calculations give values between 1.18 eV and 1.29 eV in the 1980s and 1990s [188, 194–196]. This is within 0.12 eV accuracy compared with the experimental value of 1.17 eV. However, after 2000, all-electron G_0W_0 calculations obtained 0.85 eV [160] and 0.90 eV [191]. These results triggered a debate about the reliability of both pseudopotential and all-electron based G_0W_0 results [198, 199]. In Ref. [160], this discrepancy was assigned mainly to the exclusion of the core electrons in the calculation of the self-energy, and the possible role of pseudoization was tangentially mentioned. On the other hand, the main criticism to the value of 0.85 eV in Ref. [160], put forward in Ref. [198], was the lack of convergence with respect to

Table 4.1   Comparison of band gaps from some reported all-electron and pseudopotential-based G_0W_0 calculations.

(Unit: eV)

	Expt	All-electron-G_0W_0[a]	Traditional PP-G_0W_0
C	5.50	5.49	5.54[c]
Si	1.17	0.90, 0.85[b]	1.18[c], 1.19[d], 1.24[e], 1.42[f]
AlAs($\Gamma - X$)	2.23	1.68	2.08[f]
AlAs($\Gamma - \Gamma$)	3.13	2.69	2.75[f]
AlP	2.50	2.15	2.59[g]
AlSb	1.69	1.32	1.64[g]
Ge($\Gamma - L$)	0.74	0.47, 0.51[a]	0.75[f]
Ge($\Gamma - \Gamma$)	0.89	0.79, 1.11[a]	0.71[f]
GaAs	1.63	1.42	1.29[f]
GaP($\Gamma - X$)	2.35	1.90	2.55[h]
GaP($\Gamma - \Gamma$)	2.86	2.53	2.93[h]
GaSb	0.82	0.49	0.62[g]
InP	1.42	1.25	1.44[g]
InAs	0.42	0.32	0.40[g]
InSb	0.24	0.32	0.18[g]
ZnS	3.80	3.22	3.98[i]
ZnSe	2.80	2.21	2.84[i]
CdTe	1.61	1.07	1.76[i]

Notes: [a]Ref. [191], [b]Ref. [160], [c]Ref. [194], [d]Ref. [195], [e]Ref. [196], [f]Ref. [188], [g]Ref. [190], [h]Ref. [197], [i]Ref. [189].

the number of unoccupied states included in the calculation.[g] The validity of this criticism has been confirmed by the all-electron G_0W_0 result from Friedrich *et al.* [192], who obtained, after careful convergence with respect to the number of unoccupied states, a fundamental band gap of 1.05 eV in Si. This trend was later supported by another LAPW-based G_0W_0 calculation in Refs. [162, 163], which gives a value of 1.00 eV. Meanwhile, the value of 0.90 eV in Ref. [191], based on the LMTO method, has been further increased to 0.95 eV in Ref. [200] after including local orbitals. From the pseudopotential perspective, in 2004, Tiago *et al.* [201] performed a set of pseudopotential calculations for Si in which only the $1s$ orbitals are treated as core, obtaining a fundamental band gap of 1.04 eV for Si. With these, it

---

[g]Related to other arguments put forward in this discussion, namely the cancellation of errors between lack of self-consistency and absence of vertex correction, we consider them pertinent to support the G_0W_0 approach itself, but irrelevant to an explanation of the differences between all-electron and pseudopotential-based calculations.

seems clear that the earlier reported all-electron results in Refs. [160, 191] should have some problems related to its convergence with respect to the number of unoccupied states included in the calculations of the self-energy. When this issue is taken care of, larger band gaps and smaller discrepancy with the early pseudopotential-based results should be expected. However, one notes that differences on the order of 0.1 eV still remain in many of the simple semiconductors. Physically, there is no other reason to explain such a discrepancy, except for the intrinsic approximations underlying the pseudopotential method. In Refs. [162, 163, 193], a systematic analysis on how these approximations underlying the pseudopotential-based GW calculations are given.

In spite of this clarification, we note that there is still no perfect method for calculating the single-particle excitation energy from the theoretical perspective due to the different approximations we have already made in this chapter for the simplification of the many-body perturbation theory into a practical scheme. These approximations include, most seriously, the neglect of the vertex correction. Concerning the self-consistent solution of the Dyson equation within GW approximation, we acknowledge that great progress has been made in the last few years [202–211]. From a practical perspective, the pseudopotential-based G_0W_0 method prevails nowadays due to its efficiency and well-defined physical meaning for the interpretation of the quasi-particle energies, especially in calculations of large systems. In some solids, when the all-electron method is applicable and its error is controllable, the all-electron method can be used to set a benchmark for the pseudopotential method-based calculations. However, application of this GW method to real poly-atomic systems and real problems in physics, to a large extent, still relies on a good combination of these different kinds of implementation.

4.4   Numerical Implementation of an All-Electron G_0W_0 Code: FHI-Gap

In the earlier discussions, we have explained the principles underlying the current state-of-the-art single-particle excitation theory. In the following, we take FHI-gap, an implementation of the G_0W_0 approximation in which one of the authors of this book (XZL) has participated, as an example to show how these methods are implemented in a computer code. To be simple, this introduction will focus on explaining some features which are common in an all-electron implementation of the G_0W_0 method, even with other basis sets. Therefore, the explanations of the technical details are far from being complete. For these technical details, we strongly recommend the readers to Ref. [212], where a very complete explanation of algorithms is given.[h]

The basis set with which the KS wave functions are expanded is the so-called (Linearized) Augmented Plane Wave plus local orbitals ((L)APW+lo) [167, 168], as explained in the previous chapter. Due to its combined abilities to provide the most reliable results for periodic systems within DFT and to address the widest range of materials, this method is a FP method currently considered to be a benchmark in the DFT calculations. Consequently, the development of a G_0W_0 code has constituted a demanding task. The whole process has been imbued with the compromise between computational efficiency and the numerical precision necessary to achieve the ambitious goals of reliability and wide applicability complying with the FP-(L)APW+lo standards.

Among the different existing implementations of the method, the Wien2k code [213] is used as the basis on which the G_0W_0 correction is applied, although we acknowledge that nowadays it has been extended so that other LAPW-based DFT codes such as Exciting [214–216] can also be used to generate the input KS orbitals. In Sec. 4.4.1, we present a summary of the G_0W_0 equations to be implemented. The representation of the non-local operators (polarization, dielectric function, bare and screened Coulomb potentials) requires an efficient basis set, able to address extended as well as localized valence states and core states. An optimized set of functions consisting of plane waves in the interstitial region and a spherical harmonics expansion within the MT spheres can be used [178], which will be introduced in Sec. 4.4.2. In Sec. 4.4.3, we summarize the matrix form of the G_0W_0 equations after expansion in the mixed basis.

Calculating the polarization for a given wave vector **q** requires an integration over all possible transitions from occupied to unoccupied states and vice versa which conserve the total wave number. In other words, a precise **q**-dependent Brillouin-zone integration is required. The efficiency of the linear tetrahedron method is comparable to the special points methods

---

[h]Reference [212] is the Ph.D. thesis of XZL which is available online as "All-electron G_0W_0 code based on FP-(L)APW+lo and applications, Ph.D. Thesis, Free University of Berlin, 2008". Readers may contact XZL via xzli@pku.edu.cn if the thesis is inaccessible.

for semiconductors and insulators, while it is clearly superior for metallic systems. To be able to treat the widest possible range of materials, the linear tetrahedron method is extended to the **q**-dependent case. A description of this development, together with the special requirements for the polarization, are described in Sec. 4.4.4. The frequency convolution of the correlation self-energy is presented in Sec. 4.4.5. This chapter concludes with a flowchart of this FHI-gap code in Sec. 4.4.6.

4.4.1 *Summary of the G_0W_0 Equations*

In the G_0W_0 approach, the quasi-particle energy $\epsilon_{n,\mathbf{k}}^{\mathrm{qp}}$ is obtained from a first-order correction to the KS energy eigenvalue $\epsilon_{n,\mathbf{k}}$ through

$$\epsilon_{n,\mathbf{k}}^{\mathrm{qp}} = \epsilon_{n,\mathbf{k}} + \langle \varphi_{n,\mathbf{k}}(\mathbf{r}_1) | \Re[\Sigma(\mathbf{r}_1,\mathbf{r}_2;\epsilon_{n,\mathbf{k}}^{\mathrm{qp}})] - V^{\mathrm{xc}}(\mathbf{r}_1)\delta(\mathbf{r}_1 - \mathbf{r}_2) | \varphi_{n,\mathbf{k}}(\mathbf{r}_2) \rangle, \tag{4.33}$$

where $\varphi_{n,\mathbf{k}}(\mathbf{r})$ and V^{xc} are the KS eigenfunctions and XC potential, respectively. The self-energy $\Sigma(\mathbf{r}_1,\mathbf{r}_2;\omega)$ is obtained from the Fourier transform of Eq. (4.31c):

$$\Sigma(\mathbf{r}_1,\mathbf{r}_2;\omega) = \frac{i}{2\pi}\int G_0(\mathbf{r}_1,\mathbf{r}_2;\omega+\omega')W_0(\mathbf{r}_2,\mathbf{r}_1;\omega')d\omega', \tag{4.34}$$

where G_0 is the Green's function in the KS scheme defined by

$$G_0(\mathbf{r}_1,\mathbf{r}_2;\omega) = \sum_{n,\mathbf{k}} \frac{\varphi_{n,\mathbf{k}}(\mathbf{r}_1)\varphi_{n,\mathbf{k}}^*(\mathbf{r}_2)}{\omega - \epsilon_{n,\mathbf{k}} \pm i\eta}, \tag{4.35}$$

and the dynamically screened Coulomb potential $W_0(\mathbf{r}_2,\mathbf{r}_1;\omega)$ is given by

$$W_0(\mathbf{r}_1,\mathbf{r}_2;\omega) = \int \varepsilon^{-1}(\mathbf{r}_1,\mathbf{r}_3;\omega)v(\mathbf{r}_3,\mathbf{r}_2)d\mathbf{r}_3. \tag{4.36}$$

$\varepsilon(\mathbf{r}_1,\mathbf{r}_2;\omega)$ is the dielectric function, it can be calculated from

$$\varepsilon(\mathbf{r}_1,\mathbf{r}_2;\omega) = 1 - \int v(\mathbf{r}_1,\mathbf{r}_3)P(\mathbf{r}_3,\mathbf{r}_2;\omega)d\mathbf{r}_3, \tag{4.37}$$

where the polarizability $P(\mathbf{r}_1,\mathbf{r}_2;\omega)$, in the random phase approximation (RPA), is written as

$$P(\mathbf{r}_1,\mathbf{r}_2;\omega) = -\frac{i}{2\pi}\int G_0(\mathbf{r}_1,\mathbf{r}_2;\omega+\omega')G_0(\mathbf{r}_2,\mathbf{r}_1;\omega')d\omega'. \tag{4.38}$$

The self-energy can be separated into the exchange and correlation terms. If we define

$$W_0^c(\mathbf{r}_1, \mathbf{r}_2; \omega) = W_0(\mathbf{r}_1, \mathbf{r}_2; \omega) - v(\mathbf{r}_1, \mathbf{r}_2), \qquad (4.39)$$

where $v(\mathbf{r}_1, \mathbf{r}_2) = \frac{1}{|\mathbf{r}_1 - \mathbf{r}_2|}$ is the bare Coulomb potential, the exchange and correlation parts of the self-energy can be calculated from

$$\Sigma^x(\mathbf{r}_1, \mathbf{r}_2) = \frac{i}{2\pi} \int G_0(\mathbf{r}_1, \mathbf{r}_2; \omega') v(\mathbf{r}_2, \mathbf{r}_1) d\omega'$$

$$= \sum_{n,\mathbf{k}}^{\mathrm{occ}} \varphi_{n,\mathbf{k}}(\mathbf{r}_1) v(\mathbf{r}_2, \mathbf{r}_1) \varphi_{n,\mathbf{k}}^*(\mathbf{r}_2) \qquad (4.40)$$

and

$$\Sigma^c(\mathbf{r}_1, \mathbf{r}_2; \omega) = \frac{i}{2\pi} \int G_0(\mathbf{r}_1, \mathbf{r}_2; \omega + \omega') W_0^c(\mathbf{r}_2, \mathbf{r}_1; \omega') d\omega' \qquad (4.41)$$

separately.

The required input for solving this set of equations are the eigenfunctions, $(\varphi_{n,\mathbf{k}}(\mathbf{r}))$, eigenvalues $(\epsilon_{n,\mathbf{k}})$ and the exchange–correlation potential $(V_{\mathrm{xc}}(\mathbf{r}))$ of the KS orbitals, and these data can be obtained from a self-consistent DFT calculation, and in this particular case, using the Wien2k code [213].

4.4.2   *The Mixed Basis*

For periodic systems, the reciprocal space representation improves the efficiency by exploiting explicitly the translational symmetry of the Bravais lattice. However, a direct Fourier transform of the operators, which implies taking plane waves as a basis set, is computationally inefficient for their representation in a FP, all-electron implementation (see Chapter 3). Analogous to the proposal of Kotani and van Schilfgaarde [191], an optimized set of functions satisfying Blöch's theorem can be used for the expansion of the operators as summarized in Sec. 4.4.1. This "mixed" basis set uses the space partition in the MT and interstitial regions following the APW philosophy. In this section, we introduce how this mixed basis set is defined.

By replacing the expression of Eq. (4.35) for Green's function in Eq. (4.38), one can see that the spatial dependence of the non-local polarizability on each coordinate is a product of two KS wave functions. Thus, the basis set one chooses to expand this polarizability should be efficient in representing those products of two KS wave functions. From Sec. 3.2, we know that the KS wave functions are linear combinations of (L)APW+lo basis. In other words, they are expanded in terms of spherical harmonics in the MT spheres and plane waves in the interstitial region. The product of two plane waves is still a plane wave and the product of two spherical harmonics can be expanded in spherical harmonics using the Clebsch–Gordan coefficients. Therefore, the same kind of space partition can be used. As a matter of fact, this method is exactly what has been taken in FHI-gap to define the basis set for the non-local operators used in the $G_0 W_0$ calculations. Inside the MT sphere of atom α, the basis functions are defined as

$$\gamma_{\alpha,N,L,M}\left(\mathbf{r}^\alpha\right) = v_{\alpha,N,L}\left(r^\alpha\right) Y_{L,M}\left(\mathbf{r}^\alpha\right). \tag{4.42}$$

To obtain an optimal set of radial functions $v_{\alpha,N,L}\left(r^\alpha\right)$, one proceeds as follows:

- $\dot{u}_l(r)$'s are not taken into account because $\int |\dot{u}_l(r)|^2 r^2 dr$ is typically less than 10% of $\int |u_l(r)|^2 r^2 dr$. Possible errors will be taken care of by the other basis functions.
- To truncate the expansion, one takes a maximum $l_{\max}$ for the choice of $u_l(r^\alpha)$'s.
- For each L in $v_{\alpha,N,L}(r^\alpha)$, one takes all the products of two radial functions $u_l(r^\alpha)u_{l'}(r^\alpha)$ which fulfill the triangular condition $|l - l'| \leqslant L \leqslant l + l'$.
- The overlap matrix between this set of product radial functions is calculated by

$$\mathbb{O}_{(l,l');(l_1,l_1')} = \int_0^{R_{\mathrm{MT}}^\alpha} u_{\alpha,l}(r^\alpha)u_{\alpha,l'}(r^\alpha)u_{\alpha,l_1}(r^\alpha)u_{\alpha,l_1'}(r^\alpha)(r^\alpha)^2 dr^\alpha. \tag{4.43}$$

- The matrix $\mathbb{O}_{(l,l');(l_1,l_1')}$ is diagonalized, obtaining the corresponding set of eigenvalues λ^N and eigenvectors $\{c_{l,l'}^N\}$.
- Eigenvectors corresponding to eigenvalues (λ^N) smaller than a certain tolerance $\lambda_{\min}$ are assumed to be linear-dependent and discarded.
- The remaining eigenvectors, after normalization, constitute the radial basis set: $v_{\alpha,N,L}(r^\alpha) = \sum_{l,l'} c_{l,l'}^N u_l(r^\alpha)u_{l'}(r^\alpha)$.

So defined, the functions $\{\gamma_{\alpha,N,L,M}\}$ constitute an orthonormal basis set. The translational symmetry of the lattice is imposed by taking the

Blöch summation:

$$\gamma^{\mathbf{q}}_{\alpha,N,L,M}(\mathbf{r}) = \frac{1}{\sqrt{N_c}} \sum_{\mathbf{R}} e^{i\mathbf{q}\cdot(\mathbf{R}+\mathbf{r}_\alpha)} \gamma_{\alpha,N,L,M}(\mathbf{r}^\alpha), \qquad (4.44)$$

where $\mathbf{r}_\alpha$ is the position of atom α in the unit cell, and $\mathbf{R}$ is a Bravais lattice vector.

Since the interstitial plane waves are not orthogonal, one can diagonalize the overlap matrix by solving the eigenvalue equation:

$$\sum_{\mathbf{G}'} \mathbb{O}_{\mathbf{G},\mathbf{G}'} S_{\mathbf{G}',i} = \varepsilon_i S_{\mathbf{G},i}, \qquad (4.45)$$

where $\mathbb{O}^{\mathbf{q}}_{\mathbf{G},\mathbf{G}'} \equiv \langle P^{\mathbf{q}}_{\mathbf{G}} | P^{\mathbf{q}}_{\mathbf{G}'} \rangle$ and $\langle \mathbf{r} | P^{\mathbf{q}}_{\mathbf{G}} \rangle = \frac{1}{\sqrt{V}} e^{i(\mathbf{G}+\mathbf{q})\cdot\mathbf{r}}$.

The orthogonal basis set in the interstitial region is defined by

$$\tilde{P}^{\mathbf{q}}_i(\mathbf{r}) \equiv \sum_{\mathbf{G}} \tilde{S}_{\mathbf{G},i} P^{\mathbf{q}}_{\mathbf{G}}(\mathbf{r}). \qquad (4.46)$$

where $\tilde{S}_{\mathbf{G},i} = \frac{1}{\sqrt{\varepsilon_i}} S_{\mathbf{G},i}$ so that the orthogonal interstitial plane waves are normalized. The plane wave expansion can be truncated at a certain $G_{\max}$. In fact, a parameter Q that defines $G_{\max}$ in units of $G^{\mathrm{LAPW}}_{\max}$ (the plane wave cut-off of the LAPW basis functions) can be introduced. Finally, the orthonormal mixed basis set is

$$\{\chi^{\mathbf{q}}_j(\mathbf{r})\} \equiv \{\gamma^{\mathbf{q}}_{\alpha,N,L,M}(\mathbf{r}), \tilde{P}^{\mathbf{q}}_{\mathbf{G}}(\mathbf{r})\}. \qquad (4.47)$$

4.4.3   *Matrix Form of the G_0W_0 Equations*

The basis set as introduced above was derived from the requirement of efficiency to expand products of KS eigenfunctions. The principal quantity for such expansion, and also a central quantity for the whole implementation are the matrix elements:

$$M^i_{n,m}(\mathbf{k},\mathbf{q}) \equiv \int_\Omega [\tilde{\chi}^{\mathbf{q}}_i(\mathbf{r}) \varphi_{m,\mathbf{k}-\mathbf{q}}(\mathbf{r})]^* \varphi_{n,\mathbf{k}}(\mathbf{r}) d\mathbf{r}. \qquad (4.48)$$

Following the expansion of non-local operators in periodic systems, we can write the Coulomb potential matrix elements in the mixed basis as

$$v_{i,j}(\mathbf{q}) = \int_\Omega \int_\Omega (\chi^{\mathbf{q}}_i(\mathbf{r}_1))^* \sum_{\mathbf{R}} v(\mathbf{r}_1,\mathbf{r}_2-\mathbf{R}) e^{-i\mathbf{q}\cdot\mathbf{R}} \chi^{\mathbf{q}}_j(\mathbf{r}_2) d\mathbf{r}_2 d\mathbf{r}_1. \qquad (4.49)$$

Using the matrix element defined in Eq. (4.48), the polarizability can be calculated by

$$P_{i,j}(\mathbf{q}, \omega) = \sum_{\mathbf{k}}^{BZ} \sum_{n}^{occ} \sum_{n'}^{unocc} M_{n,n'}^{i}(\mathbf{k}, \mathbf{q})[M_{n,n'}^{j}(\mathbf{k}, \mathbf{q})]^*$$

$$\cdot \left\{ \frac{1}{\omega - \epsilon_{n',\mathbf{k}-\mathbf{q}} + \epsilon_{n,\mathbf{k}} + i\eta} - \frac{1}{\omega - \epsilon_{n,\mathbf{k}} + \epsilon_{n',\mathbf{k}-\mathbf{q}} - i\eta} \right\}.$$

(4.50)

To avoid the divergence at $\mathbf{q} = 0$ of the dielectric function as defined in Eq. (4.37), one resorts to the symmetrized dielectric function[i] defined as

$$\tilde{\varepsilon}_{i,j}(\mathbf{q}, \omega) = \delta_{i,j} - \sum_{l,m} v_{i,l}^{\frac{1}{2}}(\mathbf{q}) P_{l,m}(\mathbf{q}, \omega) v_{m,j}^{\frac{1}{2}}(\mathbf{q}).$$

(4.51)

Using this equation, the correlation term of the screened Coulomb interaction can be calculated through

$$W_{i,j}^{c}(\mathbf{q}, \omega) = \sum_{l,m} v_{i,l}^{\frac{1}{2}}(\mathbf{q}) [\tilde{\varepsilon}_{l,m}^{-1} - \delta_{l,m}](\mathbf{q}, \omega) v_{m,j}^{\frac{1}{2}}.$$

(4.52)

The diagonal matrix element of the self-energy in the basis of the KS states is

$$\Sigma_{n,\mathbf{k}}^{c}(\omega) = \langle \varphi_{n,\mathbf{k}} | \Sigma^{c}(\mathbf{r}_1, \mathbf{r}_2; \omega) | \varphi_{n,\mathbf{k}} \rangle$$

$$= \frac{i}{2\pi} \sum_{\mathbf{q}}^{BZ} \sum_{i,j} \sum_{n'} [M_{n,n'}^{i}(\mathbf{k}, \mathbf{q})]^* M_{n,n'}^{j}(\mathbf{k}, \mathbf{q})$$

$$\cdot \int_{-\infty}^{\infty} \frac{W_{i,j}^{c}(\mathbf{q}, \omega')}{\omega + \omega' - \epsilon_{n',\mathbf{k}-\mathbf{q}} \pm i\eta} d\omega'$$

(4.53)

for the correlation term, and

$$\Sigma_{n,\mathbf{k}}^{x} = \langle \varphi_{n,\mathbf{k}} | \Sigma^{x}(\mathbf{r}_1, \mathbf{r}_2) | \varphi_{n,\mathbf{k}} \rangle$$

$$= -\sum_{\mathbf{q}}^{BZ} \sum_{i,j} v_{i,j}(\mathbf{q}) \sum_{n'}^{occ} [M_{n,n'}^{i}(\mathbf{k}, \mathbf{q})]^* M_{n,n'}^{j}(\mathbf{k}, \mathbf{q})$$

(4.54)

for the exchange term.

---

[i]For more details about this symmetrized dielectric function and how the divergence of the related screened Coulomb potential is treated for its contribution to the self-energy, please refer to Ref. [212].

The quasi-particle energies should be obtained by solving Eq. (4.33) self-consistently, since the self-energy depends on the quasi-particle energy $\epsilon_{n,\mathbf{k}}^{\mathrm{qp}}$. Usually, a first-order Taylor expansion of the self-energy around $\epsilon_{n,\mathbf{k}}^{\mathrm{KS}}$ is used instead. The quasi-particle energies are then given by

$$
\epsilon_{n,\mathbf{k}}^{\mathrm{QP}} = \epsilon_{n,\mathbf{k}}^{\mathrm{KS}} + Z_{n,\mathbf{k}} \left[\sum_{n,\mathbf{k}}^{\mathrm{c}} (\epsilon_{n,\mathbf{k}}^{\mathrm{KS}}) + \Sigma_{n,\mathbf{k}}^{\mathrm{x}} - \left\langle \varphi_{n,\mathbf{k}}^{\mathrm{KS}} \right| V_{\mathrm{XC}}^{\mathrm{KS}} (\mathbf{r}) \left| \varphi_{n,\mathbf{k}}^{\mathrm{KS}} \right\rangle \right],
$$

(4.55)

where

$$
Z_{n,\mathbf{k}} = \left[1 - \left(\frac{\partial}{\partial \omega} \sum_{n,\mathbf{k}}^{\mathrm{c}} (\omega) \right)_{\epsilon_{n,\mathbf{k}}^{\mathrm{KS}}} \right]^{-1}.
$$

(4.56)

In this implementation of the $G_0 W_0$ method in FHI-gap, these two equations are used to get a first set of quasi-particle energies, which are then taken as the starting point to solve Eq. (4.33) iteratively with respect to $\epsilon_{n,\mathbf{k}}^{\mathrm{qp}}$. To ensure the conservation of the quasi-particle number, the quasi-particle Fermi level is aligned to the KS one after each iteration for Eq. (4.33). The iteration is stopped when this shift is smaller than a chosen tolerance.

4.4.4   Brillouin-Zone Integration of the Polarization

Brillouin-zone integration is an important ingredient of any reciprocal space method and has been a subject of intense interest since the earliest implementation of electronic structure codes. Fundamental quantities like the total energy or the density of states require an integration over the Brillouin zone of a certain operator, e.g. the eigenvalues weighted by the Fermi distribution function for the former, the energy derivative of the Fermi distribution for the latter.

In the 1970s, a large number of studies were carried out for solving these problems, among which the special points [217–220] and the linear tetrahedron method [221–223] are the ones most used nowadays. These two methods perform equally well for insulators and semiconductors. For metals, the Brillouin-zone integration becomes more cumbersome due to the presence of the Fermi surface, which defines the integration region in the Brillouin zone. The linear tetrahedron method becomes advantageous

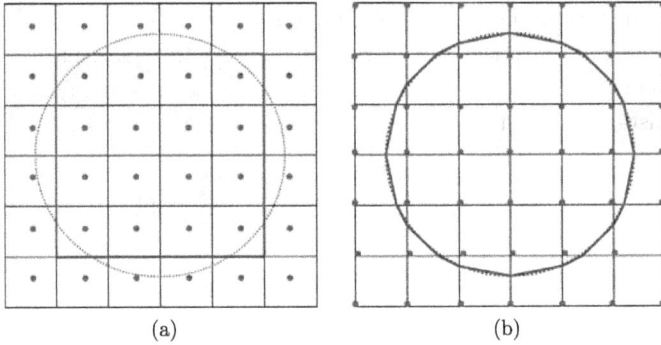

(a)   (b)

Figure 4.6 Two-dimensional sketches of the description of the Fermi surface in the special points (a) and the linear tetrahedron method (b). The red dotted line shows the exact Fermi surface, and the blue solid line the approximated one. The **k**-points grid is represented by dots. The tetrahedron method gives better description of the Fermi surface.

in these systems due to its better description of the Fermi surface (Fig. 4.6) and, therefore, of the integration region [224].

In the linear tetrahedron method, first proposed by Jepsen and Andersen [221] and Lehmann *et al.* [222], the Brillouin zone is divided into a set of tetrahedra. The energy eigenvalues ($\epsilon_{n,\mathbf{k}}$) and the integrand are calculated on the vertices of these tetrahedra and, through the procedure known as isoparametrization, linearly interpolated inside each of them. The values of the integrand can be factorized out of the integral. The remaining integrals, independent of the values at the vertices, can be integrated analytically and added to obtain integration weights dependent only on the **k**-point and the band index (Appendix C). In metallic systems, the Fermi surface is approximated, through the isoparametrization, by a plane that limits the integration region inside the tetrahedra it intersects. The occupied region of the Brillouin zone can thus be described much better than in any of the special points methods (Fig. 4.6).

The calculation of quantities like the polarizability (Eq. (4.50)) or magnetic susceptibility presents particular characteristics that require a different treatment. The integral depends on a second vector $\mathbf{q}$, it is weighted by two Fermi functions, so that the states at **k** are occupied while those at $\mathbf{k} - \mathbf{q}$ are unoccupied, and finally, the eigenvalues appear in the denominator of the integrand. The grid of **k**-points for this integration is chosen as usual in the tetrahedron method. On the other hand, the calculation of

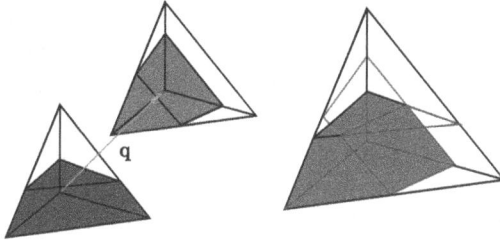

Figure 4.7   The integration region in the tetrahedron method for $\mathbf{q}$-dependent Brillouin-zone integration. The two tetrahedra on the left side are connected by the vector $\mathbf{q}$ (arrow). The right zone corresponds to the occupied region for the state $(n, \mathbf{k})$, the middle one to the unoccupied region for the state $(n', \mathbf{k} - \mathbf{q})$. The resulting integration region, determined by superimposing the two tetrahedra on the left and taking the intersection of the right and middle zones, is the left region in the tetrahedron on the right-hand side.

the self-energy (Eqs. (4.53) and (4.54)) requires a grid of $\mathbf{q}$-points to be suitable also for integration. To avoid the repeated generation of eigenvalues and eigenvectors at several different grids, the set of $\mathbf{q}$-points should be such that $\{\mathbf{k}\} = \{\mathbf{k} - \mathbf{q}\}$. For this equality to hold, the set of $\mathbf{q}$-points must include the Γ point. In FHI-gap, the same mesh for $\mathbf{k}$ and $\mathbf{q}$ is taken.

Due to the presence of the eigenvalues in the denominator of the integrand, a simultaneous isoparametrization of both, the integrand and the eigenvalues, becomes inappropriate. In 1975, Rath and Freeman proposed a solution to this problem for the calculation of the static magnetic susceptibilities in metals [223]. They approximated the numerator of the integrand by its mean value in each tetrahedron, while the denominator was included in the analytic integration to obtain the weights. In FHI-gap, two further steps were taken. First, the isoparametrization was applied not only to the eigenvalues but also to the numerator of the integrand, which significantly improves the accuracy. Besides this, we also extended the method to include the frequency dependence. In doing so, the integration inside each tetrahedron can finally be performed analytically (see Appendix C).

Since the integration runs simultaneously on two tetrahedra (at $\mathbf{k}$ and $\mathbf{k} - \mathbf{q}$), there will be situations, in metallic systems, where both tetrahedra are intersected by the Fermi surface. In this case, the integration region inside the tetrahedron is delimited by the two Fermi "planes" under the condition $\epsilon_{n\mathbf{k}} < \epsilon_F < \epsilon_{m\mathbf{k}-\mathbf{q}}$, as shown in Fig. 4.7. The complexity of the integration region is such that the integration cannot be performed analytically on the whole tetrahedron as in the standard tetrahedron method.

However, as pointed out in Ref. [223], the integration region can always be subdivided into, at most six, tetrahedra. The integration can be performed analytically inside each of these tetrahedra and then projected onto the vertices of the original tetrahedron to obtain the weights for each **k**-point. These different configurations of the distinct integration regions determined by two Fermi "planes" were then analyzed and categorized so that computer programming can be applied (see Fig. C.2 in Appendix C).

To test the accuracy and stability of this implementation, the static polarizability of the free electron gas can be calculated and compared to its well-known analytical solution (the Lindhard function). The results are shown in Fig. 4.8. This tetrahedron method performs really well for the free electron gas, which is one of the most demanding examples for the Brillouin-zone integration. Comparison of Fig. 4.8(a) with Fig. 4 in Ref. [223] shows that this implementation achieves a comparable accuracy with a coarser mesh (a $13 \times 13 \times 13$ mesh in our calculation compared to the $24 \times 24 \times 24$ mesh used in Ref. [223]).

4.4.5 *The Frequency Integration*

Then we discuss the frequency convolution for the correlation term of the self-energy in Eq. (4.53). Due to the poles of both Green's function and W^c, infinitesimally close to the real axis (Fig. 4.9), this integral is difficult to converge numerically, requiring a large number of frequencies.

Several schemes have been proposed to improve the computational efficiency in the evaluation of this convolution. One of the first methods was proposed by Godby, Schlüter, and Sham [225]. Using the idea of the well-known Matsubara summation [124, 126], which analytically continuates the integrand into the complex frequency plane and calculates the integral over the real frequency axis from the integral over the imaginary axis plus the sum of the residues corresponding to the poles of Green's function between the given frequency and the Fermi energy, they have shown that the simple form of the integrand in the imaginary axis allows a precise calculation of the integral with few frequencies only. In a different approach, proposed by Rieger *et al.* [227], the screened Coulomb potential is Fourier transformed to the imaginary-time axis. The self-energy is then obtained by direct product, according to Eq. (4.31c) and transformed back to the imaginary frequency axis. Afterwards, it is fitted by an analytic function and continued to the complex plane to obtain its dependence on the real frequency axis.

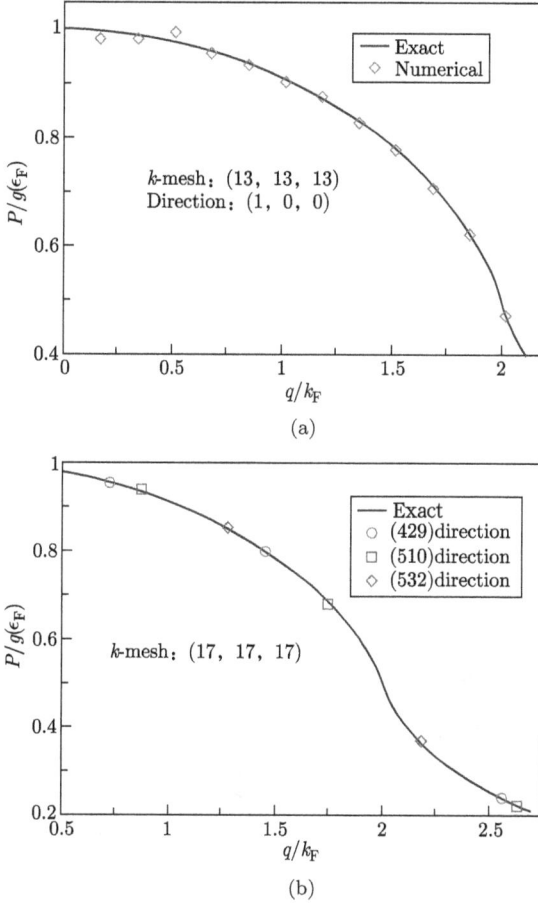

Figure 4.8 Comparison of the numerical (points) and analytical (line) results for the polarizability of the free electron gas (Lindhard function) as function of **q** (a) on the (100) direction and (b) on other directions.

In FHI-gap, the screened Coulomb potential, Green's function and the self-energy were calculated directly on the imaginary frequency axis. Equation (4.53) in this case becomes

$$\Sigma_{n,\mathbf{k}}^{c}(i\omega) = \frac{1}{\pi} \sum_{\mathbf{q}}^{BZ} \sum_{i,j} \sum_{n'} \left[M_{n,n'}^{i}(\mathbf{k},\mathbf{q}) \right]^{*}$$

$$\cdot \int_{0}^{\infty} \frac{(\epsilon_{n',\mathbf{k}-\mathbf{q}} - i\omega) W_{i,j}^{c}(\mathbf{q},i\omega')}{(i\omega - \epsilon_{n',\mathbf{k}-\mathbf{q}})^{2} + \omega'^{2}} d\omega' M_{n,n'}^{j}(\mathbf{k},\mathbf{q}), \qquad (4.57)$$

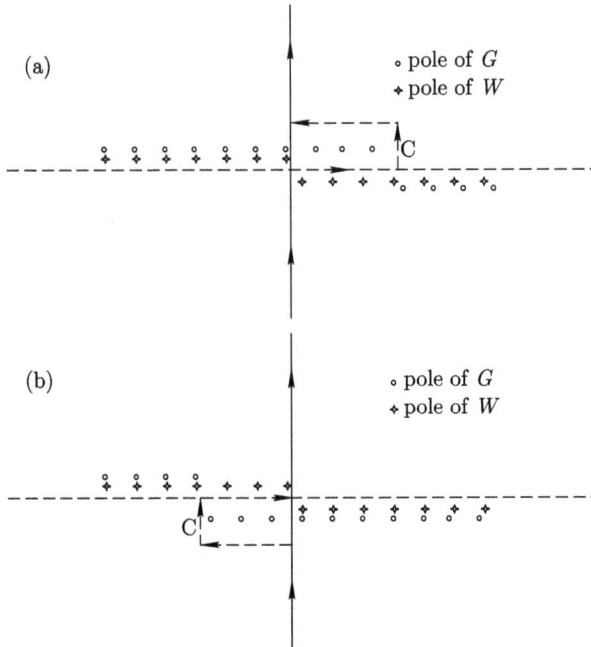

Figure 4.9   The analytic structure of $\Sigma^c = iGW^c$ for $\omega > \mu$ (a) and $\omega < \mu$ (b). Frequency integration of the self-energy along the real axis is equivalent to the integration along the imaginary axis including the path C.

where we have made use of the inversion symmetry of W^c on the imaginary frequency axis:

$$W_{i,j}^c(\mathbf{q}, i\omega) = W_{i,j}^c(\mathbf{q}, -i\omega). \tag{4.58}$$

The integrand in Eq. (4.57) is singular when $\omega = \omega'$ and $\epsilon_{n',\mathbf{k}-\mathbf{q}} = 0$. Therefore, a direct numerical integration becomes unstable for small eigenvalues. The numerical details, as well as the method to avoid this instability, are shown in Appendix 7.5. In the end, each matrix element of the self-energy is fitted with a function of the form:

$$\Sigma_{n,\mathbf{k}}^c(i\omega) = \frac{\sum_{j=0}^m a_{n,\mathbf{k},j}(i\omega)^j}{\sum_{j=0}^{m+1} b_{n,\mathbf{k},j}(i\omega)^j}, \tag{4.59}$$

which is then analytically continued onto the real frequency axis.

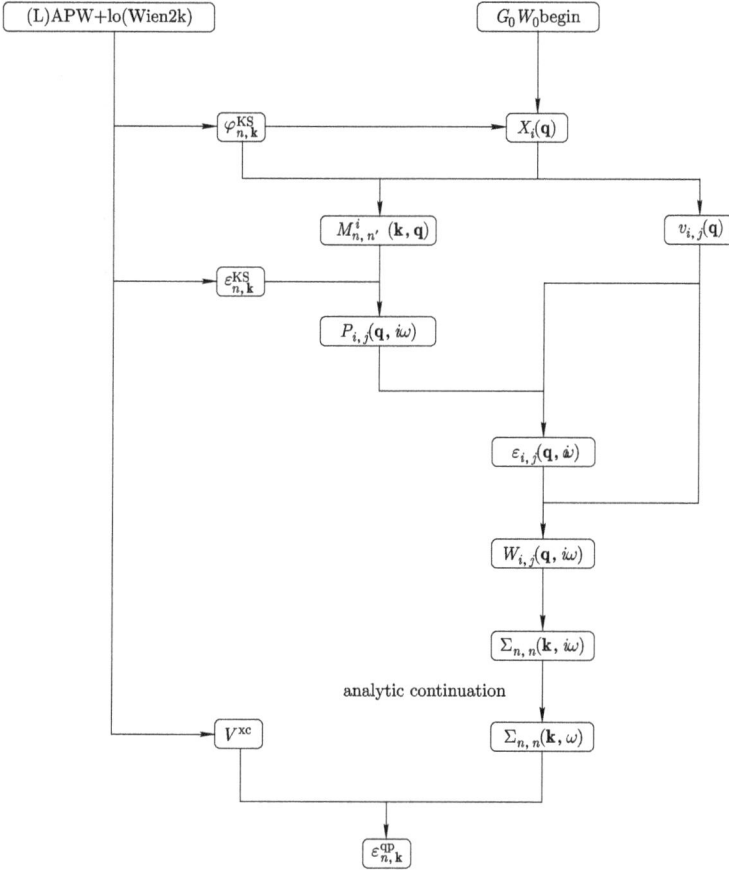

Figure 4.10   Flowchart of the FHI-gap G_0W_0 code.

4.4.6   *Flowchart*

We conclude this chapter with a short summary of the procedure carried out by the implementation of the G_0W_0 method as introduced above. The flowchart is shown in Fig. 4.10. The KS eigenvalues and eigenfunctions, as well as the XC potential are taken from Wien2k. The eigenfunctions are used to obtain the mixed basis as was described in Sec. 4.4.2. Having defined the basis functions, one can calculate the bare Coulomb matrix and the matrix elements $M^i_{n,n'}(\mathbf{k}, \mathbf{q})$. Afterwards, the KS eigenvalues are required for calculating the Brillouin-zone integration weights as described in Sec. 4.4.4 and Appendix C. Together with the matrix elements, $M^i_{n,n'}(\mathbf{k}, \mathbf{q})$, these weights are used to obtain the polarization matrix (Eq. (4.50)). The latter, together with the bare Coulomb matrix, is the input required to obtain the dielectric matrix and then the screened Coulomb potential. With these, the matrix elements of the exchange and correlation terms of the self-energy can be calculated separately. The coefficients of the expansion described in Sec. 3.2.4 of the XC potential are obtained from the Wien2k code and used to obtain its diagonal matrix elements for each eigenstate. Finally, a first set of $G_0 W_0$ quasi-particle energies are obtained by solving Eq. (4.55), which are used as the starting point for solving Eq. (4.33) self-consistently.

5
Molecular Dynamics

So far, we have discussed the electronic structures; from now on, we shift our attention to descriptions of the nuclear motion.

As introduced in Chapter 1, molecular dynamics (MD) is a technique which allows us to investigate the statistical and dynamical properties of a real poly-atomic system at the molecular level. When the Born–Oppenheimer (BO) approximation is further imposed, this simplifies into reproducing the propagation of the nuclei on their BO potential-energy surfaces (PESs). In standard treatment of such propagations, the classical equations of motion governing the microscopic evolution of a many-body entity (composed by nuclei) are often solved numerically, subject to the boundary conditions appropriate for the geometry and/or symmetry of the system. Therefore, the method is, in principle, classical and these nuclei are classical point-like particles in this treatment. We note that there are extensions of this method to the quantum (in the statistical perspective) or semiclassical (in the dynamical perspective) regime, where some of the quantum nuclear effects (QNEs) can be explicitly addressed. In this chapter, to be clear, we restrict ourselves to the classical picture that the nuclei are point-like particles for the illustration of the basic underlying principles. Its extension to the descriptions of QNEs will be described in Chapter 7.

Depending on how the interatomic interactions are calculated, the MD simulations can be separated into two categories: the traditional classical MD simulations, where the interactions between different nuclei are described by empirical potentials, and the more recent *ab initio* MD simulations, where the interactions between the nuclei are calculated *on-the-fly* using the Hellmann–Feynman theorem after the electronic structures of the system at each specific spatial configuration of the nuclei were obtained quantum mechanically in an *ab initio* manner. Throughout this book, we focus on the *ab initio* methods and consequently the *ab initio* MD, but we note that the principles underlying the propagation of the nuclei, as introduced here, also apply to those simulations where empirical potentials are used.

Within the framework of *ab initio* MD, it is also worthwhile to note that there is a scheme when the BO approximation is not rigorously done, but rather, an adiabatic separation between the electronic and the nuclear degrees of freedom has been assumed using a special treatment for the movement of the electrons and the nuclei, like the Car–Parrinello (CP) MD [228]. There is a scheme where the electronic structure optimization is rigorously performed at each spatial configuration of the nuclei along the trajectory, like the BO MD. The former treatment is an approximation to the latter one. Due to the lower computation cost which originates from the fact that the self-consistent electronic structure optimization does not have to be done at each MD step — rather, a fictitious small mass is assigned to the electronic degrees of freedom which enables an adiabatic separation between the movements of the electrons and the nuclei — the former treatment has been very popular from the 1980s till now. The BO MD method only became a standard routine after the late 1990s when the improvement of computational algorithms and the development of greater computing power made it feasible to carry out a self-consistent electronic structure optimization at each MD step. Because of this, the development of some methods within the framework of *ab initio* MD, especially those related to the thermostats, has been largely influenced by this historical trend. In later discussions, we will point this out in detail when those methods are introduced.

Our discussions in this chapter are categorized mainly into two parts. Section 5.1 presents an introduction to the ideas behind MD. With these concepts, our molecular simulation is able to sample a micro-canonical ensemble and its phase space on a constant-energy shell. In real experiments, however, it is the temperature that is kept constant. Section 5.2 then gives us an introduction to some methods which go beyond the simulations of the micro-canonical ensemble. Techniques underlying numerical controls of the molecular temperature (thermostats) are the basis for such simulations. Then, more sophisticated situations when other ensembles — such as the constant-pressure and constant-temperature (isothermal–isobaric) ones — must be simulated will also be discussed. With these, in principle, one should be able to do a proper simulation for a system under the situation to be studied and compare the results with experiments.

5.1 Introduction to Molecular Dynamics

The foundation of modern MD technique is the so-called Hamiltonian mechanics, which is often used in descriptions of a micro-canonical ensemble in Gibbs' language. Hamiltonian mechanics is, in principle, equivalent to the simplest Newton's equation of motion for the description of the nuclear propagations, with the only difference being that the equations of motion can be derived easily in non-Cartesian coordinates. Because of this advantage, properties in the phase space can be discussed in an easier way. Therefore, although it provides nothing more than what can be obtained by solving the Newton's equation of motion, we prefer using Hamiltonian mechanics for discussions in this chapter.

Within this framework of Hamiltonian mechanics, where the Hamiltonian's equation of motion for the nuclei are numerically solved, the simulations provide descriptions for the statistical properties of the system within a certain constant-energy shell in the phase space. There is neither particle exchange, energy exchange between the system under investigation and its environment, nor any change of the system's volume. In other words, most essentially, the total energy (Hamiltonian) is a conserved quantity. Later, extensions of this Hamiltonian mechanics to the non-Hamiltonian ones were proposed, where the total energy of the system is not conserved and other ensembles such as the canonical or isothermal–isobaric ones can be simulated. These different ensembles serve different purposes in simulations of the real materials. To be clear, we start our discussions from the simplest Hamiltonian simulations of the micro-canonical ensemble (or otherwise NVE ensemble, where N stands for the number of the particles, V for the volume of the cell under simulation, and E for the total energy) and then extend these techniques to the non-Hamiltonian scheme and simulations of other ensembles.

Intuitively, the three key quantities in this MD simulation of a micro-canonical ensemble within the framework of Hamiltonian mechanics should be the spatial coordinate of each nucleus ($\mathbf{r}_i$), the velocity of each nucleus ($\mathbf{v}_i$), and the force imposed by the quantum glue of the electrons and the classical internuclear Coulomb interactions on each nucleus ($\mathbf{F}_i$) at time zero. These three quantities, together with the Hamiltonian's equation of motion, determine the nuclear evolution of the system and consequently, the equilibrium and dynamical properties of the system under investigation. The entity of the interacting nuclei can be viewed as a classical many-body system, with interactions between the nuclei changing as the system evolves. For such a many-body entity, it is impossible to solve this equation of motion analytically. Therefore, one needs to resort to numerical integrations for the equation of motion.

5.1.1   The Verlet Algorithm

As mentioned above, the first step in carrying out a numerical integration for the equation of motion is to set up a proper initial state of the system. This includes setting the initial spatial configuration of the nuclei, their initial velocities, and the initial forces imposed on them. Then the trajectory of the nuclei can be integrated out numerically in principle. Seemingly simple, the integration of such an equation of motion is highly non-trivial due to the fact that a finite-time step must be used. Error accumulation ultimately results in two initially-close trajectories diverging exponentially as the dynamic evolves.

Fortunately, the MD simulations differ in principle from the trajectory predictions in classical dynamics, such as in the study of the comet's orbital (see e.g. Ref. [48, Sec. 4.3]). What we are primarily aiming at in the MD simulations is some knowledge on the statistical property of the system whose initial state we already know something about, e.g. the energy. In the language of the Gibbs' ensemble (the micro-canonical ensemble in this case), many individual microscopic configurations of a large system lead to the same macroscopic behaviors. This statement implies that it is unnecessary to know the precise propagation of every nucleus in the system in order to predict its macroscopic properties. Rather, a simple average over a large number of states visited during the simulation will give us properly-formulated macroscopic observables. As long as our integrator respects the fact that the Newton's equation of motion is time-reversible and the corresponding Hamiltonian dynamics leaves the volume of the phase space explored in the MD simulation unchanged, the energy conservation can be guaranteed if the finite-time step between movements are small enough in practical simulations [48]. This indicates that a micro-canonical ensemble is already being simulated. As it is one of the successful schemes on implementing such a propagation when the total energy and the time reversibility of the dynamics can be preserved, we take the Verlet algorithm as an example and explain the underlying principles by going through the key equations in the following [229].

In this algorithm, the spatial configuration of the system at time $t + \Delta t$ is determined by

$$\mathbf{r}_i(t + \Delta t) = \mathbf{r}_i(t) + \mathbf{v}_i(t)\Delta t + \frac{\mathbf{F}_i(t)}{2m_i}\Delta t^2 + O(\Delta t^3) + O(\Delta t^4), \qquad (5.1)$$

when the Taylor expansion is used. Here, $O(\Delta t^3)$ $(O(\Delta t^4))$ represents error on the order of Δt^3 (Δt^4), which equals $\mathbf{r}'''(t)\Delta t^3/3!$ $(\mathbf{r}''''(t)\Delta t^4/4!)$. Further expansions beyond $O(\Delta t^4)$ are neglected. We note that the quantities calculated along the trajectories are only $\mathbf{r}_i$, $\mathbf{v}_i$, and $\mathbf{F}_i$; the terms $O(\Delta t^3)$ and $O(\Delta t^4)$ will not be calculated. But it is important to write them down in Eq. (5.1) in order to show the order of accuracy for $\mathbf{r}_i$ and $\mathbf{v}_i$ calculated along the trajectory.

From Eq. (5.1), it is easy to see that in reverse time, the spatial configuration respects

$$\mathbf{r}_i(t - \Delta t) = \mathbf{r}_i(t) - \mathbf{v}_i(t)\Delta t + \frac{\mathbf{F}_i(t)}{2m_i}\Delta t^2 - O((\Delta t)^3) + O((\Delta t)^4).$$
$$(5.2)$$

If we sum up Eqs. (5.1) and (5.2), we can have

$$\mathbf{r}_i(t + \Delta t) = 2\mathbf{r}_i(t) - \mathbf{r}(t - \Delta t) + \frac{\mathbf{F}_i(t)}{m_i}\Delta t^2 + 2O(\Delta t^4). \qquad (5.3)$$

The $O(\Delta t^3)$ terms were cancelled out. Therefore, the evolution of the nuclear position obtained from this integrator contains an error on the order of Δt^4 (quantities related to the lower-order expansions needed in the analytical expression in Eq. (5.3), i.e. $\mathbf{r}_i$ and $\mathbf{F}_i$, are all calculated along the trajectory).

To obtain the new velocity, we subtract Eqs. (5.1) by (5.2). In doing so, we arrive at

$$\mathbf{r}_i(t + \Delta t) - \mathbf{r}_i(t - \Delta t) = 2\mathbf{v}_i(t)\Delta t + 2O(\Delta t^3). \qquad (5.4)$$

The velocity is then updated by

$$\mathbf{v}_i(t) = \frac{\mathbf{r}_i(t + \Delta t) - \mathbf{r}_i(t - \Delta t)}{2\Delta t} - \frac{O(\Delta t^3)}{\Delta t}. \qquad (5.5)$$

From this equation, it is clear that the velocity contains an error on the order of Δt^2.

It is worthwhile to note that Eqs. (5.1) and (5.2) are the bases of the Verlet algorithm, which is obviously time-reversible. Therefore, trajectories obtained from the Verlet algorithm respect the time reversibility of the Newton's equation of motion. Disregarding the error terms which are not calculated, in this Verlet algorithm, the trajectory is propagated through

$$\mathbf{r}_i(t + \Delta t) = 2\mathbf{r}_i(t) - \mathbf{r}_i(t - \Delta t) + \frac{\mathbf{F}_i(t)}{m_i}\Delta t^2, \qquad (5.6)$$

and

$$\mathbf{v}_i(t) = \frac{\mathbf{r}_i(t + \Delta t) - \mathbf{r}_i(t - \Delta t)}{2\Delta t}. \qquad (5.7)$$

The positions and velocities contain errors on the orders of Δt^4 and Δt^2, respectively.

5.1.2   The Velocity Verlet Algorithm

We note that in Eq. (5.6), the velocity is not used when the new position is generated. There is an algorithm equivalent to the Verlet one in which the velocity is used in the calculation of the new position. This is the so-called velocity Verlet algorithm [230].

In this algorithm, the new position is calculated from

$$\mathbf{r}_i(t + \Delta t) = \mathbf{r}_i(t) + \mathbf{v}_i(t)\Delta t + \frac{\mathbf{F}_i(t)}{2m_i}\Delta t^2. \qquad (5.8)$$

The velocity is updated by

$$\mathbf{v}_i(t + \Delta t) = \mathbf{v}_i(t) + \frac{\mathbf{F}_i(t + \Delta t) + \mathbf{F}_i(t)}{2m_i}\Delta t. \qquad (5.9)$$

We note that a key difference between this algorithm and the Euler scheme, which updates the velocity by

$$\mathbf{v}_i(t + \Delta t) = \mathbf{v}_i(t) + \frac{\mathbf{F}_i(t)}{m_i}\Delta t, \qquad (5.10)$$

is that the new velocity can only be calculated after the new position and its corresponding forces are generated in the velocity Verlet algorithm. This fine numerical difference actually results in the fundamental distinction that the trajectory obtained from the Euler algorithm is not time-reversible and the position contains an error on the order of Δt^3, whereas in the velocity Verlet algorithm, the trajectory is time-reservable and the updated position contains an error on the order of Δt^4, although this is not clear in Eqs. (5.8) and (5.9).

To prove the equality between the velocity Verlet algorithm and the original Verlet algorithm, we start from the original Verlet algorithm in Eqs. (5.6) and (5.7) and try to deduce the velocity Verlet algorithm in Eqs. (5.8) and (5.9). To this end, we first rewrite Eq. (5.6) into the following form:

$$\mathbf{r}_i(t + \Delta t) = \mathbf{r}_i(t) + \frac{\mathbf{r}_i(t + \Delta t) - \mathbf{r}_i(t - \Delta t)}{2\Delta t}\Delta t + \frac{\mathbf{F}_i(t)}{2m_i}\Delta t^2. \tag{5.11}$$

Then, by replacing the $(\mathbf{r}_i(t + \Delta t) - \mathbf{r}_i(t - \Delta t))/(2\Delta t)$ with $\mathbf{v}_i(t)$ through Eq. (5.7), we arrive at Eq. (5.8). Therefore, the updating of the position is completely equivalent in these two algorithms.

Concerning the deduction of Eq. (5.9), from Eq. (5.8), we see that

$$\mathbf{v}_i(t) = \frac{\mathbf{r}_i(t + \Delta t) - \mathbf{r}_i(t)}{\Delta t} - \frac{\mathbf{F}_i(t)}{2m_i}\Delta t. \tag{5.12}$$

This equation is equivalent to

$$\mathbf{v}_i(t + \Delta t) = \frac{\mathbf{r}_i(t + 2\Delta t) - \mathbf{r}_i(t + \Delta t)}{\Delta t} - \frac{\mathbf{F}_i(t + \Delta t)}{2m_i}\Delta t, \tag{5.13}$$

which further equals

$$\mathbf{v}_i(t + \Delta t) = \frac{\mathbf{r}_i(t + 2\Delta t) - \mathbf{r}_i(t)}{\Delta t} - \frac{\mathbf{r}_i(t + \Delta t) - \mathbf{r}_i(t)}{\Delta t} - \frac{\mathbf{F}_i(t + \Delta t)}{2m_i}\Delta t. \tag{5.14}$$

Using Eq. (5.7), it can be reduced to

$$\mathbf{v}_i(t + \Delta t) = 2\mathbf{v}_i(t + \Delta t) - \frac{\mathbf{r}_i(t + \Delta t) - \mathbf{r}_i(t)}{\Delta t} - \frac{\mathbf{F}_i(t + \Delta t)}{2m_i}\Delta t, \tag{5.15}$$

which equals

$$\mathbf{v}_i(t + \Delta t) = \frac{\mathbf{r}_i(t + \Delta t) - \mathbf{r}_i(t)}{\Delta t} + \frac{\mathbf{F}_i(t + \Delta t)}{2m_i}\Delta t. \tag{5.16}$$

From Eq. (5.12), we see that $(\mathbf{r}_i(t + \Delta t) - \mathbf{r}_i(t))/\Delta t$ equals $\mathbf{v}_i(t) + \mathbf{F}_i(t)\Delta t/(2m_i)$. Replacing this $(\mathbf{r}_i(t + \Delta t) - \mathbf{r}_i(t))/\Delta t$ term with $\mathbf{v}_i(t) + \mathbf{F}_i(t)\Delta t/(2m_i)$ in Eq. (5.16), one arrives at Eq. (5.9).

The above deduction means that starting from Eqs. (5.6) and (5.7), one easily arrives at Eqs. (5.8) and (5.9). In other words, Eqs. (5.8) and (5.9) propagate the trajectory in the same way as Eqs. (5.6) and (5.7), i.e.

the velocity Verlet and the Verlet algorithms are completely equivalent. The position (velocity) contains error on the order of Δt^4 (Δt^2). They both differ from the Euler algorithm in Eq. (5.10) in the sense that their trajectories are time-reversible and preserve a higher order of accuracy in terms of Δt.

5.1.3   The Leap Frog Algorithm

The second frequently used algorithm for the propagation of the MD trajectory, which is also equivalent to the Verlet algorithm, is the so-called Leap Frog algorithm [48, 231]. In this algorithm, the velocities and positions are updated by

$$\mathbf{v}_i(t + \Delta t/2) = \mathbf{v}_i(t - \Delta t/2) + \frac{\mathbf{F}_i(t)}{m_i}\Delta t, \tag{5.17}$$

and

$$\mathbf{r}_i(t + \Delta t) = \mathbf{r}_i(t) + \mathbf{v}_i(t + \Delta t/2)\Delta t. \tag{5.18}$$

The propagation is equivalent to the Verlet algorithm with the only difference that the velocities are updated at half-time step.

For the completeness of this introduction, we use one more paragraph to explain how these equations can be derived from Eqs. (5.6) and (5.7). From Eq. (5.7), we have

$$\mathbf{v}_i(t - \Delta t/2) = \frac{\mathbf{r}_i(t) - \mathbf{r}_i(t - \Delta t)}{\Delta t}, \tag{5.19}$$

and

$$\mathbf{v}_i(t + \Delta t/2) = \frac{\mathbf{r}_i(t + \Delta t) - \mathbf{r}_i(t)}{\Delta t}. \tag{5.20}$$

From Eq. (5.20), it is easy to obtain Eq. (5.18) for the update of the positions. Concerning the deduction of Eq. (5.17), one can rewrite Eq. (5.6) into the following form:

$$\frac{\mathbf{r}_i(t + \Delta t) - \mathbf{r}_i(t)}{\Delta t} = \frac{\mathbf{r}_i(t) - \mathbf{r}_i(t - \Delta t)}{\Delta t} + \frac{\mathbf{F}_i(t)}{m_i}\Delta t. \tag{5.21}$$

Then by inputting Eqs. (5.19) and (5.20) into Eq. (5.21), one gets

$$\mathbf{v}_i(t + \Delta t/2) = \mathbf{v}_i(t - \Delta t/2) + \frac{\mathbf{F}_i(t)}{m_i}\Delta t, \tag{5.22}$$

which is the same as Eq. (5.17). Therefore, the trajectory is equivalent to that of the Verlet algorithm. However, due to the fact that the velocities are not calculated at the same time slides as the positions, the total energy cannot be rigorously calculated.

The Verlet, velocity Verlet, and the Leap Frog algorithms are the main frequently used simple algorithms in standard implementations of the MD method. For most MD simulations, they are sufficiently accurate. From Eq. (5.1), it is easy to see that if higher orders of expansion for the updating of the position in terms of Δt are included, one may use a larger time step for the propagation of the trajectory and consequently, obtain higher efficiency of the simulations. The predictor–corrector algorithm is one such example [232]. However, these extensions may not respect the time-reversal symmetry of the Newton's equation and consequently may not lead to more reliable sampling of the corresponding phase space. In this chapter, we restrict ourselves to the simplest ones introduced and refer the readers interested in more complicated ones to Refs. [48, 233].

5.2   Other Ensembles

When a proper initial state of the system under investigation is set up and the trajectory is generated using the algorithms as introduced above, if the time step is sufficiently small, the total energy E will be conserved within a certain tolerance. On the basis of this, if the *ergodic hypothesis* is further imposed — which means that given an infinite amount of time the trajectory will go through the entire constant-energy hypersurface in the phase space, a proper micro-canonical ensemble will be simulated. This micro-canonical ensemble provides an accurate description for the statistical/dynamical behavior of the system within a certain constant-energy shell in the phase space, when there is neither particle exchange, energy exchange between the system under investigation and its environment, nor any change in the system's volume [234–238]. Most experiments, however, are performed in situations where these quantities are not conserved. The simplest situation is when the experiment is carried out at a constant-temperature T instead of the total energy E, due to the energy exchange between the system under investigation and its environment. In this case, the statistical behavior of the system at a finite temperature must be accurately taken into account, and the states associated with the system in the phase space visited during the time when the experiment is carried out are not within a certain constant-energy shell anymore. Canonical, or in other words, the constant-volume and constant-temperature (NVT) ensemble therefore needs to be simulated [239–246].

Theoretically, this can be done by introducing a proper thermostat in the simulation, through which the energy exchange between the system under investigation and the environment can be properly described. However, other thermodynamic variables such as the volume of the simulation cell at finite pressure P may also change due to fluctuations. To describe situations like this, a barostat needs to be further introduced which keeps the tensor of the system constant during the simulations. This ensemble with constant-temperature and constant-pressure is called an NPT, or isothermal–isobaric ensemble in Refs. [247–249]. In ascending order of complexity, other ensembles which simulate experiments at different situations, such as the isoenthalpic–isobaric (NPH) [238, 250–252], grand-micro-canonical (μVL, with μ standing for the chemical potential) [238, 253], etc., also exist. In this chapter, we aim at providing an introduction to the principles underlying theoretical simulations of these ensembles. Therefore, we will use the simplest case, i.e. the canonical (NVT) ensemble, as an example to explain how these principles work. Its extension to the isothermal–isobaric (NPT) ensemble will also be briefly discussed. For readers interested in those methods for simulating more complicated ensembles, please refer to their original papers [238, 250–253]. We note that in these simulations, since the total energy is not conserved and the equation of motion cannot be obtained from the Hamiltonian of the system under simulations, they are often called the non-Hamiltonian MD simulations in the literature.

5.2.1 *Andersen Thermostat*

There are several schemes for the temperature to be controlled. The underlying principle is that at a finite temperature T, the probability of finding a classical particle, e.g. the ith atom in the system, with kinetic energy $\mathbf{p}_i^2/(2m_i)$, which is labeled as $P(\mathbf{p}_i)$ throughout our discussions, respects the Maxwell–Boltzmann distribution:

$$P(\mathbf{p}_i) = \left(\frac{\beta}{2\pi m_i}\right)^{\frac{3}{2}} e^{-\beta \frac{\mathbf{p}_i^2}{2m_i}}. \tag{5.23}$$

Here, $\beta = 1/(k_B T)$. The simplest implementation of this restriction on simulation of real systems naturally is achieved through a stochastic collision of nuclei with a thermostat that imposes such a Maxwell–Boltzmann distribution on the velocities. The corresponding thermostat is the so-called Andersen thermostat [239]. When it is used, with a certain initial state, the propagator as introduced in Sec. 5.1 ensures that the system propagates on a certain constant-energy shell in the phase space between collisions. The collisions themselves, on the other hand, randomly act on a selected particle and reset its new velocity from the Maxwell–Boltzmann distribution function at the desired temperature, which effectively reduces correlation between events along the trajectory for increased sampling efficiency, at the cost of accuracy in short-time dynamics. With a reasonable choice of the collision frequency, the canonical ensemble with velocities respecting the Maxwell–Boltzmann distribution can be conveniently simulated. Reference [48, Chapter 6] gives details of how this is implemented, including how the program should be written, which we do not cover. For such details, please refer to their explanation.

Here, we note that although it is efficient and simple to implement while imposing a canonical ensemble to be sampled, the stochastic collision results in non-deterministic and discontinuous trajectories, which lead to all real-time information being lost at collision. Related physical quantities, such as the vibrational frequencies, diffusion coefficients etc., therefore cannot be calculated. In addition to this, more seriously, in spite of the fact that a fairly ergodic sampling can often be guaranteed on all degrees of freedom in such a stochastic simulation, when an adiabatic separation between the electronic and ionic degrees of freedom must be assumed, like in CP MD, a breakdown of such adiabatic separation often happens [254, 255]. As mentioned in the introduction of this chapter, *ab initio* MD has been dominated by the CP MD treatment before the late 1990s due to its lower computational cost. Therefore, deterministic methods such as the Nosé–Hoover thermostat which avoid such a numerical problem have been widely adopted since the 1980s [240–244]. It is only after the wide use of BO MD that stochastic methods such as the Andersen or Langevin thermostat became popular in *ab initio* MD simulations. Because of this historical reason, when one goes through the literature, one may realize a change of the taste in attempts to control the temperature influenced by this trend. In this chapter, to present a general introduction of these methods and be consistent with this historical trend, we will give a detailed explanation of the principles underlying the Nosé–Hoover thermostat and its related Nosé–Hoover chain. After this, we come back to another stochastic method, i.e. the Langevin thermostat, which is widely used in *ab initio* BO MD simulations nowadays. We note that extensions of this method were also designed to be compatible with CP MD recently [254].

5.2.2 Nosé–Hoover Thermostat

As one of the first successful attempts to incorporate deterministic trajectories in the MD simulation at finite temperature, Nosé and Hoover have separately used an extended system with an additional and artificial degree of freedom to control the temperature [240–243]. The corresponding thermostat is called the Nosé–Hoover thermostat. We note that this idea of extracting a canonical ensemble out of a micro-canonical ensemble on an extended system is similar to how the concept of canonical ensemble is derived from the concept of micro-canonical ensemble in standard textbooks on statistics. In that case, it is required that the degree of freedom in the thermostat is much larger than that in the canonical ensemble. Here, a very clever choice of the potential, to which the added degree of freedom particle is subjected, ensures that one single extra degree of freedom imposes a canonical ensemble on the system to be studied. In the following, we will show how this works by going through the underlying key equations.

The main assumption of this extended system method is that an additional coordinate s needs to be introduced to the classical N-nuclei system, with the Lagrangian for the extended system

$$L_{\text{Nosé}} = \sum_{i=1}^{N} \frac{m_i}{2} s^2 \dot{\mathbf{r}}_i^2 - V(\mathbf{r}_1, \ldots, \mathbf{r}_N) + \frac{Q}{2} \dot{s}^2 - \frac{g}{\beta} \ln s. \qquad (5.24)$$

Here, we use "Nosé" to denote the extended system. Q is the effective mass associated with the additional degree of freedom, g is a parameter to be fixed, and $[g\ln s]/\beta$ is the external potential acting on this extra degree of freedom. The corresponding Hamiltonian of this "Nosé" system can then be written as

$$H_{\text{Nosé}} = \sum_{i=1}^{N} \frac{1}{2m_i s^2} \mathbf{p}_i^2 + V(\mathbf{r}_1, \ldots, \mathbf{r}_N) + \frac{1}{2Q} p_s^2 + \frac{g}{\beta} \ln s, \qquad (5.25)$$

where s, p_s, $\mathbf{r}_i$ and $\mathbf{p}_i$ are variables in the phase space and $\mathbf{p}_i = \partial L_{\text{Nosé}} / \partial \dot{\mathbf{r}}_i = m_i s^2 \dot{\mathbf{r}}_i$.

From this Hamiltonian, the partition function for the micro-canonical (Nosé) ensemble is defined as

$$Z_{\text{Nosé}} = \frac{1}{N!} \int dp_s ds dp_1 \cdots dp_N dr_1 \cdots dr_N \delta(H_{\text{Nosé}} - E), \qquad (5.26)$$

Now, if we scale the momentum p_i with the additional degree of freedom by

$$\mathbf{p}_i' = \mathbf{p}_i/s, \qquad (5.27)$$

the partition function in Eq. (5.26) simplifies into

$$Z_{\text{Nosé}} = \frac{1}{N!} \int dp_s ds dp_1' \cdots dp_N' dr_1 \cdots dr_N s^{3N}$$

$$\cdot \delta \left(\left(\sum_{i=1}^{N} \frac{1}{2m_i} \mathbf{p}_i'^2 + V(\mathbf{r}_1, \ldots, \mathbf{r}_N) + \frac{1}{2Q} p_s^2 + \frac{g}{\beta} \ln s \right) - E \right). \qquad (5.28)$$

Further simplification of this equation requires use of the following equation:

$$\delta [h(s)] = \frac{\delta(s - s_0)}{h'(s_0)}, \qquad (5.29)$$

where $h(s)$ is a function of s with a single root for $h(s) = 0$ at s_0 and $h'(s_0)$ is its derivative at this point. Now, if we take

$$h(s) = \left(\sum_{i=1}^{N} \frac{1}{2m_i} \mathbf{p}_i'^2 + V(\mathbf{r}_1, \ldots, \mathbf{r}_N) + \frac{1}{2Q} p_s^2 + \frac{g}{\beta} \ln s \right) - E \qquad (5.30)$$

in Eq. (5.28) and make use of Eq. (5.29), Eq. (5.28) can be simplified into

$$Z_{\text{Nosé}} = \frac{1}{N!} \int dp_s ds dp_1' \cdots dp_N' dr_1 \cdots dr_N s^{3N} \frac{\beta s}{g}$$

$$\cdot \delta \left(s - \exp \left[-\beta \frac{\sum_{i=1}^{N} \frac{1}{2m_i} \mathbf{p}_i'^2 + V(\mathbf{r}_1, \ldots, \mathbf{r}_N) + \frac{p_s^2}{2Q} - E}{g} \right] \right)$$

$$= \frac{1}{N!} \int dp_s ds dp_1' \cdots dp_N' dr_1 \cdots dr_N \frac{\beta s^{(3N+1)}}{g}$$

$$\cdot \delta \left(s - \exp \left[-\beta \frac{\sum_{i=1}^{N} \frac{1}{2m_i} \mathbf{p}_i'^2 + V(\mathbf{r}_1, \ldots, \mathbf{r}_N) + \frac{p_s^2}{2Q} - E}{g} \right] \right). \qquad (5.31)$$

We integrate s out first. In doing so, we can get the partition function of the "Nosé" system as

$$
Z_{\text{Nosé}} = \frac{1}{N!} \int dp_s dp_1' \cdots dp_N' dr_1 \cdots dr_N \frac{\beta}{g}
$$

$$
\cdot \exp \left[-\beta \left(\frac{\sum_{i=1}^N \frac{1}{2m_i} \mathbf{p}_i'^2 + V(\mathbf{r}_1, \ldots, \mathbf{r}_N) + \frac{p_s^2}{2Q} - E}{g} \right) (3N+1) \right]
$$

$$
= \frac{1}{N!} \frac{\beta \exp[E\beta(3N+1)/g]}{g} \int dp_s \exp \left[-\beta \frac{3N+1}{g} \frac{p_s^2}{2Q} \right]
$$

$$
\cdot \int dp_1' \cdots dp_N' dr_1 \cdots dr_N \exp
$$

$$
\cdot \left[-\beta \frac{3N+1}{g} \left(\sum_{i=1}^N \frac{1}{2m_i} \mathbf{p}_i'^2 + V(\mathbf{r}_1, \ldots, \mathbf{r}_N) \right) \right]. \tag{5.32}
$$

This equation indicates that in the extended system with a Hamiltonian as shown in Eq. (5.25), its partition function can be separated into a product of two parts, i.e. that of a subsystem with variables $\mathbf{p}_i'$ and $\mathbf{r}_i$ in the phase space and that of a constant C, as

$$
Z_{\text{Nosé}} = C \int dp_1' \cdots dp_N' dr_1 \cdots dr_N \exp \left[-\beta \frac{3N+1}{g} H(\mathbf{p}', \mathbf{r}) \right]. \tag{5.33}
$$

Here,

$$
C = \frac{1}{N!} \frac{\beta \exp(E\beta(3N+1)/g)}{g} \int dp_s \exp \left[-\beta \frac{3N+1}{g} \frac{p_s^2}{2Q} \right], \tag{5.34}
$$

and

$$
H(\mathbf{p}', \mathbf{r}) = \sum_{i=1}^N \frac{1}{2m_i} \mathbf{p}_i'^2 + V(\mathbf{r}_1, \ldots, \mathbf{r}_N). \tag{5.35}
$$

The equation of motion for this extended system, rigorously, is determined by the Hamiltonian in Eq. (5.25) through

$$
\frac{d\mathbf{r}_i}{dt} = \frac{\partial H_{\text{Nosé}}}{\partial \mathbf{p}_i} = \frac{\mathbf{p}_i}{m_i s^2}, \tag{5.36a}
$$

$$
\frac{d\mathbf{p}_i}{dt} = -\frac{\partial H_{\text{Nosé}}}{\partial \mathbf{r}_i} = -\frac{\partial V(\mathbf{r}_1, \ldots, \mathbf{r}_N)}{\partial \mathbf{r}_i}, \tag{5.36b}
$$

$$\frac{\mathrm{d}s}{\mathrm{d}t} = \frac{\partial H_{\mathrm{Nosé}}}{\partial s} = \frac{p_s}{Q}, \tag{5.36c}$$

$$\frac{\mathrm{d}p_s}{\mathrm{d}t} = -\frac{\partial H_{\mathrm{Nosé}}}{\partial s} = \left[\sum_{i=1}^{N} \frac{\mathbf{p}_i^2}{m_i s^2} - \frac{g}{\beta}\right] \Big/ s. \tag{5.36d}$$

With the quasi-ergodic hypothesis that the time average along the trajectory equals the ensemble average when the simulation time is sufficiently long, the ensemble average of a certain physical quantity A that depends on $\mathbf{p}'$ and $\mathbf{r}$ in the "Nosé" system equals

$$\langle A(\mathbf{p}', \mathbf{r}) \rangle = \lim_{\tau \to \infty} \frac{1}{\tau} \int_0^\tau A(\mathbf{p}', \mathbf{r}) \mathrm{d}t. \tag{5.37}$$

Here, $\langle \cdots \rangle$ indicates the micro-canonical ensemble average of the extended (Nosé) system, and the right-hand side is the time average over the trajectory generated by Eq. (5.36). If we put in the partition function as simplified in Eq. (5.33) for the "Nosé" system to the left-hand side of Eq. (5.37), we can have

$$\langle A(\mathbf{p}', \mathbf{r}) \rangle = \frac{\int \mathrm{d}\mathbf{p}'_1 \cdots \mathrm{d}\mathbf{p}'_N \mathrm{d}\mathbf{r}_1 \cdots \mathrm{d}\mathbf{r}_N A(\mathbf{p}', \mathbf{r}) \exp\left[-\beta \frac{3N+1}{g} H(\mathbf{p}', \mathbf{r})\right]}{\int \mathrm{d}\mathbf{p}'_1 \cdots \mathrm{d}\mathbf{p}'_N \mathrm{d}\mathbf{r}_1 \cdots \mathrm{d}\mathbf{r}_N \exp\left[-\beta \frac{3N+1}{g} H(\mathbf{p}', \mathbf{r})\right]}. \tag{5.38}$$

Now, if we take $g = 3N + 1$, this further simplifies into

$$\langle A(\mathbf{p}', \mathbf{r}) \rangle = \frac{\frac{1}{N!} \int \mathrm{d}\mathbf{p}'_1 \cdots \mathrm{d}\mathbf{p}'_N \mathrm{d}\mathbf{r}_1 \cdots \mathrm{d}\mathbf{r}_N A(\mathbf{p}', \mathbf{r}) \exp\left[-\beta H(\mathbf{p}', \mathbf{r})\right]}{\frac{1}{N!} \int \mathrm{d}\mathbf{p}'_1 \cdots \mathrm{d}\mathbf{p}'_N \mathrm{d}\mathbf{r}_1 \cdots \mathrm{d}\mathbf{r}_N \exp\left[-\beta H(\mathbf{p}', \mathbf{r})\right]}. \tag{5.39}$$

We label the right-hand side of the above equation as $\langle A(\mathbf{p}', \mathbf{r}) \rangle_{\mathrm{NVT}}$, since it is equivalent to the canonical ensemble average of the operator $A(\mathbf{p}', \mathbf{r})$ in a system with Hamiltonian $H(\mathbf{p}', \mathbf{r})$ at T. In other words,

$$\langle A(\mathbf{p}', \mathbf{r}) \rangle = \langle A(\mathbf{p}', \mathbf{r}) \rangle_{\mathrm{NVT}}. \tag{5.40}$$

From this equality, it is clear that $\langle A(\mathbf{p}', \mathbf{r}) \rangle_{\mathrm{NVT}}$, the canonical ensemble average of $A(\mathbf{p}', \mathbf{r})$ in the subsystem, can be evaluated using the micro-canonical ensemble average of the extended system. Numerical, if the quasi-ergodic hypothesis is further imposed, by combining Eqs. (5.37) and (5.40), this canonical ensemble average further equals the time average of this physical quantity along the deterministic trajectory of the "Nosé" system, through

$$\langle A(\mathbf{p}', \mathbf{r})\rangle_{\text{NVT}} = \lim_{\tau \to \infty} \frac{1}{\tau} \int_0^\tau A(\mathbf{p}', \mathbf{r}) \mathrm{d}t. \qquad (5.41)$$

In other words, imagining that the real system we want to investigate is subjected to the Hamiltonian in Eq. (5.35), Eq. (5.40) means that the canonical ensemble of a certain physical quantity in this real system can be evaluated using the micro-canonical ensemble average of an artificial extended system, in which a deterministic trajectory is generated. When the quasi-ergodic hypothesis is imposed on the MD simulation of the micro-canonical extended system, the equality in Eq. (5.41) further means that this canonical ensemble average of the physical quantity in the subsystem is completely equivalent to the time average of this physical quantity on the deterministic trajectory of the extended system.

Till now, the principle underlying the Nosé–Hoover thermostat is clear. The micro-canonical distribution of the extended ("Nosé") system with an augmented variable s and Hamiltonian as shown in Eq. (5.25) is equivalent to a canonical distribution of the subsystem with variables $\mathbf{p}'_i$ and $\mathbf{r}_i$, where $\mathbf{p}'_i$ is the scaled momentum $\mathbf{p}_i/s$. This feature is obtained by artificially introducing an additional degree of freedom s, which is subjected to an external potential of a special form, and the property that the canonical ensemble can be sampled this way is guaranteed by the above equations. We note that the trajectory in Eq. (5.41) needs to be generated by Eq. (5.36), using an even distribution of the time interval $\mathrm{d}t$.

From Eqs. (5.27), (5.36) and (5.39), one can also see that the scaled momentum $\mathbf{p}'_i$ is the real momentum of the canonical system under investigation and the momentum $\mathbf{p}_i$ before scaling is only for the artificial extended "Nosé" system. Because of this difference, from now on, we call the variables related to $\mathbf{p}'_i$ the real variables and those related to $\mathbf{p}_i$ the virtual ones. Since Eq. (5.36) is the equation of motion from which the trajectory is determined, the ensemble average in Eq. (5.39) must follow an even distribution of the virtual time (t). Following this rigorous route that the trajectory is generated for the extended system with an even distribution of the virtual time and the propagation generated from Eq. (5.36) is on the virtual variables, while the ensemble average is done for the subsystem with real momentum $\mathbf{p}'_i$, there exists some obvious inconsistency which is cumbersome to handle.

To avoid such an inconsistency where the trajectory is calculated and the ensemble averages are performed in different ways, we can rewrite the equations of motion in Eq. (5.36) by scaling the key variables in the virtual system to the real one through: $\mathbf{r}'_i = \mathbf{r}_i$, $\mathbf{p}'_i = \mathbf{p}_i/s$, $s' = s$, $p'_s = p_s/s$, and $dt' = dt/s$. In terms of these real variables, the equation of motion in Eq. (5.36) can be rewritten as

$$\frac{d\mathbf{r}'_i}{dt'} = s\frac{d\mathbf{r}_i}{dt} = s\frac{\mathbf{p}_i}{m_i s^2} = \frac{\mathbf{p}_i}{m_i s} = \frac{\mathbf{p}'_i}{m_i}, \tag{5.42a}$$

$$\frac{d\mathbf{p}'_i}{dt'} = s\frac{d(\mathbf{p}_i/s)}{dt} = \frac{d\mathbf{p}_i}{dt} - \frac{1}{s}\frac{ds}{dt}\mathbf{p}_i = -\frac{\partial V(\mathbf{r}_1,\ldots,\mathbf{r}_N)}{\partial \mathbf{r}_i} - \frac{ds}{dt}\mathbf{p}'_i$$

$$= -\frac{\partial V(\mathbf{r}'_1,\ldots,\mathbf{r}'_N)}{\partial \mathbf{r}'_i} - \frac{s'p'_s}{Q}\mathbf{p}'_i, \tag{5.42b}$$

$$\frac{ds'}{dt'} = s\frac{ds}{dt} = \frac{sp_s}{Q} = \frac{s'^2 p'_s}{Q}, \tag{5.42c}$$

$$\frac{dp'_s}{dt'} = s\frac{d(p_s/s)}{dt} = \frac{dp_s}{dt} - \frac{1}{s}\frac{ds}{dt}p_s = \frac{1}{s'}\left[\sum_{i=1}^{N}\frac{\mathbf{p}'^2_i}{m_i} - \frac{g}{\beta}\right] - \frac{s'p'^2_s}{Q}. \tag{5.42d}$$

The deterministic trajectory of a system involving variables $\mathbf{r}'_i = \mathbf{r}_i$, $\mathbf{p}'_i = \mathbf{p}_i/s$, $s' = s$, $p'_s = p_s/s$, and $dt' = dt/s$ can be determined from Eq. (5.42). Along this trajectory, the conserved quantity is

$$H'_{\text{Nosé}} = \sum_{i=1}^{N}\frac{1}{2m_i}\mathbf{p}'^2_i + V(\mathbf{r}'_1,\ldots,\mathbf{r}'_N) + \frac{s'^2 p'^2_s}{2Q} + \frac{g}{\beta}\ln s'. \tag{5.43}$$

But we note that this is not the Hamiltonian of the system since the equations of motion cannot be obtained from it. The MD simulation performed in this manner, accordingly, is called non-Hamiltonian MD in Ref. [246].

Now, we can generate a trajectory using Eq. (5.42) which satisfies the equation of motion defined by the Hamiltonian in Eq. (5.25). Accordingly, the expectation value of a physical quantity $A(\mathbf{p}',\mathbf{r}')$ should be evaluated with it. However, we note that the application of the quasi-ergodic hypothesis as shown in Eq. (5.41), which relates the ensemble average of the extended micro-canonical ensemble to the real canonical subsystem, is achieved by using the trajectory generated with Eq. (5.36) for the extended system, where an even distribution of dt is imposed. In order for this quasi-ergodic hypothesis to be applied to the new trajectory of the extended system, where an even distribution of dt' is used, a mathematical transformation and justification must be made, which we will show in the following paragraphs.

Along the new trajectory, the time average of a certain physical quantity, e.g. $\mathbf{A}(\mathbf{p}(t')/s(t'), \mathbf{r}(t'))$, can be evaluated by

$$\lim_{\tau' \to \infty} \frac{1}{\tau'} \int_0^{\tau'} \mathbf{A}(\mathbf{p}(t')/s(t'), \mathbf{r}(t')) dt', \tag{5.44}$$

according to the quasi-ergodic hypothesis, where an even distribution on dt' is imposed. Since $dt' = dt/s$ and the trajectory is determined for the real subsystem, the virtual time interval dt varies in each time step during the MD simulation. The τ' in Eq. (5.44) is related to the virtual time τ in Eq. (5.37) through

$$\tau' = \int_0^\tau \frac{1}{s(t)} dt. \tag{5.45}$$

We note that till now, along the trajectory generated by Eq. (5.42), the time average of the physical quantity as given in Eq. (5.44) is not related to the ensemble average of any kind. In order to relate it to an ensemble average of some kind, we can use Eq. (5.45) and rewrite Eq. (5.44) in terms of the trajectory in the virtual time as

$$\frac{1}{\tau'} \lim_{\tau' \to \infty} \int_0^{\tau'} \mathbf{A}(\mathbf{p}(t')/s(t'), \mathbf{r}(t')) dt' = \frac{\tau}{\tau'} \frac{1}{\tau} \lim_{\tau \to \infty} \int_0^\tau \frac{\mathbf{A}(\mathbf{p}(t)/s(t), \mathbf{r}(t))}{s(t)} dt$$

$$= \frac{\frac{1}{\tau} \lim_{\tau \to \infty} \int_0^\tau \frac{\mathbf{A}(\mathbf{p}(t)/s(t), \mathbf{r}(t))}{s(t)} dt}{\tau'/\tau}$$

$$= \frac{\frac{1}{\tau} \lim_{\tau \to \infty} \int_0^\tau \frac{\mathbf{A}(\mathbf{p}(t)/s(t), \mathbf{r}(t))}{s(t)} dt}{\frac{1}{\tau} \lim_{\tau \to \infty} \int_0^\tau \frac{1}{s(t)} dt}. \tag{5.46}$$

Here, $\frac{1}{\tau} \lim_{\tau \to \infty} \int_0^\tau \cdots dt$ is the time average of a physical quantity along the trajectory in the virtual time (that generated by Eq. (5.36)).

Then, making use of Eq. (5.37), these quantities related to the time average of the virtual trajectory can be further simplified into

$$\frac{\left\langle \frac{\mathbf{A}(\mathbf{p}(t)/s(t), \mathbf{r}(t))}{s(t)} \right\rangle}{\left\langle \frac{1}{s(t)} \right\rangle}, \tag{5.47}$$

where $\langle \cdots \rangle$ denotes the ensemble average of the extended (Nosé) system. Now, we can make use of the partition function in Eq. (5.31) again and further simplify Eq. (5.47) into

$$\frac{\left\langle \frac{A(p(t)/s(t),r(t))}{s(t)} \right\rangle}{\left\langle \frac{1}{s(t)} \right\rangle}$$

$$= \frac{\left\{ \frac{\int d\mathbf{p}'_1 \cdots d\mathbf{p}'_N d\mathbf{r}_1 \cdots d\mathbf{r}_N A(\mathbf{p}',\mathbf{r}) \exp[-\beta \frac{3N}{g} H(\mathbf{p}',\mathbf{r})]}{\int d\mathbf{p}'_1 \cdots d\mathbf{p}'_N d\mathbf{r}_1 \cdots d\mathbf{r}_N \exp[-\beta \frac{3N+1}{g} H(\mathbf{p}',\mathbf{r})]} \right\}}{\left\{ \frac{\int d\mathbf{p}'_1 \cdots d\mathbf{p}'_N d\mathbf{r}_1 \cdots d\mathbf{r}_N \exp[-\beta \frac{3N}{g} H(\mathbf{p}',\mathbf{r})]}{\int d\mathbf{p}'_1 \cdots d\mathbf{p}'_N d\mathbf{r}_1 \cdots d\mathbf{r}_N \exp[-\beta \frac{3N+1}{g} H(\mathbf{p}',\mathbf{r})]} \right\}}$$

$$= \frac{\int d\mathbf{p}'_1 \cdots d\mathbf{p}'_N d\mathbf{r}_1 \cdots d\mathbf{r}_N A(\mathbf{p}',\mathbf{r}) \exp[-\beta \frac{3N}{g} H(\mathbf{p}',\mathbf{r})]}{\int d\mathbf{p}'_1 \cdots d\mathbf{p}'_N d\mathbf{r}_1 \cdots d\mathbf{r}_N \exp[-\beta \frac{3N}{g} H(\mathbf{p}',\mathbf{r})]}$$

$$= \frac{\frac{1}{N!} \int d\mathbf{p}'_1 \cdots d\mathbf{p}'_N d\mathbf{r}_1 \cdots d\mathbf{r}_N A(\mathbf{p}',\mathbf{r}) \exp[-\beta \frac{3N}{g} H(\mathbf{p}',\mathbf{r})]}{\frac{1}{N!} \int d\mathbf{p}'_1 \cdots d\mathbf{p}'_N d\mathbf{r}_1 \cdots d\mathbf{r}_N \exp[-\beta \frac{3N}{g} H(\mathbf{p}',\mathbf{r})]}. \tag{5.48}$$

From this equation, it is clear that if we take $g = 3N$, this quantity equals the ensemble average of a canonical system simulated by Eq. (5.42), which we denote as $\langle \cdots \rangle_c$, and by linking Eqs. (5.44), (5.46), (5.47), with (5.48), we can get

$$\frac{1}{\tau'} \lim_{\tau' \to \infty} \int_0^{\tau'} A(\mathbf{p}(t')/s(t'), \mathbf{r}(t')) dt' = \langle A(\mathbf{p}', \mathbf{r}) \rangle_c. \tag{5.49}$$

It is worthwhile to note that in Eq. (5.46), the quantity to be integrated has a denominator $s(t)$, which cancels one $s(t)$ in the $s(t)^{3N+1}$ in Eq. (5.31). It is this tiny difference that results in the fundamentally different ways for the canonical ensemble to be sampled by the trajectories from Eqs. (5.36) and (5.42). In the former case, the "Nosé" thermostat simulates a canonical ensemble using trajectory generated with an evenly distributed virtual time, and the parameter g needs to be set as $3N + 1$. In the latter case, the refined form of the "Nosé" thermostat simulates a canonical ensemble using trajectory generated with an evenly distributed real time, and the parameter g is set as $3N$. In most of the current implementations of this thermostat, the latter form is used.

We note that in real implementations of Eq. (5.42), it can be further simplified by replacing the variable s' by ξ' and p'_s by $p'_{\xi'}$, where $\xi' = \ln s'$

and $p'_{\xi'} = s'p_{s'}$. In doing so, Eq. (5.42) can be rewritten as

$$\frac{dr'_i}{dt'} = \frac{\mathbf{p}'_i}{m_i}, \tag{5.50a}$$

$$\frac{d\mathbf{p}'_i}{dt'} = -\frac{\partial V(\mathbf{r}'_1, \ldots, \mathbf{r}'_N)}{\partial \mathbf{r}'_i} - \frac{p'_{\xi'}}{Q}\mathbf{p}'_i, \tag{5.50b}$$

$$\frac{d\xi'}{dt'} = \frac{p'_{\xi'}}{Q}, \tag{5.50c}$$

$$\frac{dp'_{\xi'}}{dt'} = \sum_{i=1}^{N} \mathbf{p}'^2_i/m_i - g/\beta. \tag{5.50d}$$

Here, Q is still a parameter we need to set in our simulations which is associated with the mass of the particle in the extra degree of freedom. From Eq. (5.50b), it is clear that $p'_{\xi'}$ is associated with the friction coefficient in controlling the temperature. Since the variables $\mathbf{r}'_i$, $\mathbf{p}'_i$, dt' are the real variables, g equals $3N$. The energy which is conserved (Eq. (5.43)) is now reformed to

$$H'_{\text{Nosé}} = \sum_{i=1}^{N} \frac{\mathbf{p}'^2_i}{2m_i} + V(\mathbf{r}'_1, \ldots, \mathbf{r}'_N) + \frac{p'^2_{\xi'}}{2Q} + \frac{g}{\beta}\xi'. \tag{5.51}$$

Again, we note that this is not a real Hamiltonian since the equations of motion cannot be generated from it. It is a quantity which should be conserved during the MD simulation.

5.2.3 Nosé–Hoover Chain

As simple as it is, the Nosé–Hoover thermostat succeeded in sampling the canonical ensemble in a series of systems. There are, however, many situations when the quasi-ergodic hypothesis fails. A 1D system when the external potential is a harmonic oscillator is such an example. In this case, the Andersen thermostat still works well on sampling the Maxwell–Boltzmann distribution of the velocities. The Nosé–Hoover thermostat, however, only samples a micro-canonical ensemble [242]. Now, it has been realized that the reason for this breakdown is due to the fact that in the deduction of the equations from Eq. (5.28) to Eq. (5.33), it is essential to identify all the conserved quantities [246].

In the 1990s, a scheme in which the Nosé–Hoover thermostat is coupled to other thermostats to take these additional conservation laws into account has been proposed by Martyna *et al.* to get over this deficiency [245, 246]. In this improved version of the Nosé–Hoover thermostat, a system of N nuclei is coupled to a chain of thermostats (say, the chain number is M) through

$$\frac{d\mathbf{r}'_i}{dt'} = \frac{\mathbf{p}'_i}{m_i}, \tag{5.52a}$$

$$\frac{d\mathbf{p}'_i}{dt'} = -\frac{\partial V(\mathbf{r}'_1, \ldots, \mathbf{r}'_N)}{\partial \mathbf{r}'_i} - \frac{p'_{\xi'_1}}{Q_1}\mathbf{p}'_i, \tag{5.52b}$$

$$\frac{d\xi'_1}{dt'} = \frac{p'_{\xi'_1}}{Q_1}, \tag{5.52c}$$

$$\frac{dp'_{\xi'_1}}{dt'} = \left[\sum_{i=1}^{N} \frac{\mathbf{p}'^2_i}{m_i} - g/\beta\right] - \frac{p'_{\xi'_2}}{Q_2}p'_{\xi'_1}, \tag{5.52d}$$

$$\frac{d\xi'_j}{dt'} = \frac{p'_{\xi'_j}}{Q_j} \ldots\ldots\ldots\ldots\ldots 2 \leqslant j \leqslant M-1, \tag{5.52e}$$

$$\frac{dp'_{\xi'_j}}{dt'} = \left[\frac{p'^2_{\xi'_{j-1}}}{Q_{j-1}} - 1/\beta\right] - \frac{p'_{\xi'_{j+1}}}{Q_{j+1}}p'_{\xi'_j}, \tag{5.52f}$$

$$\frac{d\xi'_M}{dt'} = \frac{p'_{\xi'_M}}{Q_M}, \tag{5.52g}$$

$$\frac{dp'_{\xi'_M}}{dt'} = \left[\frac{p'^2_{\xi'_{M-1}}}{Q_{M-1}} - 1/\beta\right], \tag{5.52h}$$

where $g = 3N$. The conserved total energy out of these equations of motion is

$$H'_{\text{NHC}} = \sum_{i=1}^{N} \frac{\mathbf{p}'^2_i}{2m_i} + V(\mathbf{r}'_1, \ldots, \mathbf{r}'_N) + \sum_{j=1}^{M} \frac{p'^2_{\xi'_j}}{2Q_j} + \frac{g}{\beta}\xi'_1 + \sum_{j=2}^{M} \frac{1}{\beta}\xi'_j. \tag{5.53}$$

Now, it has been shown that the Nosé–Hoover chain is very efficient in sampling ergodically the statistical property of many systems. It is also of practical use that dynamic properties such as the vibrational frequency can be extracted from their trajectories, although a rigorous justification of such applications is absent. As mentioned at the end of Sec. 5.2.1, after the BO MD method became a standard routine in practical *ab initio* MD simulations since the late 1990s, other stochastic methods such as the Andersen and Langevin thermostats were also widely used. In Sec. 5.2.1, we have introduced the Andersen thermostat. In the following, we will give a brief introduction to the method of temperature control using Langevin dynamics.

5.2.4　Langevin Thermostat

The Langevin dynamics is a simple and efficient scheme to control the temperature of a canonical ensemble, which is often used in *ab initio* MD simulations nowadays. The equations of motion for the particles are

$$\dot{\mathbf{r}}_i = \mathbf{p}_i/m_i, \tag{5.54a}$$

$$\dot{\mathbf{p}}_i = \mathbf{F}_i - \gamma\mathbf{p}_i + \mathbf{R}_i. \tag{5.54b}$$

Here, $\mathbf{r}_i$ and $\mathbf{p}_i$ are the position and velocity of the ith particle in the simulation respectively. $\mathbf{F}_i$ is the force imposed by the other particles explicitly treated in the simulation. The two parameters which are essential to control the temperature in this scheme are γ and $\mathbf{R}_i$. The first one (γ) represents a viscous damping due to fictitious "heat bath" particles, while the second one ($\mathbf{R}_i$) represents the effect of collisions with these particles. A balance between them, which decelerates or accelerates the particles to be simulated, respectively, keeps the average kinetic energy and consequently, the temperature of the system constant.

In MD simulations where all atoms are explicitly treated, these "heat bath" particles are purely fictitious, originating from the thermal fluctuations. In computer simulations of large molecules in a solvent composed of smaller ones (though these solvent molecules will not be explicitly simulated), this "heat bath" represents thermal effects originating from real collisions between the smaller molecules in the solvent which are implicitly treated and the larger ones which are explicitly treated under simulation. Therefore, in the limiting case for very large friction coefficients, the Brownian movement of the large molecules in solvent can also be described by this equation, which simplifies into

$$\dot{\mathbf{r}}_i = \gamma^{-1}m_i^{-1}\left[\mathbf{F}_i + \mathbf{R}_i\right] \tag{5.55}$$

because $\dot{\mathbf{p}}_i = 0$. In this book, we focus on the MD simulations where all nuclei are explicitly treated. Therefore, the Brownian movement of the large molecules in solvent is unrelated to the main problem we want to solve. But it is worthwhile to note that the principles underlying this dynamics also apply to these problems.

There are four properties for the stochastic forces $\mathbf{R}_i$ to comply with in order to control the temperature in simulations using Langevin dynamics: (i) they are uncorrelated with $\mathbf{r}_i$ and $\mathbf{p}_i$, (ii) the autocorrelation function of $\mathbf{R}_i(t)$ has a form of δ-function, (iii) their mean values are zero, and (iv) their distribution is Gaussian-like and the unit variance of this Gaussian function is $\sqrt{2k_\mathrm{B}T\gamma m/\Delta t}$. For a clarification of these four properties, we note that the first one is easy to understand since it just implies that the random force imposed by the collisions of the nuclei and the fictitious "heat bath" particles is uncorrelated with the position and momentum of the nuclei. The second point needs some explanation. It originates from the fact that the timescale of the collisions between the particles under simulations and the fictitious "heat bath" ones is much smaller than that of the ionic movement. Therefore, the autocorrelation function of the random force $\mathbf{R}_i$ reads

$$\langle \mathbf{R}_i(t)\mathbf{R}_i(t')\rangle = 2k_\mathrm{B}T\gamma m\delta(t - t'). \tag{5.56}$$

These t and t' only take values with integer time step along the trajectory. Besides this, it is also assumed that during an MD movement, there are many collisions between the particles under simulation and the "heat bath". Therefore, one can impose a Gaussian distribution on $\mathbf{R}_i$ based on the central limit theorem [256]. Then the Stokes–Einstein relation for the diffusion coefficient can be used to show that the unit variance associated with this Gaussian distribution is $\sqrt{2k_\mathrm{B}T\gamma m/\Delta t}$ and its mean value is zero.

With this, the principles underlying the choice of $\mathbf{R}_i$ for the simulation of a canonical ensemble using the Langevin dynamics is clear. We note that this choice of $\mathbf{R}_i$ limits this thermostat to the equilibrium simulations only. For its extension to the non-equilibrium state simulations, one may refer to Ref. [257]. The last point which needs to be explained for the simulation of a canonical ensemble using Langevin dynamics is that the equations of motion in Eq. (5.54) are equivalent to a Fokker–Planck equation in description of the phase space probability density distribution function $\rho(\mathbf{r}_1,\ldots,\mathbf{r}_N,\mathbf{p}_1,\ldots,\mathbf{p}_N)$, of the form

$$\frac{\partial\rho}{\partial t} + \sum_{i=1}^{N}\left[\frac{\mathbf{p}_i}{m_i}\cdot\nabla_{\mathbf{r}_i}\rho + \mathbf{F}_i\cdot\nabla_{\mathbf{p}_i}\rho\right] = \gamma\sum_{i=1}^{N}\nabla_{\mathbf{p}_i}\cdot\left[\mathbf{p}_i\rho + m_ik_\mathrm{B}T\nabla_{\mathbf{p}_i}\rho\right],$$

$$\tag{5.57}$$

which has the canonical phase space probability distribution function as a stationary solution. Therefore, this method is justified for the simulation of the canonical ensemble. From these principles, it is clear that the choice of γ is the key issue in this simulation, which actually determines a compromise between the statistical sampling efficiency of the preservation of the accuracy in short-term dynamics. As a stochastic method, one should not expect the dynamics to be comparable to those from the simulations using deterministic trajectories, like those of Nosé–Hoover thermostat simulations. But the stochastic feature normally guarantees the ergodicity to be sampled in a more efficient way.

In the end, it is worthwhile to note that although traditional stochastic thermostats often cause breakdown of the adiabatic separation in the CP MD simulations, a recent development of the Langevin dynamics actually showed that using correlated noise, it is possible to tune the coupling of a stochastic thermostat with the various degrees of freedom for the nuclear motion [254]. This development not only enables the use of the Langevin dynamics in the CP MD simulations, it also improves the canonical sampling efficiency because this thermostat can be tailored in a predictable and controlled fashion. For more details about this recent development, please see Ref. [254].

5.2.5   Andersen and Parrinello–Rahman Barostats

In simulations of an NVT ensemble, a non-flexible unit cell is used. Therefore, although the thermal fluctuation of the nuclei can be properly described using thermostats, the fluctuation of the simulation cell is completely absent. This fluctuation of the unit cell is a consequence of the mechanical contact of the system with its environment, which is inevitable in most experiments. Therefore, similar to the control of temperatures using thermostats in the NVT simulations, it is also highly desirable to introduce a barostat to account for this fluctuation of the unit cell, so that the environment under simulation is as close to real experiments as possible.

Based on this consideration, there have been many methods proposed to control this quantity in the MD simulations from the earliest Andersen scheme [239, 258] and Parrinello–Rahman scheme [247, 248], to the more recent work using Langevin dynamics [256, 259]. The corresponding simulations are called NPT, or isobaric–isothermal simulations in the literature. In this section, we will use the Andersen and Parrinello–Rahman schemes as examples to illustrate how this is done. For principles underlying the scheme using Langevin dynamics, please see Refs. [256, 259].

In the Andersen scheme [239], a constant-pressure simulation method for a fluid system is proposed by introducing the instantaneous volume of the simulation cell as an environmental variable in the effective Lagrangian. Due to the isotropic feature of the system (fluid) to be simulated, a cubic cell is used with volume Ω. The side length of this cubic cell equals $\Omega^{1/3}$. Then, the Cartesian coordinates of the nuclei ($\mathbf{r}_i$) can be scaled with respect to the instantaneous size of the simulation cell by $\mathbf{x}_i = \mathbf{r}_i/(\Omega^{1/3})$. In terms of these scaled variables and the instantaneous Ω, the Lagrangian can be written as

$$L_{\text{Andersen}} = \sum_{i=1}^{N} \frac{m_i}{2} \Omega^{2/3} \dot{\mathbf{x}}_i^2 - \sum_{i>j} V(\Omega^{1/3} \mathbf{x}_{i,j}) + \frac{M}{2} \dot{\Omega}^2 - P_0 \Omega. \tag{5.58}$$

Here, M is a parameter which can be viewed as an effective inertia for the expansion and contraction of the volume. P_0 is the external potential. $\mathbf{x}_{i,j}$ is the scaled distance (with no unit) between the ith atom and the jth atom in the contracted or expanded cell.

There is a physical interpretation of the $P_0 \Omega$ term in Eq. (5.58), in the original words of Andersen: "Suppose the fluid to be simulated is in a container of variable volume. The fluid can be compressed by a piston. Thus, Ω, whose value is the volume, is the coordinate of the piston, $P_0 \Omega$ is a PV potential derived from an external pressure P_0 acting on the piston, and M is the mass of the piston. The piston is not of the usual cylindrical type that expands or contracts the system along only one direction; instead, a change in Ω causes an isotropic expansion or contraction." But we note that this interpretation is not entirely consistent with the Lagrangian in Eq. (5.58), since from $\mathbf{x}_i = \mathbf{r}_i/(\Omega^{1/3})$, one easily derives

$$\dot{\mathbf{r}}_i = \Omega^{1/3} \dot{\mathbf{x}}_i + \frac{1}{3} \Omega^{-2/3} \dot{\Omega} \mathbf{x}_i. \tag{5.59}$$

This equation means that the kinetic energy of the atoms should also contain contributions from $\dot{\Omega}$, which are absent in Eq. (5.58). In spite of this, we acknowledge that Eq. (5.58) still presents a well-defined Lagrangian.

Now, we analyze the dynamics generated from this Lagrangian. The momentum conjugate to $\mathbf{x}_i$ is denoted as π_i, which equals

$$\pi_i = \frac{\partial L_{\text{Andersen}}}{\partial \dot{\mathbf{x}}_i} = m_i \Omega^{2/3} \dot{\mathbf{x}}_i. \tag{5.60}$$

The momentum conjugate to Ω is denoted as Π, which equals

$$\Pi = \frac{\partial L_{\text{Andersen}}}{\partial \dot{\Omega}} = M \dot{\Omega}. \tag{5.61}$$

The Hamiltonian corresponding to the Lagrangian in Eq. (5.58) is then written as

$$H_{\text{Andersen}}(\mathbf{x}_1, \ldots, \mathbf{x}_N, \pi_1, \ldots, \pi_N, \Omega, \Pi) = \sum_{i=1}^{N} \dot{\mathbf{x}}_i \cdot \pi_i + \Omega \Pi - L_{\text{Andersen}}$$

$$= (2\Omega^{2/3})^{-1} \sum_{i=1}^{N} \frac{1}{m_i} \pi_i \cdot \pi_i + \sum_{i>j} V(\Omega^{1/3} \mathbf{x}_{i,j}) + (2M)^{-1} \Pi^2 + P_0 \Omega.$$

$$(5.62)$$

With this Hamiltonian, one can derive the equations of motion as

$$\frac{d\mathbf{x}_i}{dt} = \frac{H_{\text{Andersen}}}{\partial \pi_i} = \frac{\pi_i}{m_i \Omega^{2/3}},$$

$$\frac{d\pi_i}{dt} = -\frac{H_{\text{Andersen}}}{\partial \mathbf{x}_i} = -\Omega^{1/3} \sum_{i>j} \frac{\mathbf{x}_{i,j}}{|\mathbf{x}_{i,j}|} \frac{\partial V(\Omega^{1/3} \mathbf{x}_{i,j})}{\partial \mathbf{x}_i},$$

$$\frac{d\Omega}{dt} = \frac{H_{\text{Andersen}}}{\partial \Pi} = \frac{\Pi}{M},$$

$$\frac{d\Pi}{dt} = \frac{H_{\text{Andersen}}}{\partial \Omega}$$

$$= -(3\Omega)^{-1} \left(-2(2\Omega^{2/3})^{-1} \sum_{i=1}^{N} \frac{1}{m_i} \pi_i \cdot \pi_i + \Omega^{1/3} \right.$$

$$\left. \times \sum_{i<j} \mathbf{x}_{i,j} \frac{\partial V(\Omega^{1/3} \mathbf{x}_{i,j})}{\partial \mathbf{x}_i} + 3P_0 \Omega \right). \quad (5.63)$$

With these equations of motion, one can use the MD method as introduced in Sec. 5.1 for the micro-canonical system to generate a trajectory for $\mathbf{x}_1, \ldots, \mathbf{x}_N, \pi_1, \ldots, \pi_N, \Omega$, and Π. We note that the volume is not a conserved quantity in real space. However, the scaled extended system with variables $\mathbf{x}_1, \ldots, \mathbf{x}_N, \pi_1, \ldots, \pi_N, \Omega$, and Π can still be regarded as a micro-canonical ensemble in which the volume is unity since the coordinates $\mathbf{x}_i$ are all scaled to lie within a dimensionless unit cell. Under the ergodic hypothesis, the time average over this trajectory is equivalent to the ensemble average of the micro-canonical extended system. Therefore, for any function of $\mathbf{x}_1, \ldots, \mathbf{x}_N, \pi_1, \ldots, \pi_N, \Omega$, and Π, the time average over this trajectory will be regarded as equivalent to the ensemble average of this quantity on the extended system, and the NPT ensemble average of the real subsystem can be naturally evaluated using this trajectory too. We note that this method is one of the first methods which employed an extended Lagrangian in an MD simulation. Later development of many ensemble simulation methods, e.g. the Nosé–Hoover thermostat, has benefited a lot from this idea.

One shortcoming of the Andersen barostat is that an isotropic change of the simulation cell must be assumed. This assumption is reasonable in simulations of fluids. However, when it comes to solids, especially in the case when the unit cell is not cubic, in addition to the volume, one must also allow the shape of the simulation cell to change. This generalization is essential in studies of structural phase transitions. Based on this consideration, Parrinello and Rahman soon designed a scheme which allows the change of the simulation cell's shape in MD simulations [247, 248]. In this method, instead of using only one variable Ω, nine variables associated with three vectors in the Cartesian coordinate ($\mathbf{a}$, $\mathbf{b}$, and $\mathbf{c}$) are used, which allow both the shape and the volume of the simulation cell to be changed. These nine variables are $\mathbf{A}_{i,j}$, where i (and j) goes from 1 to 3. $\mathbf{A}$ is the matrix representation of ($\mathbf{a}$, $\mathbf{b}$, $\mathbf{c}$). The volume (Ω) of the simulation cell equals $\mathbf{a} \cdot (\mathbf{b} \times \mathbf{c})$. The Lagrangian for this extended system is

$$
L_{\mathrm{PR}} = \frac{1}{2} \sum_{i=1}^{N} m_i (\mathbf{A}\dot{\mathbf{x}}_i)^{\mathrm{T}} \mathbf{A}\dot{\mathbf{x}}_i - \sum_{i>j} V(\mathbf{A}^{-1}\mathbf{x}_{i,j}\mathbf{A}) + \frac{M}{2} \sum_{i,j=1}^{3} \dot{\mathbf{A}}_{i,j} - P_0\Omega.
$$

$$(5.64)$$

Using the same trick that we have applied to the Andersen barostat (and Nosé–Hoover thermostat), one can derive the equations of motion for $\mathbf{x}_i$ and $\mathbf{A}_{i,j}$. We do not go through this deduction here; for such details, please see Refs. [247, 248].

5.3 Examples for Practical Simulations in Real Poly-Atomic Systems

After understanding the principles of MD as introduced above, one can easily carry out a practical MD simulation using a computer code which is accessible in either a commercial or an open-source manner. Taking one of the most frequently used settings, an *ab initio* BO MD simulation using a Nosé–Hoover chain thermostat for the NVT ensemble as an example, in

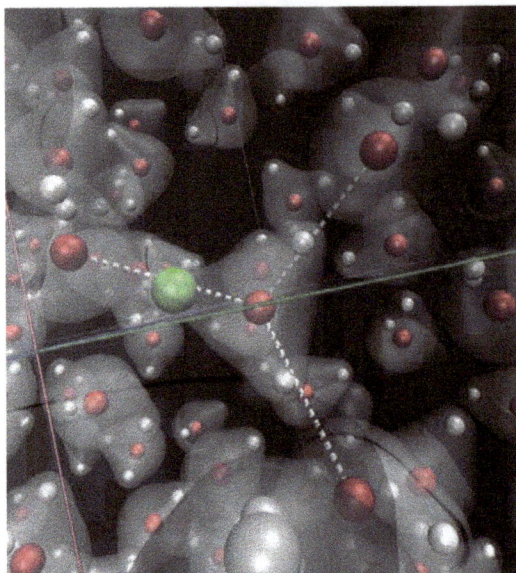

Figure 5.1   Illustration of an *ab initio* MD simulation. We show one snapshot of an *ab initio* MD simulation on a system composed of 32 water molecules and one extra proton. The extra proton is denoted by the sphere in green. The oxygen nuclei and other hydrogen nuclei are shown by spheres in red and white, respectively. The silver contour indicates the electron density distribution, which is computed in an *ab initio* manner "on-the-fly" as the system evolves.

Fig. 5.1, we show one snapshot of a system composed of 32 water molecules and one extra proton. This simulation is based on the implementation of the MD method in the VASP code [260–264]. As mentioned in the introduction, the nuclei are treated as classical point-like particles. Therefore, the ball-and-stick model is still used for the representation of the nuclei in this snapshot. The electronic structures, however, are computed "on-the-fly" in an *ab initio* manner as the system evolves. This is reflected by the silver contour representing the density distribution of the electrons at this snapshot, which describes the bond making and breaking processes during the MD simulation in a seamless manner.

From such snapshots and statistical (as well as dynamical) analysis, one can get quite some information concerning the evolution of the electronic structures and associated nuclear motion at the molecular level in such systems. For example, one can analyze the electron charge transfer between the extra proton and the bulk water solution explicitly using the electronic structure of such a snapshot. Specifically, this can be done by calculating the

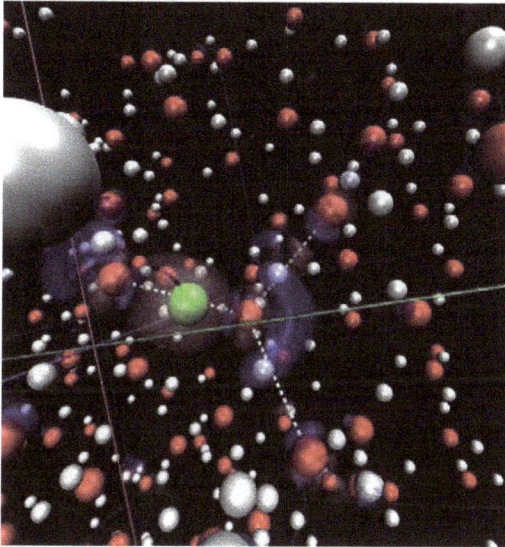

Figure 5.2   Illustration of how the electron charge transfer is analyzed in an *ab initio* MD simulation. The same simulation cell and notation of the nuclei are used in Fig. 5.1. The contour in brown shows electron charge accumulation and the contour in blue indicate electron charge depletion. This analysis shows that an extraordinarily strong hydrogen bond is associated with the extra proton. Besides this, significant electron charge redistribution also exists on the neighboring hydrogen bonds, enhancing most of them. Therefore, for a reasonable characterization of the transfer process of this extra proton, besides the dynamics of this extra proton, the hydrogen bond dynamics of the "shell" is also crucial.

electron density distribution of the system composed of the water molecules and the extra proton first, and then that for each of them (solely the water molecules and solely the proton) separately. Taking the difference between the first one and the sum over the latter two, one obtains how the electron charge density is redistributed by the interactions between the extra proton and the water solution in a very clear manner. In Fig. 5.2, we show how this electron charge transfer behaves in the neighborhood of the excess proton. Besides the electron accumulation and depletion on the water molecules and hydrogen bond associated with the extra proton, obvious electron charge redistribution exists in the neighboring region, indicating that the hydrogen bond strengths in the second shell are also significantly influenced by the presence of this extra proton.

The above analysis also indicates that the behaviors of the hydrogen bonds in the neighboring shells and the excess proton are correlated.

Because of this, for a better understanding of the proton transport mechanism in liquid water, one can investigate in more detail the hydrogen bond dynamics in the presence of this excess proton. As a matter of fact, this issue has been intensively studied from both the experimental and theoretical perspectives in the last two decades [264–274]. Now, it has been widely accepted that the behavior of this extra proton is basically determined by the dynamics of the hydrogen bonds within the neighboring shells [264–271]. This story can be understood pictorially by taking some sequential snapshots of the MD simulation, as shown in Fig. 5.3. In these snapshots, the

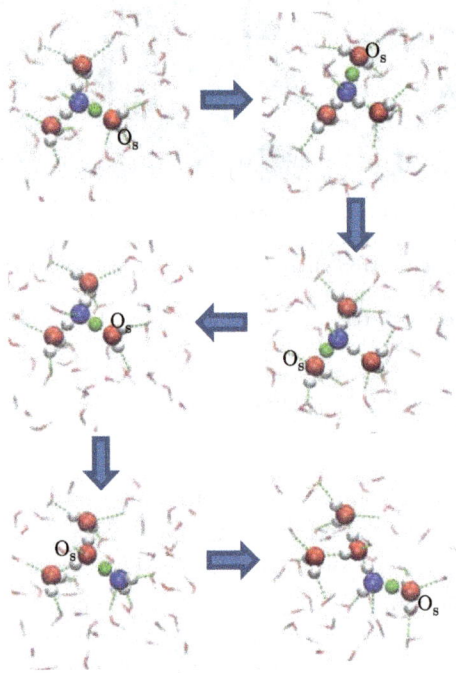

Figure 5.3   Illustration of the hydrogen bond dynamics related to the proton transport in liquid water obtained from an *ab initio* MD simulation using classical nuclei. Red spheres are oxygen nuclei and white spheres are hydrogen nuclei. The excess proton is in green and the pivot oxygen (O_p) is in blue. Hydrogen bonds (except for the most active one) are indicated by green dashed lines. O_s is the oxygen which shares the most active proton with the pivot oxygen. The first three panels show that the most active proton switches identity between the three hydrogen bonds related to O_p. Going from the third to the fourth, the coordination number of O_s reduces from 4 to 3. Then proton transport happens going from the fourth to the fifth panel. After the proton transport, the most active proton begins to exchange sites about a new pivot oxygen (the last two panels).

pivot oxygen (O_p) is defined as the oxygen nucleus which owns three hydrogen atoms simultaneously, and O_s is defined as the oxygen nucleus which shares the most active proton with the pivot one. The pair (hydrogen bond) between O_p and O_s is called "the special pair". In the MD simulation, the pivot oxygen normally stays on one site, i.e. the pivot oxygen's identity does not change, peacefully for a certain time interval (~ 1 ps). During this time, the special pair switches identity between the hydrogen bonds in which the pivot oxygen is involved frequently, which is often called the special pair dancing in Refs. [267–271]. The coordination number of O_s is normally 4 during this dancing process (the first three panels in Fig. 5.3). Upon going from the third panel to the fourth one, the coordination number of O_s reduces from 4 to 3. Then proton transport happens (the fifth panel). After the proton transport, the most active proton begins to exchange sites about a new pivot oxygen (the last two panels). In short, the proton transport happens through the well-known Grotthuss mechanism [275, 276], with the hydrogen bond dynamics determining the interconversion between different hydration states of the excess proton.

The second example we want to show briefly here for practical simulations using the *ab initio* MD method is the phase diagram of condensed matter under extreme conditions, e.g. high pressures. It is well known that experimental realization of pressures above 100 GPa is extremely difficult. For theoreticians, however, this is readily doable since an *ab initio* MD simulation using an NPT ensemble easily reproduces the atomic level evolution of the system under investigation if (i) the electronic structures are accurate, (ii) the simulation cell is large enough, and (iii) the simulation time is long enough. As a matter of fact, this simulation technique has already been widely used in studies of the high-pressure phase diagrams for a series of systems [277–280, 291]. One of the most prominent examples is the high-pressure phase diagram of hydrogen, as shown in Fig. 5.4. Below 300 GPa, it is well known that three regions exist, consisting of the molecular solid, molecular liquid, and the atomic liquid. The dissociation line between the molecular liquid and the atomic liquid as shown in the figure was just determined by the *ab initio* MD simulations in Refs. [277, 278].

We note that the accuracy of the electronic structures obtained from the present *ab initio* method, which is affordable for the MD simulations, is still a problem which determines the accuracy of the theoretical predictions of this phase diagram [282–285]. As a matter of fact, this is a problem which is vital for the accuracy of the MD simulations in general. It is also

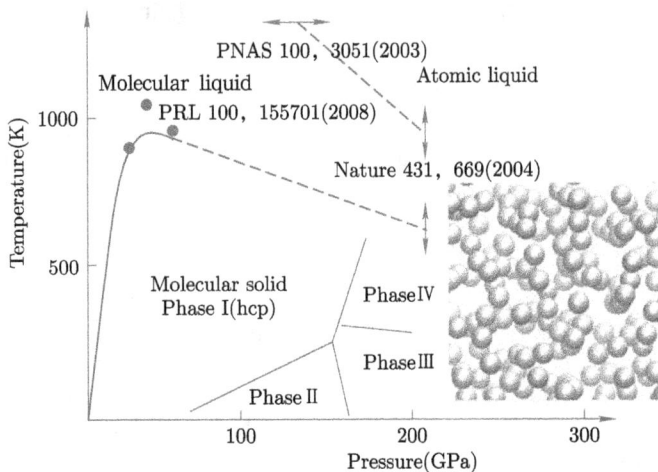

Figure 5.4   Illustration of hydrogen phase diagram under pressure. The dissociation line between the molecular liquid phase and the atomic liquid phase was obtained using the *ab initio* MD simulations in Refs. [277, 278].

why when chemical bond breaking and forming happen, non-reactive force field-based method is to be abandoned in descriptions of the interatomic interactions and *ab initio* MD method prevails in recent years. Therefore, to conclude the chapter with a reminder, when the MD simulation technique is used to interpret the experiments or make predictions, one must pay special attention to the accuracy of the interatomic interactions for all spatial configurations covered during the simulation.

6

Extension of Molecular Dynamics, Enhanced Sampling and the Free-Energy Calculations

The atomistic level MD method as introduced in the previous chapter allows us to simulate the propagation of a system using a finite-time step so that the total energy of this system is conserved and the temperature is under control. This means that the propagation needs to be done with a time interval on the order of femtoseconds, or even shorter. Processes of real chemical interest, however, happen at much longer time scales. Taking the *ab initio* MD simulation of some molecules being adsorbed on a metal surface as an example, currently, it requires a powerful computer cluster with a few hundred processors roughly one month to simulate the MD propagation of 100 ps [286]. Imagining that we want to see a rare-event chemical reaction happening at a time scale of one millisecond, we need to continue this simulation on this cluster for about 10^6 years in order to see this event happening once. Even if the *ab initio* method is renounced in descriptions of the interatomic interactions, the empirical potentials are used so that the MD simulation can propagate for, say 100 nanoseconds a day, reproducing the propagation of 1 millisecond still requires a computer simulation of about 30 years. Plus, this event needs to happen several times in order for the statistics to work. Therefore, from a statistical point of view, the brute force MD technique is completely unacceptable in practical simulations involving such rare events. A better scheme, in which rare events like this can be simulated, is highly desired.

During the past years, great effort has been made in order for this purpose to be fulfilled. One route is to give up the all-atom description and use a coarse-grained model [287, 288]. In doing so, the time step used for the MD simulation can be significantly increased so that the time scale accessible to computer simulations increases, sometimes to a value large enough to complement biological or chemical experiments. However, it is worth noting that an *a priori* detailed knowledge of the system to be simulated is required, which unfortunately is often not available.

A different route, which keeps the atomistic feature for the description of the propagation, is to develop a method so that the frequency with which a rare event happens can be significantly increased. Or, in other words, an enhanced sampling efficiency is guaranteed. Over the last few decades (almost four), great success has been achieved in this direction. A number of methods, e.g. the umbrella sampling [94, 95], the adaptive umbrella sampling [33], the replica exchange [34, 35], the metadynamics [38, 39], the multi-canonical simulations [40, 41], the conformational flooding [42], the conformational space annealing [43], the integrated tempering sampling [44–47] methods, etc. had been proposed, each having its own strength and weakness. For example, in the umbrella sampling, adaptive umbrella sampling, metadynamics, and configurational flooding methods, predetermined reaction coordinates are required before the simulations were carried out. While in replica exchange and multi-canonical simulations, different temperature simulations are needed, which make the trajectories obtained lose their dynamical information. In this chapter, we will take the umbrella sampling, the adaptive umbrella sampling, and the metadynamics methods as examples of the former class to show how they work in practice. In addition to this, the integrated tempering sampling method from the second class will also be discussed.

The enhanced sampling methods as introduced above focus on exploring the free energy profile, or otherwise often called potential of mean force (PMF), throughout the entire conformational phase space of the polyatomic system under investigation. In some applications of the molecular simulations, e.g. phase transitions, this entire PMF exploration can be avoided if the two competing phases are well-defined metastable states in standard MD simulations. In such cases, the free energy of the system at each phase can be calculated using methods like thermodynamic integration at a certain temperature and pressure. Then, by monitoring the competition of these two free energies at different temperatures and pressures, one can find out which phase is more stable at a certain condition and consequently determine the phase diagram of this substance under investigation. Since phase transition is also an important topic in molecular simulations, principles underlying the thermodynamic integration method will also be explained in this chapter. We note that although this is the motivation for our explanation of the thermodynamic integration method, the range of its applications is much wider. As a matter of fact, the principles underlying the idea of thermodynamic integration can be applied to the mapping of the PMF in a much more general sense, as long as a certain parameter can be defined to link states of interest whose free energies are to be calculated.

This chapter is organized as follows. In Sec. 6.1, we introduce principles underlying PMF exploration and the umbrella sampling, as well as the adaptive umbrella sampling methods. Then, we will explain, in brief, the metadynamics method in Sec. 6.2. The integrated tempering sampling method, as an example of enhanced sampling in which an *a priori* definition of transition coordinate is not required, is explained in Sec. 6.3. The thermodynamic integration method for the calculation of the free energy of a certain phase is introduced in Sec. 6.4. We hope this introduction can help graduate students who have started working on molecular simulations to set up some concepts before carrying out their enhanced sampling simulations.

6.1 Umbrella Sampling and Adaptive Umbrella Sampling Methods

As mentioned in the introduction, a key concept in our understanding of the enhanced sampling method is the PMF, which was first given by Kirkwood in 1935 [289]. In this definition, a reaction coordinate ξ is used, which is a function of the poly-atomic system's nuclear configuration in the Cartesian space. Here, we denote this spatial configuration of the nuclei as $\mathbf{x}$. It is a $3N$-dimensional vector composed of $\mathbf{x}^i$, with i going through 1 and N. $\mathbf{x}^i$ represents the Cartesian coordinate of the ith nucleus and ξ can be written as $\xi(\mathbf{x})$. This nomenclature for the nuclear configuration will also be used for our discussions of the path integral method in Chapter 7.

From the principles of statistical mechanics, one can first write down the free energy profile of the poly-atomic system as a function of ξ, with the variable denoted by ξ_0, as

$$F(\xi_0) = -\frac{1}{\beta} \ln P(\xi_0)$$

$$= -\frac{1}{\beta} \ln \frac{\int e^{-\beta V(\mathbf{x})} \delta(\xi(\mathbf{x}) - \xi_0) \, d\mathbf{x}}{Q}. \tag{6.1}$$

Here, β equals $1/(k_{\mathrm{B}}T)$ and Q is the canonical partition function of the system:

$$Q = \int e^{-\beta V(\mathbf{x})} d\mathbf{x}. \tag{6.2}$$

ξ_0 is the variable used to construct the PMF. In the following discussion, we use a 1D case as an example, for simplicity. The extension of the equations to higher-dimensional situations is straightforward. This PMF is a key function in descriptions of the configurational (or conformational) equilibrium properties as well as the transition rate of the dynamically activated processes in computer simulations for the behaviors of poly-atomic systems under finite temperatures.

From this definition of the PMF, one knows that it is closely related to the BO PES since the latter, together with the thermal and quantum fluctuations of the nuclei, determine the probability distribution function $P(\xi_0)$ at a finite temperature. In other words, one can think of the PMF as a revised version of the PES, with the thermal and quantum effects included when addressing the electronic and nuclear degrees of freedom. In the introduction of this chapter, one can see that in cases when there exist multiple deep local minima on the PES, the probability distribution function $P(\xi_0)$ is far from being ergodic and one needs the enhanced sampling method to map out the PMF over the entire conformational space relevant to the problem of interest. Among the various efforts carried out in the last, close-to-four decades, the umbrella sampling method first proposed by Torrie and Valleau is now believed to be one of the most influential methods [94, 95]. The mathematical basis for a treatment like this is that in order for an ergodic sampling of the conformational space relevant to the problem of interest, an additional biased potential V_{b} as a function ξ can be added to the real interatomic potential $V(\mathbf{x})$ in the molecular simulations. With this biased additional potential, a biased PMF can be constructed using

$$F_{\mathrm{b}}(\xi_0) = -\frac{1}{\beta} \ln P_{\mathrm{b}}(\xi_0)$$

$$= -\frac{1}{\beta} \ln \frac{\int e^{-\beta[V(\mathbf{x})+V_{\mathrm{b}}(\xi(\mathbf{x}))]}\delta(\xi(\mathbf{x})-\xi_0) d\mathbf{x}}{Q_{\mathrm{b}}}, \tag{6.3}$$

where

$$Q_{\mathrm{b}} = \int e^{-\beta[V(\mathbf{x})+V_{\mathrm{b}}(\xi(\mathbf{x}))]} d\mathbf{x}. \tag{6.4}$$

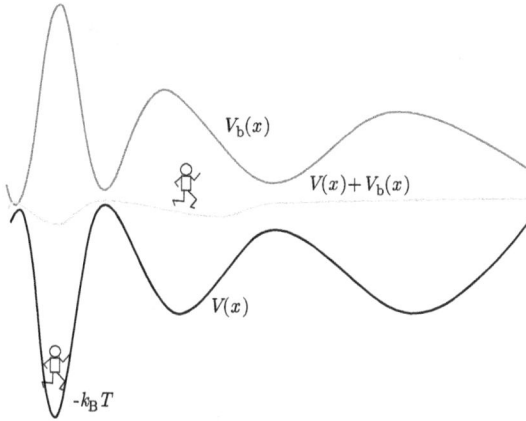

Figure 6.1   Taking the simplest case, i.e. the system is 1D and the ξ is the same as the x, as an example, the original PES $V(x)$ has a very deep valley so that the MD simulation cannot go through the whole space in practical MD simulations. However, if an additional biased potential $V_b(x)$ which roughly equals $-V(x)$ is added to the original potential, the new potential $V_b(x) + V(x)$ will be very flat so that in an MD simulation using this potential, the particle can go through the entire x axis freely and $F_b(x)$ can be sampled very well. Based on this $F_b(x)$, the original free energy profile can be reconstructed using Eq. (6.7). This principle also applies to the system with higher dimensions for the nuclear degree of freedom and more complex form of the reaction coordinate ξ.

Due to the fact that the biased potential can be chosen in a way that $V(\mathbf{x})+V_b(\xi(\mathbf{x}))$ is rather flat as a function of ξ (see Fig. 6.1), the probability distribution $P_b(\xi_0)$ can be much better-sampled over the whole conformational space relevant to the problem of interest compared with $P(\xi_0)$ in the MD simulations so that the biased PMF can be calculated efficiently using MD statistics.

Then, one relates the biased PMF $F_b(\xi_0)$ with the unbiased one $F(\xi_0)$ using the following equation:

$$F_b(\xi_0) = -\frac{1}{\beta}\ln\frac{\int e^{-\beta[V(\mathbf{x})+V_b(\xi(\mathbf{x}))]}\delta(\xi(\mathbf{x})-\xi_0)\,d\mathbf{x}}{Q_b}$$

$$= -\frac{1}{\beta}\ln e^{-\beta V_b(\xi_0)}\frac{\int e^{-\beta V(\mathbf{x})}\delta(\xi(\mathbf{x})-\xi_0)d\mathbf{x}}{Q_b}$$

$$= V_b(\xi_0) - \frac{1}{\beta}\ln\left[\frac{\int e^{-\beta V(\mathbf{x})}\delta(\xi(\mathbf{x})-\xi_0)d\mathbf{x}}{Q}\frac{Q}{Q_b}\right]$$

$$= V_{\mathrm{b}}(\xi_0) - \frac{1}{\beta} \ln \frac{\int e^{-\beta V(\mathbf{x})} \delta(\xi(\mathbf{x}) - \xi_0) d\mathbf{x}}{Q} - \frac{1}{\beta} \ln \frac{Q}{Q_{\mathrm{b}}}$$

$$= V_{\mathrm{b}}(\xi_0) + F(\xi_0) - \frac{1}{\beta} \ln \frac{Q}{Q_{\mathrm{b}}}. \tag{6.5}$$

Therefore, from the biased PMF $F_{\mathrm{b}}(\xi_0)$, which can be sampled well in practical MD simulations over the whole conformational space, one can reconstruct the unbiased PMF through

$$F(\xi_0) = F_{\mathrm{b}}(\xi_0) - V_{\mathrm{b}}(\xi_0) + \frac{1}{\beta} \ln \frac{Q_{\mathrm{b}}}{Q}. \tag{6.6}$$

We note that this $(1/\beta) \ln(Q_{\mathrm{b}}/Q)$ is a constant, which we denote as F. It is independent of ξ_0, but determined by the choice of $V_{\mathrm{b}}(\xi)$, the conformational space of the poly-atomic system relevant to the problem of interest, the original interatomic potential, and the temperature. Rewriting Eq. (6.6) also gives us

$$F(\xi_0) = F_{\mathrm{b}}(\xi_0) - V_{\mathrm{b}}(\xi_0) + F. \tag{6.7}$$

As simple as it looks, the addition of a biased potential easily solves the problem for the ergodic exploration of the conformational space for the mapping of the PMF in principle, in the MD simulations. However, we note that in practical applications, the high-dimensional PES function $V(\mathbf{x})$ is always unknown. Therefore, it is impossible to define an additional potential in its negative form in an *a priori* manner so that the system can go through the entire conformational space relevant to the problem of interest efficiently, using this biased potential. When this is the case and the chosen additional potential does not compensate for the original one, the system will still be trapped in one local minimum in the MD simulations under the biased potential (see e.g. Fig. 6.2).

One way to circumvent this problem and ensure that the whole conformational space relevant to the problem of interest is sufficiently (or even uniformly) sampled is to separate this conformational space into some boxes using the chosen reaction coordinates. When there is only one reaction coordinate, this is simplified into taking some values of ξ in the region relevant to the problem of interest, say ξ^i. Then, a series of MD simulations can be carried out using these biased potentials which constrain the system in the neighborhood of these ξ^is so that the conformational spaces of the system with reaction coordinates in the neighborhood of all these ξ^is are all well sampled. Taking the 1D PMF associated with two stable states (reactant

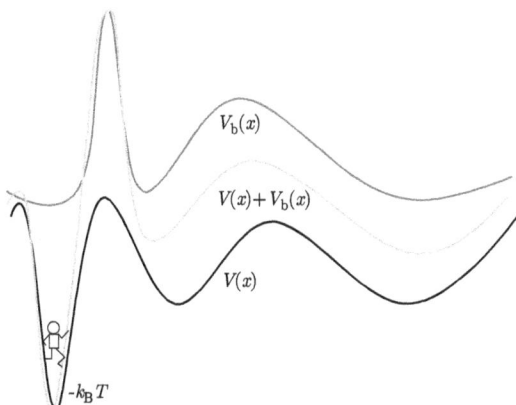

Figure 6.2 The situation is similar to Fig. 6.1. However, one does not know the form of the original potential and therefore the additional one does not compensate for it well. In this case, the sampling of the system with the biased potential will not be ergodic on the conformational space either.

and product) separated by a high energy barrier as an example (Fig. 6.3), in the language of Torrie and Valleau [94, 95], a series of bias potentials can be used along the reaction path between the reactant and the product. At each point along this path, a bias potential is used to confine the system to its neighborhood and the corresponding simulation is often called a "biased window simulation". From this simulation, one can reconstruct the PMF in this neighborhood from the MD simulation using the biased potential. Then, by linking the PMF in each region, one obtains the PMF along the reaction coordinate all over the conformational space concerned. We note that there can be many choices for this additional constraint potential. One of the most often used is the harmonic function with the form $V_b^i(\xi_0) = K(\xi_0 - \xi^i)^2/2$, as shown by the red curves in Fig. 6.3. Due to the fact that this additional bias potential looks like an umbrella (Fig. 6.3), this method is called the umbrella sampling in the literature. With this treatment, a uniform series of sample points in the region relevant to the problem of interest will ensure that the conformational space is sufficiently sampled and consequently the PMF well reconstructed.

The main technical problem related to the reconstruction of the PMF from umbrella sampling as mentioned above originates from the fact that in Eq. (6.7), there is a constant which depends on the choice of the additional potential and the region it explores. This indicates that this constant is different in each of the biased MD simulations. Consequently, one needs to

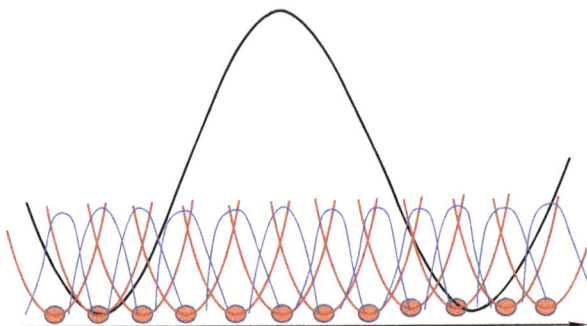

Figure 6.3   Illustration of how umbrella sampling is used in simulations of the transition between a reactant state and a product, one separated by high energy barrier. Along the reaction coordinate, a series of points were taken. At each point ξ_i, a bias potential with the form $V_{\mathrm{b}}(\xi) = K(\xi - \xi_i)^2/2$ is used to constrain the system to its neighborhood. From this biased simulation, the PMF at this neighborhood can be obtained from the corresponding probability distribution (indicated by the blue curve). These series of additional biasing potentials ensure that the whole conformation space relevant to the problem of interest is sufficiently sampled.

align the PMF reconstructed in each region of the simulation window so that the relative values of the PMF obtained from different MD simulations make sense. Traditionally, this is done by adjusting $F(\xi_0)$ of the adjacent boxes (windows) in which they overlap so that they match [94–96], using, e.g. least-square method [290]. In doing so, the PMF over the whole region of interest can be constructed. However, we note that there are serious limitations for this scheme. First, in matching the overlap region between the PMF of the neighboring windows, a significant overlap is required to ensure statistical accuracy. This indicates that a lot of sampling in the MD simulations are superfluous and consequently wasted, and more importantly, when more than one reaction coordinate is used for the construction of the PMF, the value of the constant F in one simulation window allowing the best fit with its adjacent region in one direction may not ensure the best for others [291]. Therefore, this scheme is of limited use in practical simulations of complex systems when analysis of high-dimensional PMF is required.

Among the various attempts to solve these problems [290, 292–294], the weighted histogram analysis method (WHAM) proposed by Kumar *et al.* is now the most popular, mainly due to its numerical stability and convenience in addressing PMF with multiple variables [295]. The central equation for the WHAM method includes the optimization of the unbiased distribution function as a weighted sum over its expression in different windows as

$$P(\xi_0) = \sum_{i=1}^{N_w} \left[P^i(\xi_0) \frac{n_i e^{-\beta(V_b^i(\xi_0) - F_i)}}{\sum_{j=1}^{N_w} n_j e^{-\beta(V_b^j(\xi_0) - F_j)}} \right], \tag{6.8}$$

where N_w is the number of windows and $P^i(\xi_0)$ represents the unbiased probability distribution reconstructed from the biased MD simulation in the ith window using $V_b^i(\xi_0)$. Similar to discussions above, ξ_0 denotes the variable and $V_b^i(\xi_0) = K(\xi_0 - \xi^i)^2/2$. $P(\xi_0)$ is the optimized overall distribution function obtained from the N_w biased simulations. This definition of $P(\xi_0)$ (instead of a direct sum over $P^i(\xi_0)$) is very reasonable since not only the relative weights between the number of independent data points within each window (i.e. n_i) are taken into account (more data, better statistics and consequently larger weights in the summation), but also, the small weights of $P^i(\xi_0)$ in the region with large $V_b^i(\xi_0) - F_i$ means that the error originating from worse statistics of $P^i(\xi_0)$ in this region becomes numerically well under control. This expression, together with the right values of the F_is (i goes from 1 to N_w), gives the unbiased distribution function of $P(\xi_0)$ throughout the regions on the ξ axes relevant to the conformational space of interest. What one needs to do next is to optimize the values of F_is in Eq. (6.8).

To understand how this is done in practice, we go back to Eq. (6.7) for the biased MD simulation in each window. This equation changes to

$$F^i(\xi_0) = F_b^i(\xi_0) - V_b^i(\xi_0) + F^i, \tag{6.9}$$

where $F^i(\xi_0) = -\ln P^i(\xi_0)/\beta$ and $F_b^i(\xi_0) = -\ln P_b^i(\xi_0)/\beta$. $P_b^i(\xi_0)$ represents the biased probability distribution sampled with the biased potential around ξ^i, which is directly obtained from the ith MD simulations. Statistically, it samples the region around ξ^i rather well. By inputting the expressions of $F^i(\xi_0)$ and $F_b^i(\xi_0)$ into Eq. (6.9), one gets

$$P^i(\xi_0) = P_b^i(\xi_0) e^{V_b^i(\xi_0) - F^i}. \tag{6.10}$$

Therefore, Eq. (6.8) evolves into

$$P(\xi_0) = \sum_{i=1}^{N_w} \frac{n_i P_b^i(\xi_0)}{\sum_{j=1}^{N_w} n_j e^{-\beta(V_b^j(\xi_0) - F_j)}}. \tag{6.11}$$

The free energies F_Is in Eq. (6.11), on the other hand, satisfy the following equation:

$$F_i = -\frac{1}{\beta} \ln \frac{Q_{\mathrm{b}}^i}{Q}$$

$$= -\frac{1}{\beta} \ln \frac{\int e^{-\beta[V(\mathbf{x})+V_{\mathrm{b}}^i(\xi(\mathbf{x}))]} \mathrm{d}\mathbf{x}}{Q}$$

$$= -\frac{1}{\beta} \ln \int e^{-\beta V_{\mathrm{b}}^i(\xi_0)} P(\xi_0) \mathrm{d}\xi_0. \tag{6.12}$$

Therefore, Eqs. (6.11) and (6.12) compose a set of equations which define the relationship between the F_is and $P(\xi_0)$. By solving these two equations self-consistently, one can obtain a much better and more efficient estimate of $P(\xi_0)$ compared with the traditional PMF matching method from the statistical point of view. Once again, we note that the essence of this method, in the language of Roux [96], is "constructing an optimal estimate of the unbiased distribution function as a weighted sum over the data extracted from all the simulations and determining the functional form of the weight factors that minimizes the statistical error".

So far we have discussed the umbrella sampling method. From this introduction, we know that a key point on reconstructing the PMF using the biased potential is that a sufficient sampling, in the best case a uniform sampling, of the conformational space relevant to the problem of interest is guaranteed. To this end, an *a priori* set of umbrella potentials are needed in order for this uniform sampling to be carried out in a practical manner. For simulations of relatively simple systems, in which the variable related to the umbrella potential is easy to define, this method works well. However, for complex systems where this *a priori* definition of the umbrella potential is unlikely, which unfortunately is usually true in reactions with more than one degree of freedom, this method often fails.

An alternative method within a similar scheme, which is more often used in present explorations of the PMF in complex systems, is the so-called adaptive umbrella sampling method [296–298]. A key difference between the adaptive umbrella sampling method and the umbrella sampling method is that in the former case, the umbrella potentials are chosen and updated in the simulations, while in the latter one, the umbrella potentials do not change. Plus, in the umbrella sampling method, the conformational space is separated into a series of regions and the MD simulation with bias potential centered at each region have different F_i. A key point in combining the PMFs constructed from the biased MD simulations in umbrella sampling is to align these different F_is. Conversely, in the adaptive umbrella sampling method, the umbrella potential does not focus a specific region, but rather tries to compensate the original PES. Consequently, the problem related to the treatment of the constant in Eq. (6.7) transforms into the shift of this value during the iterations so that in the end, it equals the free energy density of the system over the whole conformation space relevant to the study.

The central idea behind the adaptive umbrella sampling method is that if an umbrella potential with the form $V_b(\xi_0)$ results in a uniform biased probability distribution $P_b(\xi_0)$, the corresponding umbrella potential satisfies the following equation:

$$V_b(\xi_0) = -V(\xi_0) = \frac{1}{\beta} \ln P(\xi_0), \tag{6.13}$$

where $P(\xi_0)$ stands for the unbiased probability distribution we want to simulate. Based on this equation, the biasing potential can be adapted to the PMF which is determined using the information from the previous simulations in an iterative manner [299]. The analysis of the PMF is often carried out using WHAM [295]. To be more specific, an initial guess of the umbrella potential, which is often taken as $V_b^1(\xi_0) = 0$, can be used for the MD simulation and a biased probability distribution $P_b^1(\xi_0)$ is obtained. From this biased probability distribution, one can construct the unbiased one ($P^1(\xi_0)$). Due to the ergodic problem related to the original PES, this $P^1(\xi_0)$ will be very different from the final $P(\xi_0)$. However, one can input this unbiased probability distribution into Eq. (6.13) and get a new umbrella potential $V_b^2(\xi_0)$. Adding this biasing potential to the original potential, one can perform the second biased MD simulation and generate a new $P_b^2(\xi_0)$. From this biased probability distribution, a normalized unbiased $P^2(\xi_0)$ can be achieved. This unbiased potential will be summed up with the unbiased one in the first round subjected to some normalized weighting factors to generate the new input for Eq. (6.13). From Eq. (6.13), again, one obtains a new umbrella potential. Continue this iteration and pay attention to the fact that the input for Eq. (6.13) is always generated using weighted sum over unbiased probability distributions obtained from earlier iterations, until the conformational space relevant to the problem of interest has been sampled adequately. One can then get a well-converged unbiased PMF. This unbiased PMF can then be used for statistical or dynamical studies in the future. We note that this is just a general description of how the adaptive umbrella sampling works. There are many numerical details concerning real implementation. For these details, please see Ref. [299, 300].

6.2   Metadynamics

In the above discussions on the umbrella sampling and adaptive umbrella sampling methods, Eq. (6.7) is the basis for the analysis to be carried out and rigorous justification for the construction of the unbiased PMF exists behind the biased MD simulations. Parallel to this scheme, however, there is another algorithm which (as pointed by A. Laio, the main founder of this method) did not follow from any ordinary thermodynamic identity but was rather postulated on a heuristic basis and later verified empirically to be very successful in the enhanced sampling simulations of many complex systems [301–307]. This method is the so-called metadynamics [38, 39].

The principles underlying this metadynamics method can be understood pictorially using Fig. 6.4. The first step is to choose a sensible collective variable (labelled as CV, corresponding to the reaction coordinate used in the earlier discussion) which, in principle, should be able to distinguish the initial, intermediate, and final states and describe the slow processes relevant to the problem of interest. This is similar to the umbrella sampling and adaptive umbrella sampling methods introduced before. The difference appears afterwards. Imagine that only one CV is chosen and the BO PES looks like the black curve in Fig. 6.4, starting from one deep valley, the system will take a time which is unacceptably long for atomic simulations to escape from it using the standard MD method. In the language of metadynamics, what one can do in order to impose an efficient enhanced sampling on the PES is to add in some additional potentials so that the deep valley can be filled up quickly (Fig. 6.4(a)). This potential can be written in terms of Gaussian functions as

$$V_{\mathrm{G}}(\xi(\mathbf{x}), t) = w \sum_{t' = \tau_{\mathrm{G}}, 2\tau_{\mathrm{G}}, \dots \text{ and } t' < t} e^{-\frac{(\xi(\mathbf{x}) - \xi(t'))^2}{2(\delta s)^2}}, \qquad (6.14)$$

where $\xi(\mathbf{x})$ stands for the CV coordinate associated with the spatial configuration of the nuclei $\mathbf{x}$, and $\xi(t')$ stands for its instantaneous value at t'. It is obvious that the Gaussian height w, Gaussian width δs, and the time interval τ_{G} at which the Gaussians are added control the form of this additional potential and consequently the accuracy and efficiency of the

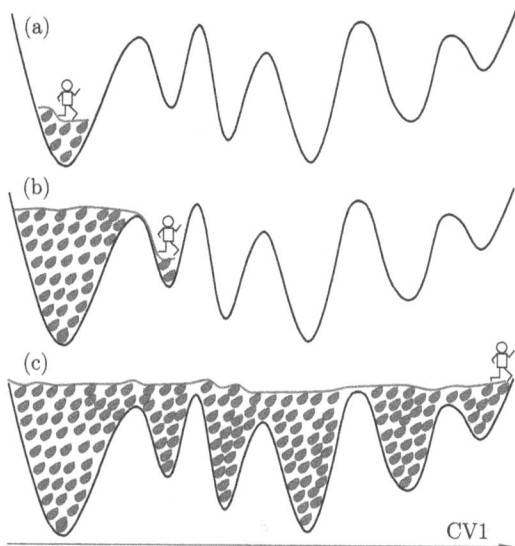

Figure 6.4   Illustration of the principles underlying the metadynamics method, taking the 1D case as an example. Imagine that the PES has multiple deep valleys. In metadynamics, some artificial potentials will be added along the chosen collective variable coordinates as the system evolves. At a certain valley, this is similar to the case that a man falls into a deep well. To get out of it, simply by jumping is unlikely, but imagine that he has an infinite amount of stones in his "magic" pocket, he can throw a stone every few steps he walks. With time going on, these stones will fill up the well and he can get out of it easily. By remembering the geometric shape of the space filled up by the stone, intuitively the PMF can be reconstructed. It is easy to understand that the size of the stone determines the accuracy and efficiency of the reconstructed PMF. As mentioned earlier, this rationalization is postulated on a heuristic basis. However, we note that it has been verified empirically to be very successful in the enhanced sampling simulations of many complex systems [301–307].

enhanced sampling. Imagine that the Gaussian potentials are like small stones (size of the stone controlled by w and δs) which fill up the valley gradually, the difference between the curve the stick man is standing on and the curve below then naturally represents the sum over the Gaussian functions as shown in Eq. (6.14), and the size of the stones determines the efficiency and accuracy for the construction of this difference between the two curves. After one valley was filled up, the walker which describes the evolution of the system in the enhanced sampling process starts to explore the neighboring one (Fig. 6.4(b)), until all the valleys are filled. After this, the system is allowed to travel between the valleys in a free manner (Fig. 6.4(c)). When the exploration of the walker is controlled using an ensemble-based method, such as the MD or the Monte-Carlo (MC) algorithm, the random walking above all the barriers will naturally include thermal fluctuations and entropy. Consequently, if a canonical ensemble is used, the difference between the curve the man treads on (which is not flat) and the curve below intuitively reflects the profile for the PMF. As a result, the following equality exists:

$$F(\xi) = -\lim_{t \to \infty} V_{\mathrm{G}}(\xi, t). \tag{6.15}$$

This equation serves as the basic assumption of metadynamics. In practice, many details exist concerning the implementation of the algorithm described above. Readers are directed to Ref. [39] for a detailed explanation. We note that at first glance, this metadynamics looks similar to the adaptive umbrella sampling, due to the fact that the sum of the Gaussian functions gradually gives us the PMF using Eq. (6.15). In the adaptive umbrella sampling method, a gradual history-dependent improvement on the assessment of the PMF also exists by going through the iterations. However, we note that the philosophy is very different. In the adaptive sampling method, the update of the PMF using Eq. (6.13) has a rigorous justification. On the other hand, in the metadynamics, the construction of the PMF using Eq. (6.15) is postulated on a heuristic basis. This postulation was later verified empirically in several systems with increasing complexity, as can be seen, for example in Refs. [38, 39, 301–307].

6.3   Integrated Tempering Sampling

In all methods introduced above, one or more reaction coordinates are required in order to describe the atomistic level evolution of the system in its conformational space. By increasing the complexity of the poly-atomic system, however, it can easily happen that the definition of such reaction coordinates becomes difficult or even impossible. In these cases, another kind of methods, which effectively avoid an *a priori* selection of the reaction coordinates, serve as an alternative. In these methods, a commonly used technical trick is to alter the potential energy landscape using the temperature or energy itself, so that the exploration of the conformational space can be accelerated and the preselection of the transition coordinates is avoided. One of the earliest examples is the Tsallis statistical method proposed in the late 1980s [308]. In this method, a high temperature sampling, which can be viewed as an exploration of the "potential energy surface scaled by temperature", does enforce an enhanced sampling of the rare events. However, due to the relatively primitive algorithm, the high temperature regions, which do not play an important role at the targeting temperature of the canonical ensemble under investigation, can be easily oversampled [309]. To improve on this, a series of methods were developed in the last two decades, with the Wang–Landau [310, 311], replica exchange [34–37], integrated tempering sampling (ITS) [44–46], and selective integrated tempering sampling (SITS) [312] methods being the most prominent successful examples. Among them, the Wang–Landau method possesses a uniform distribution on the energy scale, which is very suitable for MC simulations. The replica exchange, ITS, and SITS are presently very frequently used methods in the MD simulations. A thorough description for all of them is beyond the scope of this chapter. Here, we will take the ITS method as an example to show how such enhanced samplings are realized when the exploration of the PMF focuses on the energy space in the MD simulations.

The ITS method is intrinsically temperature-based, in which a generalized non-Boltzmann ensemble is used. This generalized non-Boltzmann ensemble allows enhanced sampling in a desired broad energy and temperature range. The key quantity is a distribution function of the interatomic potential. It is defined as an integral or summation of the Boltzmann terms over temperature through

$$p(U) = \sum_{k=1}^{M} n_k e^{-\beta_k U}, \tag{6.16}$$

where U stands for the physical interatomic interaction potential and $\beta_k = 1/(k_B T_k)$. M is the number of temperatures used for the summation over the Boltzmann distribution. T_k means the temperature used in each of them, which increases with k from T_1 to T_M. The highest temperature used (T_M) is determined by the height of the barrier and the associated time scale one wants to simulate. n_k is a weight. It is determined through the requirement that each term in the summation in Eq. (6.16) contributes a desired fraction of the system's Boltzmann distribution (at a given temperature T_k) to the total non-Boltzmann one.

In order to allow an MD-based method to sample the distribution function in Eq. (6.16), one must impose an equality between a Boltzmann-like distribution function and the non-Boltzmann one in Eq. (6.16), at a targeting temperature T (with the corresponding β equals $1/(k_B T)$). To this end, the interatomic potential must be revised. Let us assume that this revised form of the interatomic potential is U'. Then, the requirement that the distribution function $p(U)$ in Eq. (6.16) can be reproduced by a finite temperature MD simulation imposes

$$e^{-\beta U'} = p(U) = \sum_{k=1}^{M} n_k e^{-\beta_k U}. \tag{6.17}$$

From this equality, one easily obtains

$$U' = -\frac{1}{\beta} \ln \left[\sum_{k=1}^{M} n_k e^{-\beta_k U} \right], \tag{6.18}$$

and consequently, the biased force $\mathbf{F}_b$ on the ith atom can be calculated from

$$\mathbf{F}_b^i = \frac{\sum_{k=1}^{M} n_k \beta_k e^{-\beta_k U}}{\beta \sum_{k=1}^{M} n_k e^{-\beta_k U}} \mathbf{F}^i, \tag{6.19}$$

where $\mathbf{F}^i$ stands for the Hellmann–Feynman force on the ith nuclei when the inter-atomic potential is the physical one (namely U).

From Eqs. (6.17)–(6.19), it is clear that an MD simulation using a Boltzmann ensemble (but with an "artificial" interatomic potential) can be employed to reproduce the distribution function associated with the non-Boltzmann ensemble in Eq. (6.16). Due to the fact that a wide temperature range is included in the summation over the Boltzmann distribution function in Eq. (6.16) for the non-Boltzmann ensemble, it is expected that the probability of the rare event, which hardly happens in a normal MD simulation at the targeting physical temperature, can be increased. However, one notes that there are still some weighting factors to be determined in Eq. (6.19). These factors are determined through the requirement that each term in the summation of Eq. (6.16) contributes to the total distribution with a desired fraction. In other words, if one defines

$$P_k = n_k \int e^{-\beta_k U(\mathbf{x})} d\mathbf{x}, \tag{6.20}$$

where $\mathbf{x}$ stands for a $3N$-dimensional vector in the conformation space (N means the number of nuclei in the poly-atomic system), each term in the summation of Eq. (6.16) will contribute to the total distribution with a fraction

$$p_k = \frac{P_k}{\sum_{k=1}^{M} P_k}.$$ (6.21)

These p_ks should be aimed at some predetermined quantities with which we want our finite temperature (T_k) Boltzmann distribution to contribute to the non-Boltzmann one. Let us label such a fraction as p_k^0. In practice, the fraction that all these p_k^0s equal, $1/M$ is often used.

With these predetermined p_k^0s, an initial guess for the numbers n_k will be employed in the first, a certain number of, say N_τ, MD steps using the forces generated by Eq. (6.19). The targeting temperature is T. At the end of these N_τ steps, the values of p_k will be calculated using Eqs. (6.20) and (6.21). It is often true that these p_ks are different from their targeting values p_k^0s. To ensure that they fluctuate around and approach their targeting values with the simulation going on, in the next N_τ steps, the value of n_k will be changed into its original value multiplied by p_k^0/p_k. At the end of these N_τ steps, the values of p_k will be calculated again and the values of n_k are updated by the same relation. With the simulation going on, a sufficient sampling of the non-Boltzmann in Eq. (6.16) will be guaranteed. We note that this algorithm reflects only the principles underlying the ITS method. In practical applications, there are more reliable algorithms used. For these technical details, please refer to Gao, Yang, Fan, and Shao's works in Refs. [44, 46].

6.4 Thermodynamic Integration

In the earlier sections, we have introduced some extensions of the MD method as presented in Chapter 5 by mainly focusing on the mapping of the PMF. Another problem which can be studied using the MD simulation technique concerns the phase behavior of a given substance, in particular, transitions between two competing phases obtained from either random structure searching or enhanced sampling molecular dynamics simulations. Melting from a solid to a liquid phase is one example, and evaluation of the relative stability between two competing solid phases is another one. In these examples, the transition between the two competing phases is first-order and their transition curve can be calculated from the principle that at coexistence, the Gibbs free energies of the two phases are equal. Here, we will introduce a method to calculate the free energy of a given phase at a finite temperature and pressure so that such phase transitions can be studied in terms of molecular simulations. We note that continuous transitions, which will not be discussed here, are by no means simpler. Interested readers are directed to Ref. [313] for more details.

The method we want to introduce here is called the thermodynamic integration method [48–58, 280]. To be clear, we start by interpreting the principles underlying this method and then explain some technical details. The key point of this method is that it is a general method used to determine the free energy difference $F_1 - F_0$ between two systems, whose total energy functions are known, denoted as U_1 and U_0. By total energy function, we mean that for a specific spatial configuration of the nuclei $(\mathbf{R}_1, \ldots, \mathbf{R}_N)$, the total energy of the first system is $U_1(\mathbf{R}_1, \ldots, \mathbf{R}_N)$ while the total energy for the second system is $U_0(\mathbf{R}_1, \ldots, \mathbf{R}_N)$. As per its definition, the Helmholtz free energy is determined by these total energy functions through

$$
F_1 = -k_{\mathrm{B}}T \ln \left\{ \frac{1}{N! \Lambda^{3N}} \int \mathrm{d}\mathbf{R}_1 \cdots \mathrm{d}\mathbf{R}_N \mathrm{e}^{-\beta U_1(\mathbf{R}_1, \ldots, \mathbf{R}_N)} \right\},
$$
$$
F_0 = -k_{\mathrm{B}}T \ln \left\{ \frac{1}{N! \Lambda^{3N}} \int \mathrm{d}\mathbf{R}_1 \cdots \mathrm{d}\mathbf{R}_N \mathrm{e}^{-\beta U_0(\mathbf{R}_1, \ldots, \mathbf{R}_N)} \right\},
$$

$$(6.22)$$

where $\Lambda = h/(2\pi M k_{\mathrm{B}}T)^{1/2}$ is the thermal wavelength and M is the nuclear mass. For a simple nomenclature, we have assumed that all the nuclei have the same mass, but we note that the extension of Eq. (6.22) to systems with nuclei of different kinds is straightforward for the illustration of the principles to be discussed below.

By thermodynamic integration, we mean that if one imposes a series of fictitious systems between the ones with free energy F_1 and F_0, with the total energy function

$$
U_\lambda(\mathbf{R}_1, \ldots, \mathbf{R}_N) = U_0(\mathbf{R}_1, \ldots, \mathbf{R}_N) + \lambda(U_1(\mathbf{R}_1, \ldots, \mathbf{R}_N) - U_0(\mathbf{R}_1, \ldots, \mathbf{R}_N)),
$$
$$(6.23)$$

F_1 can be calculated with a thermodynamic integration treatment if F_0 is known. The mathematical foundation underlying this thermodynamic integration treatment is that for one specific value of λ, there is a total energy function $U_\lambda(\mathbf{R}_1, \ldots, \mathbf{R}_N)$ and a free energy F_λ. It is clear from the definition of $U_\lambda(\mathbf{R}_1, \ldots, \mathbf{R}_N)$ in Eq. (6.23) that F_λ equals F_0 if λ equals zero, and it equals F_1 if λ equals one. Therefore, one can plot the evolution of F_λ as a function of λ between zero and one, with the value of F_λ starting

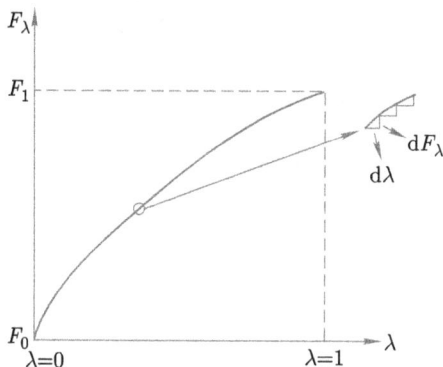

Figure 6.5   Illustration of the evolution of F_λ in the thermodynamic integration method. The x axis corresponds to the variable λ, which goes from 0 to 1. The y axis corresponds to the function F_λ, which goes from F_0 to F_1. F_0 is known while F_1 is unknown, which equals $F_0 + \int_0^1 \frac{\mathrm{d}F_\lambda}{\mathrm{d}\lambda}\mathrm{d}\lambda$.

from F_0 and ending at F_1. Using this evolution, it is clear that

$$F_1 - F_0 = \int_0^1 \frac{\mathrm{d}F_\lambda}{\mathrm{d}\lambda}\mathrm{d}\lambda. \tag{6.24}$$

This is shown in Fig. 6.5.

Now, we look at $\mathrm{d}F_\lambda/\mathrm{d}\lambda$. Similar to Eq. (6.22), F_λ is defined as

$$F_\lambda = -k_{\mathrm{B}}T\ln\left\{\frac{1}{N!\Lambda^{3N}}\int \mathrm{d}\mathbf{R}_1\cdots \mathrm{d}\mathbf{R}_N e^{-\beta U_\lambda(\mathbf{R}_1,\dots,\mathbf{R}_N)}\right\}. \tag{6.25}$$

Therefore, $\mathrm{d}F_\lambda/\mathrm{d}\lambda$ equals

$$\frac{\mathrm{d}F_\lambda}{\mathrm{d}\lambda} = \frac{\int \mathrm{d}\mathbf{R}_1\cdots \mathrm{d}\mathbf{R}_N \frac{\mathrm{d}U_\lambda(\mathbf{R}_1,\dots,\mathbf{R}_N)}{\mathrm{d}\lambda} e^{-\beta U_\lambda(\mathbf{R}_1,\dots,\mathbf{R}_N)}}{\int \mathrm{d}\mathbf{R}_1\cdots \mathrm{d}\mathbf{R}_N e^{-\beta U_\lambda(\mathbf{R}_1,\dots,\mathbf{R}_N)}}. \tag{6.26}$$

If one resorts to the definition of U_λ in Eq. (6.23), it further equals

$$\frac{\mathrm{d}F_\lambda}{\mathrm{d}\lambda} = \frac{\int \mathrm{d}\mathbf{R}_1\cdots \mathrm{d}\mathbf{R}_N[U_1(\mathbf{R}_1,\dots,\mathbf{R}_N) - U_0(\mathbf{R}_1,\dots,\mathbf{R}_n)]\,e^{-\beta U_\lambda(\mathbf{R}_1,\dots,\mathbf{R}_N)}}{\int \mathrm{d}\mathbf{R}_1\cdots\mathbf{R}_N e^{-\beta U_\lambda(\mathbf{R}_1,\dots,\mathbf{R}_N)}}. \tag{6.27}$$

In other words, by putting Eqs. (6.27) and (6.24) together, one arrives at

$$F_1 - F_0 = \int_0^1 \mathrm{d}\lambda\langle U_1(\mathbf{R}_1,\dots,\mathbf{R}_N) - U_0(\mathbf{R}_1,\dots,\mathbf{R}_N)\rangle_\lambda. \tag{6.28}$$

By $\langle\rangle_\lambda$, we mean that the thermal average of the quantity to be evaluated is calculated using an ensemble average of this quantity in a system with atomic interactions $U_\lambda(\mathbf{R}_1, \ldots, \mathbf{R}_N)$. Based on this deduction, it is clear that if the free energy of a system with total energy function $U_0(\mathbf{R}_1, \ldots, \mathbf{R}_N)$ is known, one can calculate the free energy of another system with total energy function $U_1(\mathbf{R}_1, \ldots, \mathbf{R}_N)$ by thermodynamically integrating the free energy difference between F_1 and F_0.

Now, we go to real poly-atomic systems. The definition of the thermodynamic integration as defined in Eq. (6.28) is robust, which means that as long as one has a good reference energy F_0, the free energy of the real system F_1 can be calculated through a continuous and isothermal switching of the total energy function from $U_0(\mathbf{R}_1, \ldots, \mathbf{R}_N)$ to $U_1(\mathbf{R}_1, \ldots, \mathbf{R}_N)$. However, we note that the efficiency of this thermodynamic integration sensitively depends on the similarity between the reference system and the system to be calculated. Mathematically, one can relate this similarity to the value of $U_1(\mathbf{R}_1, \ldots, \mathbf{R}_N) - U_0(\mathbf{R}_1, \ldots, \mathbf{R}_N)$ being evaluated in Eq. (6.27). The smaller this value, the fewer steps one needs to integrate for the value of λ between zero and one in Eq. (6.28).

Traditionally, this thermodynamic integration method has been extensively used in molecular simulations with empirical interatomic potentials in the 1980s and 1990s [48–54]. In studies of liquids, this reference system is often taken as the ideal gas with Lennard–Jones potential, while in studies of solids, this reference system is often chosen as the harmonic lattice. Starting from the 1990s, molecular simulations based on an *ab initio* treatment of the electronic structures have overtaken the traditional method in calculations of such free energies, due to its power in describing interatomic interactions under complex chemical environments [55–58, 280]. Since these *ab initio* calculations are much more expensive than the empirical potential calculations, it is highly recommended that one introduces an intermediate state between the idealized system with a rigorous analytical expression for the free energy and the real system [55, 280]. Due to the fact that a well-designed intermediate state held together by a well-defined empirical potential can be more similar to the real system compared with the idealized model, the thermodynamic integration between this intermediate state and the real one needs a smaller value of λ in Eq. (6.28). To compensate, more thermodynamic integration steps will be needed from the idealized model to the intermediate state. However, since the computational load for empirical potential-based MD simulations is much smaller, this treatment can help us save a lot of computation time, with errors well in control.

Up to now, the principles underlying the thermodynamic integrations are clear. In the following, we use the free energy of a crystal, with electronic structures determined by the density-functional theory (DFT) calculations, as an example to show how this is done in practice [55, 280]. We note that the treatment of the nuclear motion in this thermodynamic integration is classical. Its extensions to the quantum nuclei will be discussed in Sec. 7.3 by resorting to the path integral method as will be introduced in Chapter 7. In the *ab initio* MD-based thermodynamic-integration calculations to be discussed here, the first key quantity is the electronic free energy $U(\mathbf{R}_1, \ldots, \mathbf{R}_N; T_{el})$. Compared with the total energy function $U(\mathbf{R}_1, \ldots, \mathbf{R}_N)$ we have used in the earlier discussions, one may note two differences. The first one is that we prefer the word "free energy" here while in the earlier discussions, we use the word "total energy". The second difference is that there is one more parameter dependence of this total energy, on T_{el}, which denotes the temperature of the electrons. The physical origin of these two differences is the thermal electronic excitations in calculations of the electronic structures, which might be important in some situations [55, 280]. In many practical calculations of the DFT, a smearing factor is used to generate partially occupied orbitals so that the electronic structures can be converged faster. We note that the smearing factor used there is not necessarily related to the real electron temperature. But here, by T_{el}, we really mean the electron temperature which is determined by the environment. The corresponding electronic structure calculations should resort to the finite-temperature DFT developed by Mermin in the 1960s [30–32]. Therefore, the total energy function $U(\mathbf{R}_1, \ldots, \mathbf{R}_N)$ becomes the electronic free energy $U(\mathbf{R}_1, \ldots, \mathbf{R}_N; T_{el})$, since the electronic entropy effects are naturally included in this finite-temperature DFT. Accordingly, T_{el}-dependency of this quantity enters. We note that this electronic free energy equals

$$U(\mathbf{R}_1, \ldots, \mathbf{R}_N; T_{el}) = E(\mathbf{R}_1, \ldots, \mathbf{R}_N; T_{el}) - T_{el}S, \qquad (6.29)$$

where

$$E(\mathbf{R}_1, \ldots, \mathbf{R}_N; T_{el}) = \sum_i f_i \int d\mathbf{r}\psi_i^*(\mathbf{r}) \left(-\frac{1}{2}\nabla^2\right)\psi_i(\mathbf{r}) + \int n(\mathbf{r})V_{ext}(\mathbf{r})d\mathbf{r}$$

$$+ \frac{1}{2} \iint \frac{n(\mathbf{r})n(\mathbf{r}')}{|\mathbf{r} - \mathbf{r}'|}d\mathbf{r}d\mathbf{r}'$$

$$+ E^{xc}[n(\mathbf{r})] + E_{ion\text{-}ion}(\mathbf{R}_1, \ldots, \mathbf{R}_N), \qquad (6.30)$$

and

$$S = -2k_{\mathrm{B}}T \sum_i [f_i \ln f_i + (1 - f_i) \ln(1 - f_i)]. \qquad (6.31)$$

In Eqs. (6.30) and (6.31), f_i is the Fermi–Dirac occupation number of the Kohn–Sham (KS) orbital ψ_i at T_{el} and $n(\mathbf{r}) = \sum_i f_i |\psi_i(\mathbf{r})|^2$. Equation (6.30) can be thought of as a finite-temperature version of the total energy functional for the electronic system as introduced in Eq. (2.30), with nuclei interactions added. Such a treatment ensures that $U(\mathbf{R}_1, \ldots, \mathbf{R}_N; T_{\mathrm{el}})$ includes the electronic entropy effects and corresponds to the free energy of the whole system with static nuclei at a certain spatial configuration $(\mathbf{R}_1, \ldots, \mathbf{R}_N)$. Therefore, for the nuclear system, it is just the total energy. Since this energy depends on T_{el}, we use $U(\mathbf{R}_1, \ldots, \mathbf{R}_N; T_{\mathrm{el}})$ to label this total energy function of the nuclear configuration, with the free energy of the electronic system completely included at the finite-temperature DFT level.

With this total energy function in hand, we can calculate the Helmholtz free energy. The first thing we do is to separate the total energy function into two parts, i.e. the static energy at the perfect-lattice positions $U(\mathbf{R}_1^0, \ldots, \mathbf{R}_N^0; T_{\mathrm{el}})$ and the remainder, through

$$U(\mathbf{R}_1, \ldots, \mathbf{R}_N; T_{\mathrm{el}}) = U(\mathbf{R}_1^0, \ldots, \mathbf{R}_N^0; T_{\mathrm{el}}) + U^{\mathrm{vib}}(\mathbf{R}_1, \ldots, \mathbf{R}_N; T_{\mathrm{el}}). \qquad (6.32)$$

$U(\mathbf{R}_1^0, \ldots, \mathbf{R}_N^0; T_{\mathrm{el}})$ is a constant which does not depend on $(\mathbf{R}_1, \ldots, \mathbf{R}_N)$. Therefore, the free energy of the crystal, as determined by

$$F = -k_{\mathrm{B}}T \ln$$
$$\times \left\{ \frac{1}{N! \Lambda^{3N}} \int d\mathbf{R}_1 \cdots d\mathbf{R}_N e^{-\beta(U(\mathbf{R}_1^0, \ldots, \mathbf{R}_N^0; T_{\mathrm{el}}) + U^{\mathrm{vib}}(\mathbf{R}_1, \ldots, \mathbf{R}_N; T_{\mathrm{el}}))} \right\}, \qquad (6.33)$$

equals

$$F = U(\mathbf{R}_1^0, \ldots, \mathbf{R}_N^0; T_{\mathrm{el}}) - k_{\mathrm{B}}T \ln$$
$$\times \left\{ \frac{1}{N! \Lambda^{3N}} \int d\mathbf{R}_1 \cdots d\mathbf{R}_N e^{-\beta U^{\mathrm{vib}}(\mathbf{R}_1, \ldots, \mathbf{R}_N; T_{\mathrm{el}})} \right\}. \qquad (6.34)$$

From this equation, it is clear that the key issue relates to determining the second term in Eq. (6.34), which we label as F^{vib}. This quantity, again, can be separated into two terms, i.e. F^{harm} which represents the free energy of a harmonic lattice and the deviation of F^{vib} from it. In terms of thermodynamic integration, the reference state is the harmonic lattice, whose free energy can be calculated analytically using

$$F^{\text{harm}} = \frac{3k_{\text{B}}T}{N_{\mathbf{k},s}} \sum_{\mathbf{k},s} \ln \left[e^{\frac{1}{2}\beta\hbar\omega_{\mathbf{k},s}} - e^{-\frac{1}{2}\beta\hbar\omega_{\mathbf{k},s}} \right], \qquad (6.35)$$

where $\omega_{\mathbf{k},s}$ means the phonon frequency of branch s at reciprocal space point $\mathbf{k}$. The deviation of F^{vib} from F^{harm}, which we label as F^{anharm}, needs to be calculated using thermodynamic integration. From our discussions, this term equals

$$F^{\text{anharm}} = \int_0^1 d\lambda \langle U^{\text{vib}}(\mathbf{R}_1, \dots, \mathbf{R}_N) - U^{\text{harm}}(\mathbf{R}_1, \dots, \mathbf{R}_N) \rangle_\lambda. \qquad (6.36)$$

Here, $U^{\text{harm}}(\mathbf{R}_1, \dots, \mathbf{R}_N)$ refers to the harmonic potential associated with the Hessian matrix of the crystal. In some cases, to decrease computational load, an intermediate state can also be used, but the principles underlying these calculations for the free energy of a crystal is already there. For more details, please see Ref. [55].

7
Quantum Nuclear Effects

In Chapter 5, we have introduced the principles underlying standard molecular dynamics (MD) simulations nowadays. It was clearly pointed out in this introduction that the nuclei are treated as classical point-like particles. This classical treatment of the nuclei is normally a good approximation since the nuclear masses are much larger than that of the electron. However, it needs to be pointed out that these values are still far from being large enough so that the classical treatment is rigorous, especially for the lightest element hydrogen, where it is only ~1836 times of that of the electron. The quantum nuclear effects (QNEs) might still be important in reality. As a matter of fact, it has long been realized that the statistical properties of hydrogen-bonded systems, such as water, heavily depend on the isotope of hydrogen. Taking the melting/boiling temperatures of normal water (composed of H_2O) and heavy water (composed of D_2O) as an example, this value in heavy water is ~3.8°C/1.4°C higher than that of normal water under the same ambient pressure. In statistical mechanics, it is well known that the classical thermal effects of different isotopes are rigorously the same at the same temperature. Therefore, the difference between these statistical properties in the two materials must originate from the QNEs.

Another example where the quantum feature of the nuclei plays an important role is derived from the studies of proton tunneling, which has been well characterized in both hydrogen diffusion on metal surface [314, 315] and proton transfer in biosystems [316]. The phenomena in the latter system have an important influence on enzyme catalysis [316–318]. In order to be able to account for such QNEs at the atomic level in the simulations, it is highly desired that one can have a scheme in statistical mechanics where, in addition to the thermal effects, the QNEs can be equally addressed. As a matter of fact, due to the development of the path integral representation of the quantum mechanics starting from the late 1940s, the foundation of such a scheme has been rigorously and systematically presented by Feynman and Hibbs in Ref. [67]. Based on this foundation, the MD simulation technique as introduced in Chapter 5 was combined with the framework of this path integral representation of the statistical mechanics and a series of path integral molecular dynamics (PIMD) simulations had been performed in the 1980s (see, e.g., Refs. [68, 69, 74]). Parallel to these PIMD simulations, the Monte-Carlo (MC) sampling technique was also used and properties of liquid helium including its superfluidity were systematically studied using this path integral Monte-Carlo (PIMC) method (see, e.g., Refs. [70–73]). In these PIMD/PIMC simulations, the QNEs are addressed on the same footing as the thermal ones when the statistical properties of the system to be simulated are evaluated, as will be explained in detail soon in this chapter. Therefore, when comparing their results with the ones where the nuclei are treated as classical particles, such as the MD simulation method we have introduced in Chapter 5, the differences account for the impact of the QNEs on the statistical properties in a very clear manner.

This comparison sets up a rigorous framework for the QNEs to be analyzed, which is still used nowadays. However, it needs to be pointed out that in both these early PIMD and PIMC simulations, the empirical potentials were used to account for the interatomic interactions in the simulation. These potentials are simple and very good in describing the statistical property of many solids and liquids. However, they can easily fail when chemical reaction happens, due to a serious reconstruction of the electronic structures, which needs to be addressed "on-the-fly" as the dynamics of the system evolves. To address problems like this, where the impact of the QNEs is often more interesting, people started trying a combination of the PIMD and PIMC simulation techniques with the *ab initio* method for the description of the electronic structures after the 1990s, first within the framework of Car–Parrinello (CP) MD (see e.g. Refs. [319–321]) and then directly on the Born–Oppenheimer (BO) MD or MC schemes [282, 286, 322, 323]. These methods really allow the bond making and bond breaking events to happen, as well as the thermal and quantum nuclear effects to be accounted for in a seamless manner based on the forces computed "on-the-fly" as the dynamics of the system evolve. Now, it is fair to say that they have come to such a mature stage that not only different functionals with the density-functional theory (DFT) can be used in descriptions of the electronic structures (see, e.g., Refs. [324, 325]), but also traditional quantum chemistry methods such as the MP2 method can be used (see, e.g., Refs. [326, 327]). With these choices of the electronic structures, when the interatomic interactions are accurate enough and the sampling over the high-dimensional phase space is complete (ergodicity is satisfied, in the language of statistics), one can safely rely on results obtained from such simulations, even under low temperatures when the classical description of the nuclei fails. Therefore, on the statistical level, a scheme in which the thermal nuclear effects and QNEs are accounted for on the same footing in the atomic simulations is already there. The only thing we need to be attentive to is in its use the choice of the electronic structures for the description of the interatomic interactions and the ergodicity issue in the PIMD/PIMC sampling (which, however, is non-trivial at all).

Besides these statistical properties, another kind of property where the QNEs may also play an important role relates to the dynamics, especially when the chemical reaction rate is evaluated. This chemical reaction rate is a key parameter in chemistry which is very hard to simulate rigorously. One theory underlying descriptions of this quantity is the so-called transition-state theory (TST) [328–331]. Since the probability of finding the system close to the transition state (TS) is much smaller than that of the reactant or product state, this theory is intrinsically both statistical and dynamical. The term "statistical" means that this chemical reaction rate is proportional to the ratio of the equilibrium density of the system at the TS to its value at the reaction state, which is a statistical property. Conversely, the term "dynamical" indicates that since the TS is defined by a dividing surface separating the reaction and product states, after the system at the TS falls into either the reactant or the product region, it stays at this state for a longer time than it spends at the TS. Therefore, what happens dynamically at the TS is of crucial importance to its behavior in the future, and theories underlying descriptions of such processes should be dynamical. Following the principles of scientific research, i.e. from the easy and idealized models to the more difficult and realistic ones, the earliest methods within the TST usually assume a classical treatment of the nuclei. Later, when events like quantum tunneling were found to be crucial in describing the chemical reaction behaviors, a quantum version of this was also proposed. This development is also associated with the development of the purely statistical PIMD method to the dynamical regime. Two of the schemes most often used in descriptions of this dynamics within the scheme of path integral molecular dynamics are the centroid molecular dynamics

(CMD) and the ring-polymer molecular dynamics (RPMD). This extension of the statistical PIMD method to the dynamical regime will also be discussed.

In Chapter 6, a scheme in which the free energy with anharmonic contributions from the nuclei is calculated using the sampling method of MD was introduced. In our discussions there, the assumption that the nuclei are classical particles results in the fact that although the anharmonic contribution from the thermal fluctuations of the nuclei are accounted for in the thermodynamic integrations with MD sampling, the anharmonic effect associated with the quantum feature of the nuclei is completely neglected. This anharmonic contribution originating from the QNEs is often believed to be unimportant at moderate and high temperatures. However, when the mass of the particle is small, this assumption might fail, starting from the moderate temperature regime to low temperatures. As a matter of fact, this anharmonic correction is recently found to be very important in describing the phase diagram of hydrogen and neon (see e.g. Refs. [332, 333]). Therefore, from both the methodology and the practical simulation point of view, it is highly desired that a scheme in which the thermodynamic integration is combined with the *ab initio* PIMD method is developed to address such problems. This treatment of the anharmonic quantum nuclear correction to the free energy will also be discussed [333].

The chapter is organized as follows. In Sec. 7.1, we introduce the PIMD and related methods for statistics, where it is rigorously justified. Then, some extensions beyond these statistical studies will be briefly discussed in Sec. 7.2. After these, an introduction to how this PIMD method is combined with the thermodynamic integration method for the calculation of the free energy will be presented in Sec. 7.3. We end this chapter with some examples in Sec. 7.4 and a brief summary in Sec. 7.5.

7.1   Path Integral Molecular Simulations

7.1.1   *Path Integral Representation of the Propagator*

For a theoretical description of the QNEs, it is crucial to start our discussions from their origin, i.e. the intrinsic quantum nature of the nuclei. The development of quantum mechanics in the last century shows us that the fundamental difference between the classical world and the quantum world lies in the fact that in the quantum world, things must be described in terms of "probability". In understanding this principle, we can make use of a scene that many of us might have experienced during our primary school time, as an analogy. We remember that when a naughty boy makes a loud noise during the class and irritates his teacher, the teacher will throw a piece of chalk he is holding toward this guy to remind him to be quiet. In most cases, it works. In the language of classical/quantum mechanics, we would like to say that it works because the piece of chalk is heavy enough so that it behaves like a classical particle, and the classical particles move according to their trajectories. Therefore, as long as you control the trajectory, you control the consequence.

Now, imagine that this chalk is a particle small enough so that its quantum feature is important in descriptions of the phenomena related to it, what happens after the chalk is thrown out will then be completely out of control due to the principle of quantum mechanics. Every person in the classroom might be hit at a later time. From our textbook of quantum mechanics, we know that in order to quantitatively describe behaviors like these, a propagator needs to be used. This propagator is a function of two events, with each event representing something happening at a certain time and a certain position. Still taking the classroom with a naughty boy and his teacher as an example, we can label the event when the teacher throws the piece of chalk on the stage (position labeled as $\mathbf{x}_a$) at a certain time t_a as event a, and the event when "someone" got hit by the chalk at his/her position (labeled as $\mathbf{x}_b$) at a later time t_b as event b.

If we forget about the person who did this and only focus on the particle, these two events can be rephrased as "the generation of a particle (piece of chalk) at a and the annihilation of this particle at b". As mentioned in Chapter 4, in a many-body quantum entity, the correlation between these two events should be described using the propagator, which is also known as Green's function. After event a happens, the probability of event b happening equals the square of the propagator's absolute value. In the Schrödinger representation of the quantum mechanics, this propagator is written as

$$G\left(\mathbf{x}_b, t_b; \mathbf{x}_a, t_a\right) = \sum_j \psi_j(\mathbf{x}_b)\psi_j^*(\mathbf{x}_a)e^{(-i/\hbar)E_j(t_b-t_a)}, \qquad (7.1)$$

where j runs over all eigenstates of the quantum system. This equation indicates that if one knows the eigenstate wave functions and eigenvalues of this quantum system, this propagator can be expressed analytically and consequently, the correlation between any two events is accurately described. However, it is well known that the many-body Schrödinger equation is difficult to solve and it has a notorious scaling problem. Nowadays, many studies on the QNEs still resort to such a method, where the high-dimensional *ab initio* potential energy surfaces (PESs) are mapped out first and then the Schrödinger equation is solved directly [59–62]. This method is rigorous because not only nuclear exchange but also real-time propagation can be described rigorously in this framework [60]. But due to the scaling problem associated with both mapping the *ab initio* high-dimensional PESs and solving the Schrödinger equation, its application is seriously limited to systems less than ~6 atoms. When the system gets bigger, a practical scheme for descriptions of this quantity must be used.

Thanks to the development of the path integral representation of the quantum mechanics starting from the late 1940s by Feynman [63–67], a framework where this propagator is calculated using a method we would like to use in this chapter was systematically presented by Feynman and Hibbs in their seminal book in 1965 (Ref. [67]). In this book, it was clearly explained that this propagator can be calculated not only from Eq. (7.1) but also in terms of a numerical path integral, in which contributions from all paths between events a and b are taken into account. The trick is to divide the time interval between these two events, i.e. $t_b - t_a$, into P slides. Draw a line between t_a and t_b which intersects with t_i on $\mathbf{x}_i$. Then, on each time slide t_i, move the spatial coordinates $\mathbf{x}_i$ through the whole Cartesian space. When P equals infinity, all paths between these two events will be taken into account. The propagator is calculated by adding contributions from all these paths into one quantity through

$$G\left(\mathbf{x}_b, t_b; \mathbf{x}_a, t_a\right) = \lim_{P \to \infty} \frac{1}{A} \int_V \int_V \cdots \int_V e^{(i/\hbar)S[b,a]} \frac{d\mathbf{x}_1}{A} \frac{d\mathbf{x}_2}{A} \cdots \frac{d\mathbf{x}_{P-1}}{A}. \tag{7.2}$$

Here, A is a renormalization factor which equals $[2\pi i\hbar(t_b - t_a)/(Pm)]^{\frac{1}{2}}$. $S[b, a]$ is the action of the path linking events a and b, defined by the spatial coordinates $\mathbf{x}_1, \mathbf{x}_2, \ldots$, and $\mathbf{x}_{P-1}$ on $t_1, t_2, \ldots, t_{P-1}$. For one specific path, as shown in Eq. (7.2), its contribution to the propagator is determined by the action of this path, which is calculated from

$$S[b, a] = \int_{t_a}^{t_b} L(\dot{\mathbf{x}}, \mathbf{x}, t) dt. \tag{7.3}$$

To be more precise, taking the choice of path $\mathbf{x}_1, \mathbf{x}_2, \ldots, \mathbf{x}_{P-1}$ happening at times $t_1, t_2, \ldots, t_{P-1}$, as an example, the action of this path as defined by Eq. (7.3) can be written as

$$S[b, a] = \left[\frac{m}{2} \left(\frac{\mathbf{x}_1 - \mathbf{x}_a}{t_1 - t_a} \right)^2 - \frac{1}{2} \left(V(\mathbf{x}_1) + V(\mathbf{x}_a) \right) \right] (t_1 - t_a)$$

$$+ \sum_{i=2}^{P-1} \left[\frac{m}{2} \left(\frac{\mathbf{x}_i - \mathbf{x}_{i-1}}{t_i - t_{i-1}} \right)^2 - \frac{1}{2} \left(V(\mathbf{x}_i) + V(\mathbf{x}_{i-1}) \right) \right] (t_i - t_{i-1})$$

$$+ \left[\frac{m}{2} \left(\frac{\mathbf{x}_b - \mathbf{x}_{P-1}}{t_b - t_{P-1}} \right)^2 - \frac{1}{2} \left(V(\mathbf{x}_b) + V(\mathbf{x}_{P-1}) \right) \right] (t_b - t_{P-1}).$$

$$(7.4)$$

With this definition, it is clear that the integration in Eq. (7.2) can be calculated numerically through such a procedure, as shown in Fig. 7.1 for the 1D case. In practice, a finite number of time intervals must be chosen, one often tests the convergence of the quantity to be calculated with respect to the number of slides till a reasonable accuracy can be obtained.

With this, we hope that we have made our point clear. To put it simply, the same quantity, i.e. the propagator, can be obtained by using either (i) all eigenstate wave functions of the quantum system, or (ii) a sum over contributions from all paths between the events to be investigated. The first option looks elegant, but it is difficult to handle for large many-body systems. In cases when it is not feasible, the path integral approach provides a numerically simple alternative, which we will make use of in studies of QNEs to be discussed in the following sections.

7.1.2   Path Integral Representation of the Density Matrix

So far we have discussed the propagator; in statistical mechanics, the key quantity we are interested in is actually not the propagator, but the density matrix. Therefore, in order to understand how the concept of the path integral is used in studies on statistical mechanics, the key point is to understand how this quantity (density matrix) is expressed in terms of path integral. As the basis, the first thing we explain here is how this key quantity is expressed in quantum mechanics in general.

Imagine that the quantum system we want to study has exact eigenstate wave functions ψ_i, eigenvalues E_i, and a Hamiltonian $\hat{H}$. In the operator notation, the density operator is $e^{-\beta \hat{H}}$, where $\beta = 1/(k_B T)$ and T is the temperature. The trace of this operator is the so-called partition function

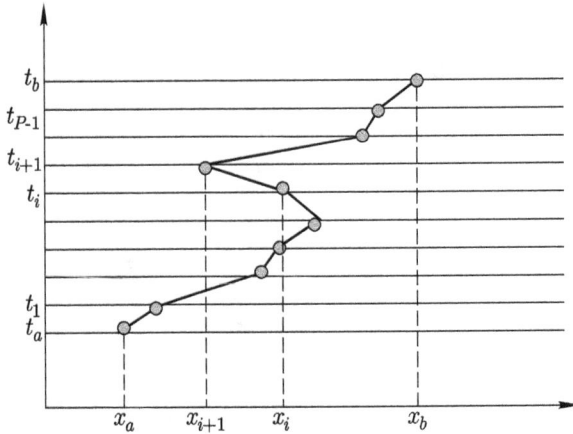

Figure 7.1   Illustration of how a propagator can be calculated using path integral. The first step is to divide the time interval into some slides. At the starting and ending points (t_a and t_b), the spatial coordinates (x_a and x_b in this 1D case) must be fixed since they represent the events whose correlation is to be investigated. With the spatial coordinates on the time slices between t_a and t_b chosen, we have one specific "path". Contribution from this path to the propagator is then calculated using Eq. (7.4). Then one further moves the spatial coordinates on each time slide between t_a and t_b through the whole space (x axis in this 1D case). In doing so, we can take account of contributions from all paths between a and b. If all these contributions are calculated numerically using Eq. (7.4) and the time interval number approaches infinity, we have the propagator rigorously defined which equals the number it gives through Eq. (7.1). Calculations of the eigenstate wave functions are avoided accordingly.

Z^Q. The expectation value of any physical observable $\hat{O}$ equals

$$\langle \hat{O} \rangle = (Z^Q)^{-1} \mathrm{Tr}[\hat{O} e^{-\beta \hat{H}}]. \tag{7.5}$$

Now, we discuss how these quantities are represented in two spaces, i.e. the system's eigenstate wave functions' Hilbert spaces and the position space. Such a comparison can help us in setting up a link between the density matrix and the propagator we have discussed in Sec. 7.1.1, which you will see soon. We first look at the system's eigenstate wave functions' Hilbert space, where the probability of finding the quantum system at its ith eigenstate equals $e^{-\beta E_i}$ at thermal equilibrium. From this probability, it is clear that for an operator $\hat{O}$, its expectation value at the thermal equilibrium equals

$$\langle \hat{O} \rangle = (Z^Q)^{-1} \sum_i \langle \psi_i | \hat{O} | \psi_i \rangle e^{-\beta E_i}. \tag{7.6}$$

The partition function Z^Q itself has the simple form

$$Z^Q = \sum_i e^{-\beta E_i}. \tag{7.7}$$

Then we come to the position space. In this space, the density matrix is written as

$$\rho\left(\mathbf{x}_a, \mathbf{x}_b, \beta\right) = \langle \mathbf{x}_b | e^{-\beta \hat{H}} | \mathbf{x}_a \rangle, \tag{7.8}$$

whose diagonal part is the so-called density function. To arrive at the wave-function-based expression of this density matrix, one can insert an identity matrix $\hat{I} = \sum_j |\psi_j\rangle\langle\psi_j|$ into the right-hand side of this equation. In doing so, we have

$$
\begin{aligned}
\rho\left(\mathbf{x}_a, \mathbf{x}_b, \beta\right) &= \left\langle \mathbf{x}_b \middle| e^{-\beta \hat{H}} \left(\sum_j |\psi_j\rangle\langle\psi_j| \right) \middle| \mathbf{x}_a \right\rangle \\
&= \sum_j \langle \mathbf{x}_b | \psi_j \rangle e^{-\beta E_j} \langle \psi_j | \mathbf{x}_a \rangle \\
&= \sum_j \psi_j(\mathbf{x}_b) \psi_j^*(\mathbf{x}_a) e^{-\beta E_j}.
\end{aligned}
\tag{7.9}
$$

Similar to the propagator, to obtain this quantity, in the Schrödinger representation of the quantum mechanics, we need to know the eigenstate wave functions. However, as mentioned, calculating the wave function of nuclei is not feasible for systems with more than ~6 atoms. An alternative approach must be adopted.

Now, we apply the concept of the path integral as introduced in Sec. 7.1.1 to the description of this density matrix, the non-local function as shown in Eq. (7.9). The trick is to compare it with Eq. (7.1) for the propagator. From this comparison, it is easy to see that these two equations for the propagator and the density matrix share a very strong similarity. The only difference is that in Eq. (7.1) for the propagator, the index of the exponential function is imaginary and it is determined by the time interval $t_b - t_a$, while in Eq. (7.9) for the density matrix, the index of the exponential function is real and it is determined by the temperature T. In Sec. 7.1.1, we have shown that for the propagator in Eq. (7.1), we can avoid using the wave functions by resorting to integrations over paths between the events to be described. Similarly, with this tiny difference between Eq. (7.9) and Eq. (7.1) in mind, we can also resort to the path integral representation of the quantum mechanics and rewrite the density matrix as shown in Eq. (7.9) in terms of the path integral. The only thing we need to do is to treat the temperature-dependent factor β as an imaginary-time interval and replace the $i(t_b - t_a)/\hbar$ term in Eq. (7.1) by this temperature-dependent imaginary-time interval. To be more precise, we should replace the time interval $t_b - t_a$ as used in Eq. (7.1) by $-i\hbar\beta$. Then, using the same trick as that mentioned for the numerical representation of the propagator in Eqs. (7.2)–(7.4), we can rewrite the density matrix in terms of path integral without resorting to the nuclear wave functions.

Following this routine, the first thing we need to do is to divide the time interval $-i\hbar\beta$ into P slices. Such a treatment results in a time step of $-i\hbar\beta/P$ along the imaginary time path to be integrated in the density matrix. Now, imagine that one path is defined by a certain choice of x_1, x_2, ..., x_{P-1} between x_a and x_b, the action which determines the weight of this path in the path integral scheme then equals

$$S[b,a] = -\left[\frac{mP^2}{2\beta^2\hbar^2}(x_1 - x_a)^2 + \frac{1}{2}(V(x_1) + V(x_a))\right]\frac{-\hbar i \beta}{P}$$

$$-\sum_{i=2}^{P-1}\left[\frac{mP^2}{2\beta^2\hbar^2}(x_i - x_{i-1})^2 + \frac{1}{2}(V(x_i) + V(x_{i-1}))\right]\frac{-\hbar i \beta}{P}$$

$$-\left[\frac{mP^2}{2\beta^2\hbar^2}(x_b - x_{P-1})^2 + \frac{1}{2}(V(x_b) + V(x_{P-1}))\right]\frac{-\hbar i \beta}{P}$$

$$(7.10)$$

from Eq. (7.4). We note that because imaginary time is used, the positive sign of the kinetic energy becomes negative, resulting in the action (which is originally defined as an integral over the Lagrangian) as a term above which looks like an integral over a Hamiltonian with a minus sign. The $(i/\hbar)S[b,a]$ term on the index of the exponential functional in Eq. (7.2) consequently becomes

$$(i/\hbar)S[b,a] = -\left[\frac{mP}{2\beta\hbar^2}(x_1 - x_a)^2 + \frac{\beta}{2P}(V(x_1) + V(x_a))\right]$$

$$-\sum_{i=2}^{P-1}\left[\frac{mP}{2\beta\hbar^2}(x_i - x_{i-1})^2 + \frac{\beta}{2P}(V(x_i) + V(x_{i-1}))\right]$$

$$-\left[\frac{mP}{2\beta\hbar^2}(x_b - x_{P-1})^2 + \frac{\beta}{2P}(V(x_b) + V(x_{P-1}))\right].$$

$$(7.11)$$

To clarify the labeling in the above equations further, we can relabel the $\mathbf{x}_a$ and $\mathbf{x}_b$ in the density matrix as $\mathbf{x}_0$ and $\mathbf{x}_P$, respectively. The imaginary-time interval is still divided into P slices and the path is determined by a consequence of spatial coordinates from $\mathbf{x}_1$ to $\mathbf{x}_{P-1}$, with $\mathbf{x}_0$ and $\mathbf{x}_P$ keeping fixed. Then, Eq. (7.11) can be rewritten in a simple form as

$$
\begin{aligned}
(i/\hbar)S[b,a] &= -\sum_{i=1}^{P}\left[\frac{mP}{2\beta\hbar^2}\left(\mathbf{x}_i - \mathbf{x}_{i-1}\right)^2 + \frac{\beta}{2P}\left(V(\mathbf{x}_i) + V(\mathbf{x}_{i-1})\right)\right] \\
&= -\beta\sum_{i=1}^{P}\left[\frac{1}{2}m\omega_P^2\left(\mathbf{x}_i - \mathbf{x}_{i-1}\right)^2 + \frac{1}{2P}\left(V(\mathbf{x}_i) + V(\mathbf{x}_{i-1})\right)\right],
\end{aligned}
$$

$$(7.12)$$

where $\omega_P = \sqrt{P}/(\beta\hbar)$. If we put this exponential index to the path integral representation of propagator as shown in Eq. (7.2), but using the imaginary time for the density matrix, the equation this density matrix ends up with will be

$$
\begin{aligned}
&\rho\left(\mathbf{x}_0, \mathbf{x}_P, \beta\right) \\
&= \lim_{P\to\infty}\frac{1}{A}\int_V\int_V\cdots\int_V e^{-\beta\sum_{i=1}^{P}\left[\frac{1}{2}m\omega_P^2(\mathbf{x}_i-\mathbf{x}_{i-1})^2+\frac{1}{2P}(V(\mathbf{x}_i)+V(\mathbf{x}_{i-1}))\right]} \\
&\quad \cdot \frac{d\mathbf{x}_1}{A}\frac{d\mathbf{x}_2}{A}\cdots\frac{d\mathbf{x}_{P-1}}{A}.
\end{aligned}
$$

$$(7.13)$$

In the case of the real-time propagator, we have mentioned that A is a renormalization factor which equals $[2\pi i\hbar(t_b - t_a)/(Pm)]^{1/2}$. Here, for the imaginary-time density matrix, $t_b - t_a$ should be replaced by $-i\hbar\beta$. Accordingly, A becomes $[2\pi\beta\hbar^2/(Pm)]^{1/2}$. Replacing the A in Eq. (7.13) with this value, we finally arrive at

$$
\begin{aligned}
&\rho\left(\mathbf{x}_0, \mathbf{x}_P, \beta\right) \\
&= \lim_{P\to\infty}\left(\frac{mP}{2\beta\pi\hbar^2}\right)^{\frac{P}{2}}\int_V\int_V\cdots\int_V \\
&\quad \cdot e^{-\beta\sum_{i=1}^{P}\left[\frac{1}{2}m\omega_P^2(\mathbf{x}_i-\mathbf{x}_{i-1})^2+\frac{1}{2P}(V(\mathbf{x}_i)+V(\mathbf{x}_{i-1}))\right]}d\mathbf{x}_1 d\mathbf{x}_2\cdots d\mathbf{x}_{P-1}.
\end{aligned}
$$

$$(7.14)$$

Till now, the numerical representation of the density matrix in terms of the path integral is already clear. However, there is still a key concept whose physical meaning needs some further explanation. In the above discussions, the variables $\mathbf{x}_0$ to $\mathbf{x}_P$ are defined as points in the Cartesian space for the nucleus to be studied for simplicity. However, we note that they can also be used to represent a poly-atomic system's spatial configuration of the nuclei under investigation. Suppose that this poly-atomic system is a molecule containing N nuclei, this $\mathbf{x}_i$ is then a $3N$-dimensional vector, representing a spatial configuration of this molecule's nuclei. To understand this, one just needs to imagine that the wave functions as used in Eqs. (7.1) and (7.9) are many-body wave functions of this N-nuclei system whose square magnitude represents the probability of finding the poly-atomic system at this specific spatial configuration. Accordingly, the path from $\mathbf{x}_0$, through $\mathbf{x}_1, \ldots, \mathbf{x}_{P-1}$, to $\mathbf{x}_P$ represents a path in the $3N$-dimensional configuration space linking $\mathbf{x}_0$ and $\mathbf{x}_P$, two spatial configurations of the nuclei. All discussions above about the position space extend to this $3N$-dimensional configuration space. The only impact on the equations above is that in Eq. (7.14), an extra iteration over the atoms needs to be included. With this extra iteration included, Eq. (7.14) becomes

$$\rho\left(\mathbf{x}_0, \mathbf{x}_P, \beta\right)$$

$$= \lim_{P \to \infty} \left[\prod_{j=1}^{N} \left(\frac{m_j P}{2\beta\pi\hbar^2} \right)^{\frac{P}{2}} \right] \int_V \int_V \cdots \int_V$$

$$\cdot \mathrm{e}^{-\beta \sum_{i=1}^{P} \left[\sum_{j=1}^{N} \frac{1}{2} m_j \omega_P^2 \left(\mathbf{x}_i^j - \mathbf{x}_{i-1}^j \right)^2 + \frac{1}{2P} \left(V\left(\mathbf{x}_i^1, \ldots, \mathbf{x}_i^N\right) + V\left(\mathbf{x}_{i-1}^1, \ldots, \mathbf{x}_{i-1}^N\right) \right) \right]}$$

$$\cdot \mathrm{d}\mathbf{x}_1 \mathrm{d}\mathbf{x}_2 \cdots \mathrm{d}\mathbf{x}_{P-1}. \tag{7.15}$$

Here, $\mathbf{x}_i^j$ means the 3D position space associated with the jth atom's ith bead. We note that this understanding of $\mathbf{x}_i$ and Eq. (7.15) associated are the frequently used theoretical foundations for discussions of the density matrix in the path integral molecular simulations. Again, P is a parameter which represents the number of slices sampled along the path. The convergence of the property under investigation with respect to this parameter must be tested in practical simulations.

7.1.3   Statistical Mechanics: Path Integral Molecular Simulations

Equation (7.15) gives an expression for the density matrix in terms of the path integral in the $3N$-dimensional configuration space, where N is the number of nuclei in the poly-atomic system. The diagonal part of this density matrix is the density function, which describes the probability of finding the poly-atomic system under investigation at a certain spatial configuration. This function can be obtained from Eq. (7.15) by setting $\mathbf{x}_0 = \mathbf{x}_P$ as

$$\rho\left(\mathbf{x}_P, \mathbf{x}_P, \beta\right)$$

$$= \lim_{P \to \infty} \left[\prod_{j=1}^{N}\left(\frac{m_j P}{2\beta\pi\hbar^2}\right)^{\frac{P}{2}}\right] \int_V \int_V \cdots \int_V$$

$$\cdot e^{-\beta \sum_{i=1}^{P}\left[\sum_{j=1}^{N} \frac{1}{2}m_j\omega_P^2\left(\mathbf{x}_i^j - \mathbf{x}_{i-1}^j\right)^2 + \frac{1}{2P}\left(V(\mathbf{x}_i^1,...,\mathbf{x}_i^N) + V(\mathbf{x}_{i-1}^1,...,\mathbf{x}_{i-1}^N)\right)\right]}$$

$$\cdot d\mathbf{x}_1 d\mathbf{x}_2 \cdots d\mathbf{x}_{P-1}. \tag{7.16}$$

In Sec. 7.1.2, we stated that the density matrix is $e^{-\beta\hat{H}}$ in the operator notation, whose trace gives us the partition function. Therefore, in the configuration space, this partition function is easily obtainable from the density matrix. What we need to do is to perform an extra integration on $\mathbf{x}_P$ over the configuration space in Eq. (7.16). This integration gives us the partition function Z^Q for the quantum canonical system in the limit of P approaching infinity through

$$Z^Q = \lim_{P \to \infty} \left[\prod_{j=1}^{N}\left(\frac{m_j P}{2\beta\pi\hbar^2}\right)^{\frac{P}{2}}\right] \int_V \int_V \cdots \int_V$$

$$\cdot e^{-\beta \sum_{i=1}^{P}\left[\sum_{j=1}^{N} \frac{1}{2}m_j\omega_P^2\left(\mathbf{x}_i^j - \mathbf{x}_{i-1}^j\right)^2 + \frac{1}{2P}\left(V(\mathbf{x}_i^1,...,\mathbf{x}_i^N) + V(\mathbf{x}_{i-1}^1,...,\mathbf{x}_{i-1}^N)\right)\right]}$$

$$\cdot d\mathbf{x}_1 d\mathbf{x}_2 \cdots d\mathbf{x}_P. \tag{7.17}$$

This quantum canonical partition function is a function of temperature. When it is known, in principle, all thermodynamic quantities of the quantum system under investigation are obtainable.

Now, we look at the partition function in Eq. (7.17). In practice, a finite P is always used. We label the partition function represented in Eq. (7.17) for a finite P as Z_P. We note that this Z_P can be understood as the "configurational" partition function for a fictitious $3N \times P$-particle system in an effective potential V^{eff}, in the form of

$$V^{\text{eff}}(\mathbf{x}_1, \mathbf{x}_2, \ldots, \mathbf{x}_P)$$

$$= \sum_{i=1}^{P} \left[\sum_{j=1}^{N} \frac{1}{2} m_j \omega_P^2 (\mathbf{x}_i^j - \mathbf{x}_{i-1}^j)^2 + \frac{1}{2P} (V(\mathbf{x}_i^1, \ldots, \mathbf{x}_i^N) \right.$$

$$\left. + V(\mathbf{x}_{i-1}^1, \ldots, \mathbf{x}_{i-1}^N)) \right], \tag{7.18}$$

where $\mathbf{x}_0 = \mathbf{x}_P$. Again, $\mathbf{x}_i$ is a $3N$-dimensional vector representing the spatial configuration of the nuclei in the ith image and $\mathbf{x}_i^j$ is a 3D vector representing the position of the jth nucleus in this image. P is the number of slices we have chosen for the sampling of the path integral along the imaginary-time interval. Because of the similarity between the cyclic path and a necklace, these sampling points are also called "beads", "images", or "replicas" in the literature. In terms of this V^{eff}, Z_P in Eq. (7.17) can be rewritten as

$$Z_P = \left[\prod_{j=1}^{N} \left(\frac{m_j P}{2\beta \pi \hbar^2} \right)^{\frac{P}{2}} \right] \int_V \int_V \cdots \int_V e^{-\beta V^{\text{eff}}(\mathbf{x}_1, \mathbf{x}_2, \ldots, \mathbf{x}_P)} \mathrm{d}\mathbf{x}_1 \mathrm{d}\mathbf{x}_2 \cdots \mathrm{d}\mathbf{x}_P.$$

$$\tag{7.19}$$

The relationship between this configuration partition function (Z_P) over a $3N \times P$ configuration space and the quantum canonical partition function (Z^Q) of the real poly-atomic system is that Z_P equals Z^Q when P goes to infinity.

Pictorially, this relationship can be understood from the comparison as shown in Fig. 7.2. We use H_2, the simplest molecule, as an example. The canonical partition function of the quantum system is what we want to simulate. From Eq. (7.9), we know that one needs the eigenstate wave functions and the eigenvalues of the nuclei, which is not feasible in the studies of most poly-atomic systems, except for descriptions of some simple gas phase small molecules [59–62]. As an alternative, one can resort to Eq. (7.17) and construct a fictitious polymer for the real system under investigation. This polymer is composed of P replicas of the real poly-atomic system. In between the replicas, the same atoms are linked by spring interactions, whose spring constant (defined as $m_j \omega_P^2$) is determined by m_j and ω_P, with $\omega_P = \sqrt{P}/(\beta \hbar)$. Within one replica, the interatomic potential is calculated by either force fields or *ab initio* methods upon which the

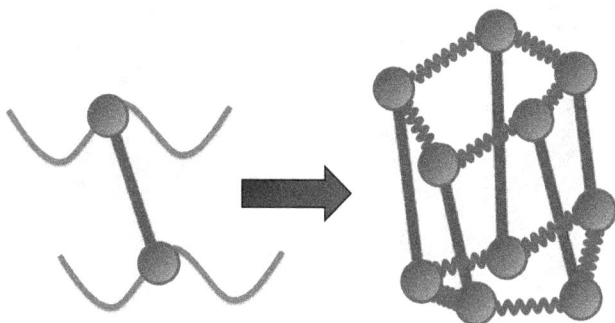

Figure 7.2 Illustration of how the mapping from the canonical quantum system to a classical polymer is done in path integral statistical mechanics. The simplest molecule, H_2, is taken as an example. In principle, one needs to calculate the nuclear wave functions, as shown on the left. Using path integral, this calculation of the wave function can be avoided. What one needs to do is to set up a fictitious polymer. This polymer is composed by P replicas of the real molecule. In each replica, the potential is determined by the real potential of the system at the specific spatial configuration of this replica. In between the replicas, the neighboring images (beads) of the same atoms are linked by springs. The spring constant is determined by m_j and ω_P as $m_j \omega_P^2$, where $\omega_P = \sqrt{P}/(\beta\hbar)$. Therefore, the higher the temperature, the heavier the nucleus, the stronger the interaction between the beads. In the limit of $T \to \infty$ and $m_j \to \infty$, one arrives at the classical limit when all images overlap with each other. The partition function of the quantum system as shown on the left equals the configurational partition function of the polymer on the right as $P \to \infty$.

molecular simulation is based. These two terms, i.e. the spring interaction and the intra-replica potential, correspond to the kinetic and the potential energies of the path integral, respectively. From the form of the spring constant as shown in Eq. (7.17), it is clear that as the temperature and m_j go to infinity, the spring constant becomes so large that all the replicas overlap in configuration space, resulting in a simulation in the classical limit. When the temperature is low and m_j is reasonably small, it is reasonable to expect that a molecular simulation based on this polymer gives results very different from the one with $P = 1$, i.e. the classical simulation. In other words, from $P = 1$ (the classical simulation), when P approaches infinity, one approaches the quantum limit of this canonical ensemble. In practice, a finite P must be taken. The statistical results obtained from these simulations with a finite P should always be converged with respect to this P. The difference between results obtained from the $P = 1$ simulation and this path integral converged simulation tells us the impact of QNEs on the statistical results.

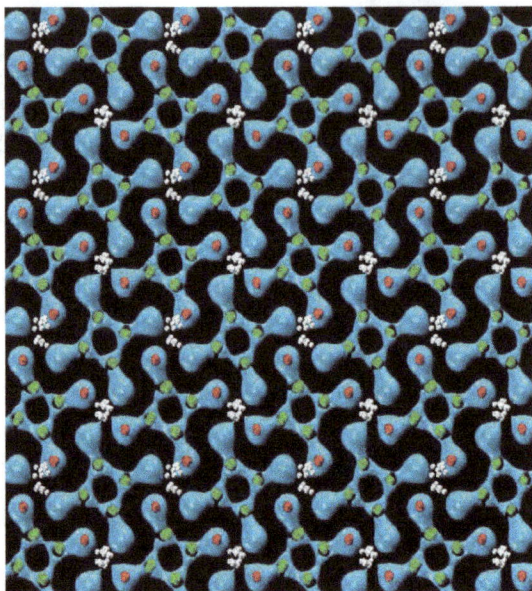

Figure 7.3   An example of the artificial polymer in a real path integral molecular simulation, courtesy of Dr. Brent Walker from our joint paper [322]. The system is a layer of squaric acid, a molecular crystal held together with intermolecular hydrogen bonds. The quantum nature of the nuclei is addressed using the path integral treatment by generating a series of images for the real poly-atomic system and connecting the same atom in neighboring images with spring potential. The electronic structures are calculated quantum mechanically for each image, and the blue contour denotes the density distribution of the electrons in one image. For more details, please see Ref. [322].

In order to provide a more vivid explanation of this artificial polymer in path integral simulation (with electronic structures calculated "on-the-fly" in an *ab initio* manner as the dynamics of the system evolves), we take a real system, i.e. a layer of squaric acid, a hydrogen-bonded molecular crystal, as an example and show a schematic scheme in Fig. 7.3. P is set as 16 and the temperature is 100 K. Around 16 images of the real system are generated and the same atoms of the neighboring images are connected by artificial spring interactions. As said, for the electronic structures of the system in each image, they are calculated quantum mechanically using the *ab initio* method. This is shown by blue contours designating the density distribution of the electrons. Since the mass of the hydrogen is small, the spring interaction with which the hydrogen nuclei is connected is weak and consequently the images of the hydrogen nuclei are delocalized in real space.

The mass of the oxygen nucleus is larger and consequently, their dispersions are smaller, but still observable. If we compare results of simulations using this polymer and the classical system ($P = 1$), it is obvious that the QNEs are small for O and large for H. With the increase of the temperature, it is also easy to expect that the difference between simulations with $P = 1$ and larger P decreases. For more details on this path integral simulation, readers are directed to Ref. [322].

Then we come back to the mathematics. With this relationship between the "configurational" partition function of the polymer and the quantum partition function as introduced in the early paragraphs in mind, we now shift our attention to this "configurational" partition function. Equation (7.19) for this "configurational" partition function of a polymer may look a little complicated to be handled numerically at first sight. However, for people working on molecular simulations, this is an equation which cannot be more simple. To understand what we mean by this statement, one just needs to rewrite Z_P using a proper partition function in a fictitious phase space, composed of variables $\mathbf{x}_1, \ldots, \mathbf{x}_P, \mathbf{p}_1, \ldots, \mathbf{p}_P$, as

$$Z_P = C \int\!\!\int \cdots \int e^{-\beta H(\mathbf{x}_1, \ldots, \mathbf{x}_P, \mathbf{p}_1, \ldots, \mathbf{p}_P)} d\mathbf{x}_1 \cdots d\mathbf{x}_P d\mathbf{p}_1 \cdots d\mathbf{p}_P. \quad (7.20)$$

The Hamiltonian in this equation is designed using the effective potential V^{eff} through

$$H(\mathbf{x}_1, \ldots, \mathbf{x}_P, \mathbf{p}_1, \ldots, \mathbf{p}_P) = \sum_{i=1}^{P} \sum_{j=1}^{N} \frac{(\mathbf{p}_i^j)^2}{2M_i^j} + V^{\text{eff}}(\mathbf{x}_1, \mathbf{x}_2, \ldots, \mathbf{x}_P).$$

$$(7.21)$$

Since the kinetic energy and the potential energy terms are separable in this Hamiltonian, one can easily replace the P-dependent constant

$$\prod_{j=1}^{N} \left(\frac{m_j P}{2\beta\pi\hbar^2} \right)^{\frac{P}{2}}$$

in Eq. (7.19) by a product of $N \times P$ uncoupled Gaussian integrals, which originate from the integrals over momentum $\mathbf{p}_i^j$, as in Eq. (7.20). The constant C in front of the integration over configurational and momentum space in Eq. (7.20) ensures that Z_P is unchanged compared to Eq. (7.19). It is determined by the choice of the artificial mass M_i^j associated with the beads of each nucleus.

We note that such an equality means that for a finite P, the path integral of a quantum system is isomorphic to a classical polymer composed of P replicas of the real poly-atomic system under investigation, subject to a classical Hamiltonian given by Eq. (7.21) [68]. In doing so, the molecular simulation techniques as introduced in the earlier chapters can be directly used. These molecular simulation techniques include both MD and MC methods, with which the partition function as given in Eq. (7.20) can be evaluated. In the following, we discuss how these path integral-based molecular simulation techniques are used, taking the MD-based implementations as the guiding example. The only thing we need to take care in is to design the fictitious polymer in a proper way, so that the contributions from the quantum nature of the nuclei to the statistical properties of the system are not miscounted. Readers interested in the PIMC method are directed to peruse Refs. [70–73].

Besides the partition function and the density matrix, other quantities, such as the expectation values of any physical observable $\hat{O}$, can also be described using this path integral. The definition of such an expectation value is already given in Eq. (7.5). From this definition, one can obtain a path integral-based representation for the expectation value of this observable using the Trotter factorization. To understand this statement, we first go back to the path integral representation of the propagator as shown in Eq. (7.2). As a matter of fact, its form originates and is equivalent to a Trotter factorization-based equation. What one does in such a Trotter factorization is to insert $P - 1$ identity matrices $\int d\mathbf{x}_i |\mathbf{x}_i\rangle\langle\mathbf{x}_i| = \mathbf{I}$ on the time slices between t_a and t_b. With this treatment, one arrives at such an equation for the propagator

$$
G\left(\mathbf{x}_b, t_b; \mathbf{x}_a, t_a\right)
$$

$$
= \lim_{P\to\infty} \frac{1}{A} \int_V \int_V \cdots \int_V \langle\mathbf{x}_a|e^{(i/\hbar)\hat{L}\Delta t}|\mathbf{x}_1\rangle\langle\mathbf{x}_1|e^{(i/\hbar)\hat{L}\Delta t}|\mathbf{x}_2\rangle \cdots
$$

$$
\cdot\langle\mathbf{x}_{P-1}|e^{(i/\hbar)\hat{L}\Delta t}|\mathbf{x}_b\rangle \frac{d\mathbf{x}_1}{A}\frac{d\mathbf{x}_2}{A} \cdots \frac{d\mathbf{x}_{P-1}}{A}. \tag{7.22}
$$

We note that this equation is equivalent to the path integral treatment in terms of action in Eqs. (7.2)–(7.4).

With this equality in mind, one can rewrite the expectation value of $\hat{O}$ in the configuration space as

$$\langle \hat{O} \rangle = (Z^Q)^{-1} \mathrm{Tr} \left[\hat{O} e^{-\beta \hat{H}} \right]$$

$$= \lim_{P \to \infty} \frac{1}{Z_P} \int_V \int_V \cdots \int_V \langle \mathbf{x}_0 | \hat{O} e^{-\frac{1}{P\beta}\hat{H}} | \mathbf{x}_1 \rangle \langle \mathbf{x}_1 | e^{-\frac{1}{P\beta}\hat{H}} | \mathbf{x}_2 \rangle \cdots$$

$$\cdot \langle \mathbf{x}_{P-1} | e^{-\frac{1}{P\beta}\hat{H}} | \mathbf{x}_P \rangle d\mathbf{x}_1 d\mathbf{x}_2 \cdots d\mathbf{x}_P$$

$$= \lim_{P \to \infty} \frac{1}{Z_P} \int_V \int_V \cdots \int_V O(\mathbf{x}_1) \langle \mathbf{x}_0 | e^{-\frac{1}{P\beta}\hat{H}} | \mathbf{x}_1 \rangle \langle \mathbf{x}_1 | e^{-\frac{1}{P\beta}\hat{H}} | \mathbf{x}_2 \rangle \cdots$$

$$\cdot \langle \mathbf{x}_{P-1} | e^{-\frac{1}{P\beta}\hat{H}} | \mathbf{x}_P \rangle d\mathbf{x}_1 d\mathbf{x}_2 \cdots d\mathbf{x}_P, \tag{7.23}$$

where $\mathbf{x}_0 = \mathbf{x}_P$ and $O(\mathbf{x}_1)$ denotes the expectation value of $\hat{O}$ at $\mathbf{x}_1$. Due to the fact that $\hat{H}/(P\beta)$ is an infinitely small Trotter factor, the same treatment also applies to $\mathbf{x}_2$, as

$$\langle \hat{O} \rangle = \lim_{P \to \infty} \frac{1}{Z_P} \int_V \int_V \cdots \int_V \langle \mathbf{x}_0 | e^{-\frac{1}{P\beta}\hat{H}} | \mathbf{x}_1 \rangle \langle \mathbf{x}_1 | \hat{O} e^{-\frac{1}{P\beta}\hat{H}} | \mathbf{x}_2 \rangle \cdots$$

$$\cdot \langle \mathbf{x}_{P-1} | e^{-\frac{1}{P\beta}\hat{H}} | \mathbf{x}_P \rangle d\mathbf{x}_1 d\mathbf{x}_2 \cdots d\mathbf{x}_P$$

$$= \lim_{P \to \infty} \frac{1}{Z_P} \int_V \int_V \cdots \int_V O(\mathbf{x}_2) \langle \mathbf{x}_0 | e^{-\frac{1}{P\beta}\hat{H}} | \mathbf{x}_1 \rangle \langle \mathbf{x}_1 | e^{-\frac{1}{P\beta}\hat{H}} | \mathbf{x}_2 \rangle \cdots$$

$$\cdot \langle \mathbf{x}_{P-1} | e^{-\frac{1}{P\beta}\hat{H}} | \mathbf{x}_P \rangle d\mathbf{x}_1 d\mathbf{x}_2 \cdots d\mathbf{x}_P. \tag{7.24}$$

Then we can continue such a cycling to $\mathbf{x}_P$ and make an average over all the images. Such a treatment gives us an expression for the expectation value of $\hat{O}$ as

$$\langle \hat{O} \rangle = \lim_{P \to \infty} \frac{1}{Z_P} \int_V \int_V \cdots \int_V \left[\frac{1}{P} \sum_{i=1}^{P} O(\mathbf{x}_i) \right]$$

$$\times \langle \mathbf{x}_0 | e^{-\frac{1}{P\beta}\hat{H}} | \mathbf{x}_1 \rangle \langle \mathbf{x}_1 | e^{-\frac{1}{P\beta}\hat{H}} | \mathbf{x}_2 \rangle \cdots$$

$$\cdot \langle \mathbf{x}_{P-1} | e^{-\frac{1}{P\beta}\hat{H}} | \mathbf{x}_P \rangle d\mathbf{x}_1 d\mathbf{x}_2 \cdots d\mathbf{x}_P. \tag{7.25}$$

In doing so, the equation from which the expectation value of any physical observable is calculated in the frequently used path integral scheme is arrived at.

We note that most of the studies in this field focus on the expectation value of local physical quantities [282, 286, 321, 322], such as the density function. In recent years, however, simulations for some non-local physical quantities such as the momentum distribution of a proton in water have also attracted much attention from the theoretical perspective, especially after the deep inelastic neutron scattering (DINS) experiment became available [334–336]. In these cases, rigorously speaking, the so-called open-path integral molecular dynamics method should be resorted to (see e.g. Morrone, Lin and Car's work in Refs. [337–339]), where the constraint $x_0 = x_P$ is released. This open path allows the density matrix itself to be simulated. However, we note that compared to the normal PIMD simulations with close path, special care must be taken in the open-path simulations concerning its stability. To deal with the problem, some new methods on simulating the momentum distribution using the conventional close-path PIMD method were also proposed; see e.g. Refs. [340–342]. Readers interested in theoretical simulations concerning this quantity may refer to two sets of works in particular: Morrone, Lin, Car, and Parrinello's in Refs. [337–341], and Ceriotti and Manolopoulos', in Ref. [342].

7.1.4 *Staging and Normal-Mode Transformations*

In terms of molecular dynamics, the canonical sampling associated with the partition function in Eq. (7.20), the density function in Eq. (7.15), and the expectation value of a physical observable in Eq. (7.25) can be obtained using the equations of motion of the fictitious polymer resulting from the Hamiltonian in Eq. (7.21).

A number of well-known numerical difficulties, however, exist in such a straightforward implementation of the PIMD method. These difficulties mainly arise from three aspects. First, from the definition of ω_P, which equals $\sqrt{P}/(\beta\hbar)$, it is clear that the stiffness of the spring constant increases with P. Therefore, with the increase in the number of beads, if the masses of the beads are independent of it, the spring interaction requires smaller and smaller time steps for the PIMD simulations to be carried out in order to characterize the interbead vibrations. Second, the stiffness of this spring interaction results in the external potential generated from the *ab initio*

(or force field) calculations within each image serving only as a small perturbation to the spring interaction. Consequently, the trajectories for the
beads of the artificial polymer will remain close to "invariant tori" in the
Cartesian space and efficient sampling of its entire configuration space is
seriously hindered. Third, even if this multi-time scale and the "invariant
tori" problems are solved, a sufficient number of thermostats should still
be incorporated into the dynamical scheme to ensure ergodicity of a simple
polymer in the PIMD simulations. Therefore, the ergodicity for the configuration space sampling of the fictitious polymer in the PIMD simulation is
technically much trickier than a simple implementation of the equation of
motion generated from Eq. (7.21). To the best of our knowledge, this non-
ergodic sampling in direct implementations of the PIMD method was first
pointed out by Hall and Berne in 1984 [343], using water and liquid neon as
examples. To a certain extent, this is also why the MC methods are prefer-
entially used in the path integral molecular simulations in the 1980s [70–73].
An MD-based method, in which an efficient sampling over the configuration
space of the artificial polymer was guaranteed, was highly desired.

These difficulties had been largely solved in the late 1980s and the
early 1990s. In order to explain these efficient sampling methods in a clear
manner, we first rewrite the Hamiltonian in Eq. (7.21) in the following form:

$$H(\mathbf{x}_1, \ldots, \mathbf{x}_P, \mathbf{p}_1, \ldots, \mathbf{p}_P)$$

$$= \sum_{j=1}^{N} \sum_{i=1}^{P} \frac{(\mathbf{p}_i^j)^2}{2M_i^j} + \sum_{j=1}^{N} \sum_{i=1}^{P} \frac{1}{2} m_j \omega_P^2 \left(\mathbf{x}_i^j - \mathbf{x}_{i-1}^j \right)^2 + \sum_{i=1}^{P} \frac{1}{P} V(\mathbf{x}_i^1, \ldots, \mathbf{x}_i^N),$$

$$(7.26)$$

where the close-path feature $\mathbf{x}_0 = \mathbf{x}_P$ is used to simplify the $V(\mathbf{x}_i^1, \ldots, \mathbf{x}_i^N)$
term, as compared to Eqs. (7.21) and (7.18). For a transparent nomencla-
ture in the following discussions, between the summation over beads and
nuclei, we take the sum over nuclei (N) as the outer loop since the form
of coordinate transformation, which serves as the key step in solving this
ergodicity problem, does not depend on the nuclei. In addition to this simplification of the potential energy term, since the first term in Eqs. (7.21)
and (7.26) is just introduced in the molecular dynamics simulations to
sample the configuration space, the mass M_i^j in this term, in principle,
is arbitrary for calculations of the statistical properties. It is allowed to
have one value for each artificial particle in the polymer (in total $N \times P$
artificial particles), as long as the spring constant is physical. With this in

mind, Eq. (7.26) can be further reformed into

$$H(\mathbf{x}_1, \ldots, \mathbf{x}_P, \mathbf{p}_1, \ldots, \mathbf{p}_P)$$

$$= \sum_{j=1}^{N} \sum_{i=1}^{P} \frac{(\mathbf{p}_i^j)^2}{2M_i^j} + \sum_{j=1}^{N} \frac{1}{2} m_j \omega_P^2 \mathbf{x}^j \mathbf{A} \mathbf{x}^j + \frac{1}{P} \sum_{i=1}^{P} V(\mathbf{x}_i^1, \ldots, \mathbf{x}_i^N),$$

$$(7.27)$$

where

$$\mathbf{A} = \begin{pmatrix} 2 & -1 & 0 & 0 & \cdots & 0 & 0 & 0 & -1 \\ -1 & 2 & -1 & 0 & \cdots & 0 & 0 & 0 & 0 \\ 0 & -1 & 2 & -1 & \cdots & 0 & 0 & 0 & 0 \\ \vdots & \vdots & \vdots & \vdots & \vdots & \vdots & \vdots & \vdots & \vdots \\ 0 & 0 & 0 & 0 & \cdots & -1 & 2 & -1 & 0 \\ 0 & 0 & 0 & 0 & \cdots & 0 & -1 & 2 & -1 \\ -1 & 0 & 0 & 0 & \cdots & 0 & 0 & -1 & 2 \end{pmatrix}. \qquad (7.28)$$

Here, $\mathbf{A}$ is a $P \times P$ matrix. For the matrix multiplication term $(1/2)m_j\omega_P^2 \mathbf{x}^j \mathbf{A} \mathbf{x}^j$ in Eq. (7.27), $\mathbf{x}^j$ can be viewed as a P-dimensional vector composed of $(\mathbf{x}_1^j, \mathbf{x}_2^j, \ldots, \mathbf{x}_P^j)$, with $\mathbf{x}_i^j$ representing the position of the ith image of the jth nucleus. When this position is a 3D vector in the Cartesian space, this matrix multiplication term goes through the coordinate of x, y, and z one by one. We note that M_i^j is the artificial mass we set for the ith image of the jth nucleus, while m_j is the physical mass of the jth nucleus.

From Eq. (7.27), it is clear that the smallest time step required in the PIMD simulations to describe the motion of the polymer, which originates from the spring interaction between neighboring beads, is associated with the largest eigenvalue of the matrix $\mathbf{A}$. However, we note that an artificial mass can be attributed to each fictitious particle in the polymer. Based on this advantage which is intrinsic in the principles of the PIMD method, a coordinate transformation can be employed to solve the infamous multi-time scale problem originating from the interbead vibrations. This coordinate transformation first decouples the harmonic interactions between neighboring beads in Eq. (7.27). Then, the mass associated with the different renormalized degrees of freedom (after coordinate transformation) can be artificially chosen so that the spring interaction results in vibrations of the same frequency, and it is advisable to choose those masses so that the resulting frequency does not depend on P. In doing so, the problems associated with the interimage vibrations can be avoided so that an efficient sampling of the polymer's configuration space can be carried out in a much easier manner.

In order to carry out such a decoupling of the spring interactions between neighboring beads, there are currently two popular schemes for the PIMD simulations to follow. The first one is the so-called "staging" method. It was originally proposed in a PIMC algorithm [71] and then employed in the PIMD simulations [319]. In this method, taking the jth nucleus as an example, the coordinate transformation from the original ones $(\mathbf{x}_1^j, \mathbf{x}_2^j, \ldots, \mathbf{x}_P^j)$ to the transformed ones $(\mathbf{u}_1^j, \mathbf{u}_2^j, \ldots, \mathbf{u}_P^j)$ is

$$\mathbf{u}_1^j = \mathbf{x}_1^j,$$
$$\mathbf{u}_i^j = \mathbf{x}_i^j - \frac{(i-1)\mathbf{x}_{i+1}^j + \mathbf{x}_1^j}{i}, \quad \text{for } i \geqslant 2,$$
(7.29)

whose form does not depend on the nuclear index j. The inverse of this relation is

$$\mathbf{x}_1^j = \mathbf{u}_1^j,$$
$$\mathbf{x}_i^j = \mathbf{u}_1^j + \sum_{l=i}^{P} \frac{(i-1)}{l-1}\mathbf{u}_l^j, \quad \text{for } i \geqslant 2,$$
(7.30)

which can also be obtained recursively by

$$\mathbf{x}_1^j = \mathbf{u}_1^j,$$
$$\mathbf{x}_i^j = \mathbf{u}_i^j + \frac{(i-1)}{i}\mathbf{x}_{i+1}^j + \frac{1}{i}\mathbf{x}_1^j, \quad \text{for } i \geqslant 2.$$
(7.31)

Since $\mathbf{x}_{P+1}^j = \mathbf{x}_1^j$, one often carries out the recursion in Eq. (7.31) in the order $\mathbf{x}_1^j, \mathbf{x}_P^j, \mathbf{x}_{P-1}^j, \ldots, \mathbf{x}_2^j$.

One advantage of such transformed coordinates is that the interbead spring interaction can be normalized through

$$\sum_{i=1}^{P}(\mathbf{x}_i^j - \mathbf{x}_{i+1}^j)^2 = \sum_{i=2}^{P} \frac{i}{i-1}(\mathbf{u}_i^j)^2.$$
(7.32)

To understand such a relationship from a practical perspective, we take the $P = 4$ case as an example. The relationship between the original coordinates and the transformed ones in Eq. (7.29) indicates

$$\mathbf{u}_1^j = \mathbf{x}_1^j,$$

$$\mathbf{u}_4^j = \mathbf{x}_4^j - \frac{3\mathbf{x}_1^j + \mathbf{x}_1^j}{4} = \mathbf{x}_4^j - \mathbf{x}_1^j,$$

$$\mathbf{u}_3^j = \mathbf{x}_3^j - \frac{2\mathbf{x}_4^j + \mathbf{x}_1^j}{3},$$

$$\mathbf{u}_2^j = \mathbf{x}_2^j - \frac{\mathbf{x}_3^j + \mathbf{x}_1^j}{2}.$$

(7.33)

Putting this relation into $\sum_{i=2}^{4} \frac{i}{i-1}(\mathbf{u}_i^j)^2$, one easily obtains

$$\sum_{i=2}^{4} \frac{i}{i-1}(\mathbf{u}_i^j)^2 = 2(\mathbf{u}_2^j)^2 + \frac{3}{2}(\mathbf{u}_3^j)^2 + \frac{4}{3}(\mathbf{u}_4^j)^2$$

$$= 2\left(\mathbf{x}_2^j - \frac{\mathbf{x}_3^j + \mathbf{x}_1^j}{2}\right)^2 + \frac{3}{2}\left(\mathbf{x}_3^j - \frac{2\mathbf{x}_4^j + \mathbf{x}_1^j}{3}\right)^2 + \frac{4}{3}(\mathbf{x}_4^j - \mathbf{x}_1^j)^2$$

$$= 2(\mathbf{x}_1^j)^2 + 2(\mathbf{x}_2^j)^2 + 2(\mathbf{x}_3^j)^2 + 2(\mathbf{x}_1^j)^2 - 2\mathbf{x}_1^j\mathbf{x}_2^j - 2\mathbf{x}_2^j\mathbf{x}_3^j$$
$$- 2\mathbf{x}_3^j\mathbf{x}_4^j - 2\mathbf{x}_4^j\mathbf{x}_1^j$$

$$= \sum_{i=1}^{4}(\mathbf{x}_i^j - \mathbf{x}_{i+1}^j)^2.$$

(7.34)

Therefore, the coordinate transformation decouples the relative motion of the beads in the position space.

Another point which is implied in Eqs. (7.29) and (7.30) is that there is a one-to-one correspondence between $(\mathbf{x}_1^j, \mathbf{x}_2^j, \ldots, \mathbf{x}_P^j)$ and $(\mathbf{u}_1^j, \mathbf{u}_2^j, \ldots, \mathbf{u}_P^j)$. Therefore, by setting

$$m_1^j = 0,$$

$$m_i^j = \frac{i}{i-1}m_j, \quad \text{for } i \geqslant 2.$$

(7.35)

Equation (7.27) can be rewritten as

$$H(\mathbf{x}_1, \ldots, \mathbf{x}_P, \mathbf{p}_1, \ldots, \mathbf{p}_P)$$

$$= H(\mathbf{u}_1, \ldots, \mathbf{u}_P, \mathbf{P}_1, \ldots, \mathbf{P}_P)$$

$$= \sum_{j=1}^{N}\sum_{i=1}^{P}\left[\frac{(\mathbf{P}_i^j)^2}{2M_i'^j} + \frac{1}{2}m_i^j\omega_P^2(\mathbf{u}_i^j)^2\right] + \sum_{i=1}^{P}\frac{1}{P}V(\mathbf{u}_i^1, \ldots, \mathbf{u}_i^N).$$

(7.36)

Here, $\mathbf{P}_i$ is assumed to be conjugate to $\mathbf{u}_i$ in the transformed coordinates, and $M_i'^j$ is the associated mass which is related to the M_i^j in Eq. (7.26) by the coordinate transition in Eq. (7.29). We do not need to care about the specific form of this relation. There is no potential originating from the harmonic spring interaction for the first artificial particle in the transformed coordinate of the polymer ($m_1^j = 0$). For the artificial particles associated with the other degrees of freedom, this potential is determined by m_i^j ($i \geqslant 2$) in Eq. (7.35). Therefore, as long as we keep $M_i'^j$ ($i \geqslant 2$) as a constant multiple of m_i^j, the vibrations originating from the spring interactions associated with the other renormalized degrees of freedom will share the same time scale. In practice, these $M_i'^j$s are often chosen as $M_1'^j = m_j$ (physical mass of the jth nucleus) and $M_i'^j = m_i^j$ for $i \geqslant 2$. The equations of motion in the transformed coordinates obtained from the Hamiltonian in Eq. (7.36) is

$$
\dot{\mathbf{u}}_i^j = \frac{\mathbf{P}_i^j}{M_i'^j},
$$
$$
\dot{\mathbf{P}}_i^j = -m_i^j \omega_P^2 \mathbf{u}_i^j - \frac{1}{P}\frac{\partial V}{\partial \mathbf{u}_i^j},
\tag{7.37}
$$

where the only term which needs to be obtained from calculations in the original Cartesian space is $\partial V/\partial \mathbf{u}_i^j$.

In order to obtain such forces in the transformed coordinate space, one needs to go back to the original Cartesian space and calculate the Hellmann–Feynman forces of the corresponding nuclear configuration, namely, $\partial V/\partial \mathbf{x}_i^j$. Then a transformation must be made to link these $\partial V/\partial \mathbf{x}_i^j$s with the $\partial V/\partial \mathbf{u}_i^j$s to be used for the simulation of the propagation in Eq. (7.37). Such a transformation is defined from Eqs. (7.29) and (7.30). The starting point is the following equation:

$$
\frac{1}{P}\frac{\partial V}{\partial \mathbf{u}_k^j} = \frac{1}{P}\sum_{i=1}^{P}\frac{\partial V}{\partial \mathbf{x}_i^j}\frac{\partial \mathbf{x}_i^j}{\partial \mathbf{u}_k^j},
$$
$$
\frac{1}{P}\frac{\partial V}{\partial \mathbf{x}_k^j} = \frac{1}{P}\sum_{i=1}^{P}\frac{\partial V}{\partial \mathbf{u}_i^j}\frac{\partial \mathbf{u}_i^j}{\partial \mathbf{x}_k^j},
\tag{7.38}
$$

for a certain k. When $k = 1$, from Eq. (7.30), it is clear that $\partial \mathbf{x}_i^j/\partial \mathbf{u}_k^j$ equals one for any i. Therefore, using the first equation in Eq. (7.38), one easily obtains

$$
\frac{1}{P}\frac{\partial V}{\partial \mathbf{u}_1^j} = \frac{1}{P}\sum_{i=1}^{P}\frac{\partial V}{\partial \mathbf{x}_i^j}.
\tag{7.39}
$$

When $k \geqslant 2$, from Eq. (7.29), it is clear that $\partial u_1^j / \partial x_k^j = 0$ and $\partial u_i^j / \partial x_k^j \neq 0$ for $i \geqslant 2$. Therefore, the second equation in Eq. (7.38) becomes

$$\frac{1}{P}\frac{\partial V}{\partial x_k^j} = \frac{1}{P}\sum_{i=1}^{P}\frac{\partial V}{\partial u_i^j}\frac{\partial u_i^j}{\partial x_k^j} = \frac{1}{P}\sum_{i=2}^{P}\frac{\partial V}{\partial u_i^j}\frac{\partial u_i^j}{\partial x_k^j}. \tag{7.40}$$

Then, we can make use of the second equation in Eq. (7.29) and rewrite it into

$$\frac{1}{P}\frac{\partial V}{\partial x_k^j} = \frac{1}{P}\sum_{i=2}^{P}\frac{\partial V}{\partial u_i^j}\frac{\partial u_i^j}{x_k^j}$$

$$= \frac{1}{P}\sum_{i=2}^{P}\frac{\partial V}{\partial u_i^j}\left[\frac{\partial x_i^j}{\partial x_k^j} - \frac{i-1}{i}\frac{\partial x_{i+1}^j}{\partial x_k^j}\right]$$

$$= \frac{1}{P}\sum_{i=2}^{P}\frac{\partial V}{\partial u_i^j}\left[\delta_{i,k} - \frac{i-1}{i}\delta_{i+1,k}\right]$$

$$= \frac{1}{P}\frac{\partial V}{\partial u_k^j} - \frac{1}{P}\frac{k-2}{k-1}\frac{\partial V}{\partial u_{k-1}^j}. \tag{7.41}$$

Therefore, the $\partial V / \partial u_k^j$ can be calculated recursively from

$$\frac{1}{P}\frac{\partial V}{\partial u_k^j} = \frac{1}{P}\frac{\partial V}{\partial x_k^j} + \frac{1}{P}\frac{k-2}{k-1}\frac{\partial V}{\partial u_{k-1}^j}. \tag{7.42}$$

Combining Eqs. (7.39) and (7.42), the $\partial V / \partial u_k^j$s to be used in the numerical simulation of the propagation in Eq. (7.37) should be calculated recursively from

$$\frac{1}{P}\frac{\partial V}{\partial u_1^j} = \frac{1}{P}\sum_{i=1}^{P}\frac{\partial V}{\partial x_i^j},$$

$$\frac{1}{P}\frac{\partial V}{\partial u_k^j} = \frac{1}{P}\frac{\partial V}{\partial x_k^j} + \frac{1}{P}\frac{k-2}{k-1}\frac{\partial V}{\partial u_{k-1}^j}, \qquad \text{for } k \geqslant 2. \tag{7.43}$$

It is worth noting that, different from the recursion for the transformation of the coordinates between x_i^j and u_i^j in Eq. (7.31), the force transformation here starts from $k = 2$, after $\partial V / \partial u_1^j$ is obtained using the first expression in Eq. (7.43).

    With these, the coordinate transformation which decouples the movement of the neighboring images and imposes one single frequency for all these interbead vibrations has been introduced. The next thing one needs to do, in order to carry out an ergodic sampling of the polymer naturally, is a massive thermostat sampling for all the degrees of freedom for the artificial polymer in the transformed coordinates. This thermostat sampling can be performed using the thermostats introduced in Chapter 5. Taking the Andersen thermostat as an example, in this case, the momentum $\mathbf{P}_i^j$ for $1 \leqslant i \leqslant P$ and $1 \leqslant j \leqslant N$ should be rescaled for the desired temperature during each collision between the artificial particle and the thermostat. In the case when Nosé–Hoover chain thermostat of a certain length is used, the equation of motion in Eq. (7.37) for the ith bead of the jth nucleus should be linked with this Nosé–Hoover chain according to Eq. (5.52). One notes that the computational cost of these thermostats is much lower than that of the force calculation in real poly-atomic systems, especially when *ab initio* methods for the electronic structures are used. Therefore, picking up a thermostat which is efficient in pushing the polymer into its equilibrium state is essential in practical PIMD simulations.

So far, we have discussed the "staging" transformation. As mentioned, the coordinate transformation aims at decoupling the relative motion between neighboring beads in the transformed coordinates. Therefore, as an alternative to the "staging" method, the most direct way to fulfill such a task is to diagonalize the matrix $\mathbf{A}$ in Eq. (7.28) and then use its eigenvectors to perform the coordinate transformation. In doing so, the Hamiltonian in Eq. (7.27) will have diagonal spring interactions in its second term in the transformed coordinates, with the constant of these diagonal spring interactions determined by the eigenvalues of the matrix $\mathbf{A}$. This method is the so-called "normal-mode" method in Refs. [81, 319]. Parallel to the "staging" method, it is the other frequently used coordinate transformation method in practical PIMD simulations.

Now, we go into the details of how such a "normal-mode" transformation is performed. Since the matrix $\mathbf{A}$ is the same for all the nuclei, a single unitary orthogonal matrix $\mathbf{U}$ can be used to transform coordinates of all nuclei from the Cartesian to the normal-mode ones. This transformation reads

$$\mathbf{u}^j = \frac{1}{\sqrt{P}}\mathbf{U}\mathbf{x}^j \quad \text{and} \quad \mathbf{x}^j = \sqrt{P}\mathbf{U}^\mathrm{T}\mathbf{u}^j, \tag{7.44}$$

where j goes through all the nuclear index from 1 to N. From our knowledge of linear algebra, we know that the matrix $\mathbf{U}$ can also be used to diagonalize the matrix $\mathbf{A}$ through $P\mathbf{U}\mathbf{A}\mathbf{U}^T$, and the resulting matrix $\mathbf{\Gamma}$ is diagonal. If we choose the number of P to be even (which equals $2n + 2$) and align Γ as

$$\Gamma = \begin{pmatrix} \lambda_{-n} & 0 & 0 & 0 & 0 & 0 \\ 0 & \lambda_{-n+1} & 0 & 0 & 0 & 0 \\ 0 & 0 & \cdots & 0 & 0 & 0 \\ 0 & 0 & 0 & \lambda_0 & 0 & 0 \\ 0 & 0 & 0 & 0 & \cdots & 0 \\ 0 & 0 & 0 & 0 & 0 & \lambda_{n+1} \end{pmatrix}, \tag{7.45}$$

the unitary orthogonal matrix $\mathbf{U}$ will have the following simple form:

$$U_{k,i} = \begin{cases} \sqrt{2/P}\sin\left(\frac{2\pi ki}{2n+2}\right), & -n \leqslant k < 0, \\ \sqrt{1/P}, & k = 0, \\ \sqrt{2/P}\cos\left(\frac{2\pi ki}{2n+2}\right), & 0 < k \leqslant n, \\ (-1)^i/\sqrt{P}, & k = n+1. \end{cases} \tag{7.46}$$

Here, the row index k goes from $-n$ to $n+1$ and the column index l goes from 1 to P [344]. In other words, we label the beads from 1 to P in the Cartesian coordinates and from $-n$ to $n+1$ in the normal-mode coordinates. The diagonal elements of Γ are $\lambda_0 = 0$, $\lambda_{\pm i} = 4P\sin^2(i\pi/P)$ for $0 < i \leqslant n$, and $\lambda_{n+1} = 4P$. In the transformed "normal-mode" coordinates, the Hamiltonian in Eq. (7.27) then reads

$$H(\mathbf{u}_{-n}, \ldots, \mathbf{u}_{n+1}, \mathbf{P}_{-n}, \ldots, \mathbf{P}_{n+1})$$

$$= \sum_{j=1}^{N} \sum_{i=-n}^{n+1} \left[\frac{(\mathbf{p}_i^j)^2}{2M_i'^j} + \frac{1}{2} m_j \omega_P^2 \lambda_i (\mathbf{u}_i^j)^2 \right] + \sum_{i=-n}^{n+1} \frac{1}{P} V(\mathbf{u}_i^1, \ldots, \mathbf{u}_i^N). \tag{7.47}$$

From the expression of the transformation matrix $\mathbf{U}$ in Eq. (7.46), it is clear that in principle, it is equivalent to a Fourier transform of a periodic path. For the eigenvector associated with the eigenvalue $\lambda_0 = 0$, from Eqs. (7.44) and (7.46), we see that the transformed coordinate should be

$$\mathbf{u}_0^j = \frac{1}{\sqrt{P}} \sum_{i=1}^{P} \mathbf{U}_{0,i} \mathbf{x}_i^j = \frac{1}{P} \sum_{i=1}^{P} \mathbf{x}_i^j; \tag{7.48}$$

which represents the centroid of the path of the jth nucleus. Therefore, in the normal-mode coordinate, the propagation of this specific mode naturally describes the evolution of the centroid. Analogous to the "staging" method, from the other eigenvalues of the matrix Γ and the fact that an artificial mass M'^j_i can be set for each artificial particle, one can easily choose

$$M^j_0 = m_j,$$

$$M^j_i = c\lambda_i m_j, \quad \text{for } -n \leqslant i \leqslant -1 \quad \text{and} \quad 1 \leqslant i \leqslant n+1,$$

(7.49)

with c representing a constant. With these, the different modes corresponding to the interbead vibrations naturally move on the same time scale.

From the Hamiltonian in Eq. (7.48), the equation of motion is easily obtainable from

$$\dot{\mathbf{u}}^j_i = \frac{\mathbf{P}^j_i}{M'^j_i},$$

$$\dot{\mathbf{P}}^j_i = -m^j_i \omega^2_P \lambda_i \mathbf{u}^j_i - \frac{1}{P}\frac{\partial V}{\partial \mathbf{u}^j_i}.$$

(7.50)

The next thing one needs to do is the same as the above descriptions for the "staging" method, i.e. calculating $\partial V/\partial \mathbf{u}^j_i$ from $\partial V/\partial \mathbf{x}^j_i$. Compared with the "staging" method, here, due to the fact that the coordinate transformation is performed with a constant matrix determined by P only, from Eqs. (7.38) and (7.44), this transformation between $\partial V/\partial \mathbf{u}^j_i$ from $\partial V/\partial \mathbf{x}^j_i$ can be carried out in a much simpler manner.

In Eq. (7.44), we can see that the transformation between the Cartesian and the normal-mode coordinates can be carried out as follows. For a specific nucleus, e.g. the jth, we go through the indices x, y, and z one by one. For each index, e.g. the x index, the Cartesian coordinates of the P images were organized as a P-dimensional vector. Then the coordinate transformation to $\mathbf{u}^j$ is carried out using Eq. (7.44). In the force transformations, we follow the same routine. For the jth nucleus, we go through the indices x, y, and z one by one. For each index, e.g. x, the forces along the x axis on the P images of the jth nucleus were organized as a P-dimensional vector, which we label as F^j. It is composed of

$$\left(\frac{\partial V}{\partial \mathbf{x}^j_1}\bigg|_x, \ldots, \frac{\partial V}{\partial \mathbf{x}^j_P}\bigg|_x\right),$$

where $\left.\frac{\partial V}{\partial \mathbf{x}_i^j}\right|_x$ means the Hellmann–Feynman force on the ith image of the jth nucleus along the x axis. We label F'^j as the vector which represents the transformed forces, composed of

$$\left(\left.\frac{\partial V}{\partial \mathbf{u}_{-n}^j}\right|_x, \dots, \left.\frac{\partial V}{\partial \mathbf{u}_{n+1}^j}\right|_x \right).$$

Then, from Eqs. (7.38) and (7.44), one easily obtains

$$\frac{1}{P}F'^j = \frac{1}{\sqrt{P}}\mathbf{U}^T F^j. \tag{7.51}$$

These forces are then used to propagate the equation of motion in Eq. (7.50). Analogous to what we say in the "staging" method, massive thermostats must again be used, with an efficient one imposed on each degree of freedom. With these, an ergodic sampling can also be realized using this "normal-mode" method. We note that due to the advantage that the zeroth (according to our labeling from $-n$ to $n + 1$) normal mode naturally describes the propagation of the centroid, which does have a physical meaning, the "normal-mode" method is getting more and more popular nowadays, especially when extensions to real-time propagation are concerned. We will give a brief discussion to this extension in Sec. 7.2.

7.1.5 Evaluation of the Zero-Point Energy

In the above discussions, we have introduced how the PIMD method should be implemented. From the corresponding PIMD simulations, one can estimate the statistical expectation value of a physical quantity at a finite temperature using Eq. (7.25). The result of such an evaluation is the expectation value of this quantity in the poly-atomic system at a finite T, with the QNEs rigorously addressed on the same footing as the thermal ones. From our discussions in Chapter 5, we know that the expectation value of the same physical quantity can also be calculated in an MD simulation, with only the thermal nuclear effects taken into account. Therefore, by comparing the results obtained from these two simulations, one can evaluate the impact of QNEs on this physical quantity in a very clear manner.

Besides these quantities which can be evaluated using Eq. (7.25), e.g. the radial distribution function [323, 338, 345], the intra and inter-molecular bond length distributions [286, 321], etc., there are also some physical quantities in which the evaluation of their expectation values from the PIMD simulations contain some subtleties. The internal energy is such an example. As a matter of fact, this internal energy at finite temperatures is among the most relevant quantities in molecular simulations. A comparison between this quantity from the PIMD and MD simulations at different temperatures and then an extrapolation of their differences toward 0 K can give us the nuclear zero-point energy in a real poly-atomic system, beyond the frequently used harmonic approximation. This zero-point energy is of primary concern in studies of many problems, whose value otherwise must be calculated from very expensive quantum MC simulations (normally with a force field treatment of the interatomic nuclear interactions; see e.g. Refs. [346–349]). As mentioned before, for large poly-atomic systems, such quantum MC simulations might not be applicable. Therefore, the PIMD simulation, in principle, gives us a useful estimator for this key physical quantity in practical calculations of poly-atomic systems.

To understand how this purpose is fulfilled, we first go back to the original definition of the internal energy in statistical mechanics, which is

$$\langle E \rangle = -\frac{1}{Z^Q} \frac{\partial Z^Q}{\partial \beta}. \tag{7.52}$$

Here, Z^Q is the partition function of the quantum poly-atomic system, whose expression can be given in different ways, e.g. Eq. (7.20) in the limit of $P \to \infty$ and Eq. (7.17), etc. We take its expression in Eq. (7.20) and rewrite it as

$$Z^Q = \lim_{P \to \infty} C \int \cdots \int$$

$$\times e^{-\beta \left\{ \sum_{j=1}^{N} \sum_{i=1}^{P} \left[\frac{(\mathbf{p}_i^j)^2}{2M_i^j} + \frac{1}{2} m_j \omega_P^2 \left(\mathbf{x}_i^j - \mathbf{x}_{i-1}^j \right)^2 \right] + \sum_{i=1}^{P} \frac{1}{P} V(\mathbf{x}_i^1, \ldots, \mathbf{x}_i^N) \right\}}$$

$$\cdot d\mathbf{x}_1 \cdots d\mathbf{x}_P d\mathbf{p}_1 \cdots d\mathbf{p}_P$$

$$= \lim_{P \to \infty} C \int \cdots \int$$

$$\times e^{\sum_{j=1}^{N} \sum_{i=1}^{P} \left[-\beta \frac{(\mathbf{p}_i^j)^2}{2M_i^j} - \frac{1}{\beta} \frac{P}{2\hbar^2} m_j \left(\mathbf{x}_i^j - \mathbf{x}_{i-1}^j \right)^2 \right] - \sum_{i=1}^{P} \beta \frac{1}{P} V(\mathbf{x}_i^1, \ldots, \mathbf{x}_i^N)}$$

$$\cdot d\mathbf{x}_1 \cdots d\mathbf{x}_P d\mathbf{p}_1 \cdots d\mathbf{p}_P. \tag{7.53}$$

Using this expression of the quantum partition function, an expansion of Eq. (7.52) easily gives us

$$\langle E \rangle = \lim_{P \to \infty} \frac{C}{Z^Q} \int \cdots \int \left\{ \sum_{j=1}^{N} \sum_{i=1}^{P} \left[\frac{(\mathbf{p}_i^j)^2}{2M_i^j} - \frac{1}{\beta^2} \frac{P}{2\hbar^2} m_j (\mathbf{x}_i^j - \mathbf{x}_{i-1}^j)^2 \right] \right.$$

$$\left. + \sum_{i=1}^{P} \frac{1}{P} V(\mathbf{x}_i^1, \ldots, \mathbf{x}_i^N) \right\}$$

$$\cdot e^{-\beta \sum_{j=1}^{N} \sum_{i=1}^{P} \left[\frac{(\mathbf{p}_i^j)^2}{2M_i^j} + \frac{1}{2} m_j \omega_P^2 (\mathbf{x}_i^j - \mathbf{x}_{i-1}^j)^2 \right] - \sum_{i=1}^{P} \beta \frac{1}{P} V(\mathbf{x}_i^1, \ldots, \mathbf{x}_i^N)}$$

$$\times d\mathbf{x}_1 \cdots d\mathbf{x}_P d\mathbf{p}_1 \cdots d\mathbf{p}_P$$

$$= \lim_{P \to \infty} \frac{C}{Z^Q} \int \cdots \int \left\{ \sum_{j=1}^{N} \sum_{i=1}^{P} \left[\frac{(\mathbf{p}_i^j)^2}{2M_i^j} - \frac{1}{2} m_j \omega_P^2 (\mathbf{x}_i^j - \mathbf{x}_{i-1}^j)^2 \right] \right.$$

$$\left. + \sum_{i=1}^{P} \frac{1}{P} V(\mathbf{x}_i^1, \ldots, \mathbf{x}_i^N) \right\}$$

$$\cdot e^{-\beta \sum_{j=1}^{N} \sum_{i=1}^{P} \left[\frac{(\mathbf{p}_i^j)^2}{2M_i^j} + \frac{1}{2} m_j \omega_P^2 (\mathbf{x}_i^j - \mathbf{x}_{i-1}^j)^2 \right] - \sum_{i=1}^{P} \beta \frac{1}{P} V(\mathbf{x}_i^1, \ldots, \mathbf{x}_i^N)}$$

$$\times d\mathbf{x}_1 \cdots d\mathbf{x}_P d\mathbf{p}_1 \cdots d\mathbf{p}_P.$$

$$(7.54)$$

Now, if we define

$$E = \sum_{j=1}^{N} \sum_{i=1}^{P} \left[\frac{(\mathbf{p}_i^j)^2}{2M_i^j} - \frac{1}{2} m_j \omega_P^2 (\mathbf{x}_i^j - \mathbf{x}_{i-1}^j)^2 \right] + \sum_{i=1}^{P} \frac{1}{P} V(\mathbf{x}_i^1, \ldots, \mathbf{x}_i^N),$$

$$(7.55)$$

Eq. (7.54) can be further rewritten as

$$\langle E \rangle = \lim_{P \to \infty} \frac{\int \cdots \int E e^{-\beta \sum_{j=1}^{N} \sum_{i=1}^{P} \left[\frac{(\mathbf{p}_i^j)^2}{2M_i^j} + \frac{1}{2} m_j \omega_P^2 (\mathbf{x}_i^j - \mathbf{x}_{i-1}^j)^2 \right] - \sum_{i=1}^{P} \beta \frac{1}{P} V(\mathbf{x}_i^1, \ldots, \mathbf{x}_i^N)} \cdot d\mathbf{x}_1 \cdots d\mathbf{x}_P d\mathbf{p}_1 \cdots d\mathbf{p}_P}{\int \cdots \int e^{-\beta \sum_{j=1}^{N} \sum_{i=1}^{P} \left[\frac{(\mathbf{p}_i^j)^2}{2M_i^j} + \frac{1}{2} m_j \omega_P^2 (\mathbf{x}_i^j - \mathbf{x}_{i-1}^j)^2 \right] - \sum_{i=1}^{P} \beta \frac{1}{P} V(\mathbf{x}_i^1, \ldots, \mathbf{x}_i^N)} \cdot d\mathbf{x}_1 \cdots d\mathbf{x}_P d\mathbf{p}_1 \cdots d\mathbf{p}_P}.$$

$$(7.56)$$

Therefore, the instantaneous quantity E as defined in Eq. (7.55) becomes the quantity whose ensemble average should be evaluated in the PIMD simulation. From Eq. (7.56), we see that this ensemble average gives the internal energy of the poly-atomic system, in which all QNEs are included. The minus sign in front of the second term in the quantity in Eq. (7.55) to be averaged during the simulation originates from the temperature dependence of the spring constant in the Hamiltonian of the fictitious polymer. In a simpler form of the ensemble averages at a finite T in molecular dynamics, the expectation value of the internal energy can be further reformed as

$$\langle E \rangle = \frac{3NP}{2\beta} - \left\langle \sum_{j=1}^{N} \sum_{i=1}^{P} \frac{1}{2} m_j \omega_P^2 \left(\mathbf{x}_i^j - \mathbf{x}_{i-1}^j \right)^2 \right\rangle + \left\langle \sum_{i=1}^{P} \frac{1}{P} V(\mathbf{x}_i^1, \ldots, \mathbf{x}_i^N) \right\rangle.$$

$$(7.57)$$

From this estimator, it is easy to see that the kinetic energy (first term) and the spring potential (second term) both scale linearly with the number of beads P in a finite-temperature PIMD simulation. Therefore, with the increase of P, these two quantities go to large values. Fortunately, due to the minus sign in front of the second term, the difference between them converges with P and the zero-point energy of a poly-atomic system can be evaluated with Eq. (7.57) in practical PIMD simulations. However, we note that large fluctuations on these two quantities still remain at large P. As a consequence, for highly quantum systems in which a large P must be used in order to arrive at an accurate simulation of the QNEs, a loss of precision might exist [97, 350]. An alternative estimator, in which all terms involved converge with P, is highly desired.

To circumvent this problem, a path integral version of the virial theorem was introduced by Herman et $al.$ in 1982, where an estimator of the internal energy which suffers much less from these fluctuations was proposed [350, 351]. For a clear explanation of how this works, we first go back to the original quantum partition function in Eq. (7.17), where no artificial kinetic energy for the molecular dynamics simulations is introduced and the integration goes only through the configuration space (composed of the $\mathbf{x}_i^j$s) instead of the phase space (composed of the $\mathbf{x}_i^j$s and $\mathbf{p}_i^j$s). Using the periodic boundary condition of the path integral sampling, this equation can be rewritten as

$$Z^Q = \lim_{P \to \infty} \left(\frac{mP}{2\beta \pi \hbar^2} \right)^{\frac{P}{2}} \int_V \int_V \cdots \int_V e^{-\beta V^{\mathrm{eff}}(\mathbf{x}_1, \mathbf{x}_2, \ldots, \mathbf{x}_P)} \mathrm{d}\mathbf{x}_1 \mathrm{d}\mathbf{x}_2 \cdots \mathrm{d}\mathbf{x}_P,$$

$$(7.58)$$

where

$$V^{\text{eff}}(\mathbf{x}_1, \mathbf{x}_2, \ldots, \mathbf{x}_P) = \sum_{i=1}^{P} \sum_{j=1}^{N} \left[\frac{1}{2} m_j \omega_P^2 (\mathbf{x}_i^j - \mathbf{x}_{i-1}^j)^2 \right] + \sum_{i=1}^{P} \frac{1}{P} V(\mathbf{x}_i^1, \ldots, \mathbf{x}_i^N).$$

(7.59)

Similar to what we have used for the nomenclature before, $(\mathbf{x}_1, \mathbf{x}_2, \ldots, \mathbf{x}_P)$ altogether represents the spatial configuration of the artificial polymer. $\mathbf{x}_i$ is composed of $(\mathbf{x}_i^1, \ldots, \mathbf{x}_i^N)$. It means the spatial configuration of the polyatomic system at its ith image. The key point of the virial internal energy estimator is that the first two terms in Eq. (7.57) is replaced by the mean kinetic energy of the quantum system as

$$\frac{3NP}{2\beta} - \left\langle \sum_{j=1}^{N} \sum_{i=1}^{P} \frac{1}{2} m_j \omega_P^2 (\mathbf{x}_i^j - \mathbf{x}_{i-1}^j)^2 \right\rangle = \left\langle \frac{1}{2P} \sum_{j=1}^{N} \sum_{i=1}^{P} \mathbf{x}_i^j \cdot \frac{\partial V(\mathbf{x}_i^1, \ldots, \mathbf{x}_i^N)}{\partial \mathbf{x}_i^j} \right\rangle.$$

(7.60)

With this treatment, the expectation value of the internal energy in Eq. (7.57) reforms into

$$\langle E \rangle = \left\langle \frac{1}{2P} \sum_{j=1}^{N} \sum_{i=1}^{P} \mathbf{x}_i^j \cdot \frac{\partial V(\mathbf{x}_i^1, \ldots, \mathbf{x}_i^N)}{\partial \mathbf{x}_i^j} \right\rangle + \left\langle \sum_{i=1}^{P} \frac{1}{P} V(\mathbf{x}_i^1, \ldots, \mathbf{x}_i^N) \right\rangle.$$

(7.61)

Here, you can see that neither of the two terms being evaluated in the PIMD simulation scales with P. Therefore, a smaller fluctuation of the internal energy to be evaluated in the PIMD simulations should be expected.

To understand how the equality in Eq. (7.60) exists, we use the partition function in Eq. (7.58) to evaluate the quantity on the right-hand side of Eq. (7.60). From this partition function, the ensemble average of this quantity equals

$$\left\langle \frac{1}{2P} \sum_{j=1}^{N} \sum_{i=1}^{P} \mathbf{x}_i^j \cdot \frac{\partial V(\mathbf{x}_i^1, \ldots, \mathbf{x}_i^N)}{\partial \mathbf{x}_i^j} \right\rangle$$

$$= \frac{\int_V \cdots \int_V \left[\frac{1}{2} \sum_{j=1}^{N} \sum_{i=1}^{P} \mathbf{x}_i^j \cdot \frac{\partial \frac{1}{P} V(\mathbf{x}_i^1, \ldots, \mathbf{x}_i^N)}{\partial \mathbf{x}_i^j} \right] e^{-\beta V^{\text{eff}}(\mathbf{x}_1, \mathbf{x}_2, \ldots, \mathbf{x}_P)} d\mathbf{x}_1 d\mathbf{x}_2 \cdots d\mathbf{x}_P}{\int_V \cdots \int_V e^{-\beta V^{\text{eff}}(\mathbf{x}_1, \mathbf{x}_2, \ldots, \mathbf{x}_P)} d\mathbf{x}_1 d\mathbf{x}_2 \cdots d\mathbf{x}_P}.$$

(7.62)

Here, the relation between $V(\mathbf{x}_i^1, \ldots, \mathbf{x}_i^N)/P$ and $V^{\mathrm{eff}}(\mathbf{x}_1, \ldots, \mathbf{x}_P)$ is given by Eq. (7.59). Now, if we label

$$\alpha(\mathbf{x}_1, \mathbf{x}_2, \ldots, \mathbf{x}_P) = \sum_{i=1}^{P} \sum_{j=1}^{N} \frac{1}{2} m_j \omega_P^2 (\mathbf{x}_i^j - \mathbf{x}_{i-1}^j)^2, \qquad (7.63)$$

and

$$\lambda(\mathbf{x}_1, \mathbf{x}_2, \ldots, \mathbf{x}_P) = \sum_{i=1}^{P} \frac{1}{P} V(\mathbf{x}_i^1, \mathbf{x}_i^2, \ldots, \mathbf{x}_i^N), \qquad (7.64)$$

then Eq. (7.59) will be reformed into

$$V^{\mathrm{eff}}(\mathbf{x}_1, \mathbf{x}_2, \ldots, \mathbf{x}_P) = \alpha(\mathbf{x}_1, \mathbf{x}_2, \ldots, \mathbf{x}_P) + \lambda(\mathbf{x}_1, \mathbf{x}_2, \ldots, \mathbf{x}_P), \qquad (7.65)$$

and Eq. (7.62) reforms to

$$\left\langle \frac{1}{2P} \sum_{j=1}^{N} \sum_{i=1}^{P} \mathbf{x}_i^j \cdot \frac{\partial V(\mathbf{x}_i^1, \ldots, \mathbf{x}_i^N)}{\partial \mathbf{x}_i^j} \right\rangle$$

$$= \frac{\int_V \cdots \int_V \left[\frac{1}{2} \sum_{j=1}^{N} \sum_{i=1}^{P} \mathbf{x}_i^j \cdot \frac{\partial V^{\mathrm{eff}}(\mathbf{x}_1, \ldots, \mathbf{x}_P)}{\partial \mathbf{x}_i^j} \right] e^{-\beta V^{\mathrm{eff}}(\mathbf{x}_1, \mathbf{x}_2, \ldots, \mathbf{x}_P)} d\mathbf{x}_1 d\mathbf{x}_2 \cdots d\mathbf{x}_P}{\int_V \cdots \int_V e^{-\beta V^{\mathrm{eff}}(\mathbf{x}_1, \mathbf{x}_2, \ldots, \mathbf{x}_P)} d\mathbf{x}_1 d\mathbf{x}_2 \cdots d\mathbf{x}_P}$$

$$- \frac{\int_V \cdots \int_V \left[\frac{1}{2} \sum_{j=1}^{N} \sum_{i=1}^{P} \mathbf{x}_i^j \cdot \frac{\partial \alpha(\mathbf{x}_1, \ldots, \mathbf{x}_P)}{\partial \mathbf{x}_i^j} \right] e^{-\beta V^{\mathrm{eff}}(\mathbf{x}_1, \mathbf{x}_2, \ldots, \mathbf{x}_P)} d\mathbf{x}_1 d\mathbf{x}_2 \cdots d\mathbf{x}_P}{\int_V \cdots \int_V e^{-\beta V^{\mathrm{eff}}(\mathbf{x}_1, \mathbf{x}_2, \ldots, \mathbf{x}_P)} d\mathbf{x}_1 d\mathbf{x}_2 \cdots d\mathbf{x}_P}. \qquad (7.66)$$

For a further evaluation of this quantity, we first make use of an important property of $\alpha(\mathbf{x}_1, \ldots, \mathbf{x}_P)$, that it is a homogeneous function of $(\mathbf{x}_1, \ldots, \mathbf{x}_P)$ of degree 2, so that the following equation exists:

$$\sum_{j=1}^{N} \sum_{i=1}^{P} \mathbf{x}_i^j \cdot \frac{\partial \alpha(\mathbf{x}_1, \ldots, \mathbf{x}_P)}{\partial \mathbf{x}_i^j} = 2\alpha(\mathbf{x}_1, \ldots, \mathbf{x}_P). \qquad (7.67)$$

Therefore, the second term on the right-hand side of Eq. (7.66) equals

$$
\begin{aligned}
&-\frac{\int_V \cdots \int_V \left[\frac{1}{2}\sum_{j=1}^N \sum_{i=1}^P \mathbf{x}_i^j \cdot \frac{\partial \alpha(\mathbf{x}_1,\ldots,\mathbf{x}_P)}{\partial \mathbf{x}_i^j}\right]}{\int_V \cdots \int_V e^{-\beta V^{\mathrm{eff}}(\mathbf{x}_1,\mathbf{x}_2,\ldots,\mathbf{x}_P)}\,d\mathbf{x}_1 d\mathbf{x}_2 \cdots d\mathbf{x}_P} \\[2mm]
&\,e^{-\beta V^{\mathrm{eff}}(\mathbf{x}_1,\mathbf{x}_2,\ldots,\mathbf{x}_P)}\,d\mathbf{x}_1 d\mathbf{x}_2 \cdots d\mathbf{x}_P
\end{aligned}
$$

$$
= -\frac{\int_V \cdots \int_V \alpha(\mathbf{x}_1,\ldots,\mathbf{x}_P)e^{-\beta V^{\mathrm{eff}}(\mathbf{x}_1,\mathbf{x}_2,\ldots,\mathbf{x}_P)}\,d\mathbf{x}_1 d\mathbf{x}_2 \cdots d\mathbf{x}_P}{\int_V \cdots \int_V e^{-\beta V^{\mathrm{eff}}(\mathbf{x}_1,\mathbf{x}_2,\ldots,\mathbf{x}_P)}\,d\mathbf{x}_1 d\mathbf{x}_2 \cdots d\mathbf{x}_P}
$$

$$
= \langle \alpha(\mathbf{x}_1,\ldots,\mathbf{x}_P)\rangle
$$

$$
= -\left\langle \sum_{j=1}^N \sum_{i=1}^P \frac{1}{2}m_j\omega_P^2(\mathbf{x}_i^j - \mathbf{x}_{i-1}^j)^2\right\rangle, \tag{7.68}
$$

and Eq. (7.66) further evolves to

$$
\left\langle \frac{1}{2P}\sum_{j=1}^N \sum_{i=1}^P \mathbf{x}_i^j \cdot \frac{\partial V(\mathbf{x}_i^1,\ldots,\mathbf{x}_i^N)}{\partial \mathbf{x}_i^j}\right\rangle
$$

$$
= \frac{\int_V \cdots \int_V \left[\frac{1}{2}\sum_{j=1}^N \sum_{i=1}^P \mathbf{x}_i^j \cdot \frac{\partial V^{\mathrm{eff}}(\mathbf{x}_1,\ldots,\mathbf{x}_P)}{\partial \mathbf{x}_i^j}\right]}{\int_V \cdots \int_V e^{-\beta V^{\mathrm{eff}}(\mathbf{x}_1,\mathbf{x}_2,\ldots,\mathbf{x}_P)}\,d\mathbf{x}_1 d\mathbf{x}_2 \cdots d\mathbf{x}_P} \tag{7.69}
$$

$$
\;\frac{e^{-\beta V^{\mathrm{eff}}(\mathbf{x}_1,\mathbf{x}_2,\ldots,\mathbf{x}_P)}\,d\mathbf{x}_1 d\mathbf{x}_2 \cdots d\mathbf{x}_P}{}
$$

$$
- \left\langle \sum_{j=1}^N \sum_{i=1}^P \frac{1}{2}m_j\omega_P^2(\mathbf{x}_i^j - \mathbf{x}_{i-1}^j)^2\right\rangle.
$$

Comparing Eq. (7.69) with Eq. (7.60), the only equality we need to prove becomes

$$
\frac{\int_V \cdots \int_V \left[\frac{1}{2}\sum_{j=1}^N \sum_{i=1}^P \mathbf{x}_i^j \cdot \frac{\partial V^{\mathrm{eff}}(\mathbf{x}_1,\ldots,\mathbf{x}_P)}{\partial \mathbf{x}_i^j}\right]}{\int_V \cdots \int_V e^{-\beta V^{\mathrm{eff}}(\mathbf{x}_1,\mathbf{x}_2,\ldots,\mathbf{x}_P)}\,d\mathbf{x}_1 d\mathbf{x}_2 \cdots d\mathbf{x}_P}
$$

$$
\;\frac{e^{-\beta V^{\mathrm{eff}}(\mathbf{x}_1,\mathbf{x}_2,\ldots,\mathbf{x}_P)}\,d\mathbf{x}_1 d\mathbf{x}_2 \cdots d\mathbf{x}_P}{}
$$

$$
= \frac{3NP}{2\beta}. \tag{7.70}
$$

This proof is doable if we reform the left-hand side of Eq. (7.70) in the following manner:

$$\int_V \cdots \int_V \left[\frac{1}{2} \sum_{j=1}^N \sum_{i=1}^P \mathbf{x}_i^j \cdot \frac{\partial V^{\text{eff}}(\mathbf{x}_1,\ldots,\mathbf{x}_P)}{\partial \mathbf{x}_i^j} \right]$$

$$\frac{\mathrm{e}^{-\beta V^{\text{eff}}(\mathbf{x}_1,\mathbf{x}_2,\ldots,\mathbf{x}_P)} \mathrm{d}\mathbf{x}_1 \mathrm{d}\mathbf{x}_2 \cdots \mathrm{d}\mathbf{x}_P}{\int_V \cdots \int_V \mathrm{e}^{-\beta V^{\text{eff}}(\mathbf{x}_1,\mathbf{x}_2,\ldots,\mathbf{x}_P)} \mathrm{d}\mathbf{x}_1 \mathrm{d}\mathbf{x}_2 \cdots \mathrm{d}\mathbf{x}_P}$$
(7.71)

$$= -\frac{1}{\beta} \frac{\int_V \cdots \int_V \frac{1}{2} \sum_{j=1}^N \sum_{i=1}^P \mathbf{x}_i^j \cdot \left[\frac{\partial}{\partial \mathbf{x}_i^j} \mathrm{e}^{-\beta V^{\text{eff}}(\mathbf{x}_1,\mathbf{x}_2,\ldots,\mathbf{x}_P)} \right] \mathrm{d}\mathbf{x}_1 \mathrm{d}\mathbf{x}_2 \cdots \mathrm{d}\mathbf{x}_P}{\int_V \cdots \int_V \mathrm{e}^{-\beta V^{\text{eff}}(\mathbf{x}_1,\mathbf{x}_2,\ldots,\mathbf{x}_P)} \mathrm{d}\mathbf{x}_1 \mathrm{d}\mathbf{x}_2 \cdots \mathrm{d}\mathbf{x}_P}.$$

Then, making use of an integration by parts, this quantity further evolves into

$$-\frac{1}{\beta} \frac{\int_V \cdots \int_V \frac{1}{2} \sum_{j=1}^N \sum_{i=1}^P \mathbf{x}_i^j \cdot \left[\frac{\partial}{\partial \mathbf{x}_i^j} \mathrm{e}^{-\beta V^{\text{eff}}(\mathbf{x}_1,\mathbf{x}_2,\ldots,\mathbf{x}_P)} \right] \mathrm{d}\mathbf{x}_1 \mathrm{d}\mathbf{x}_2 \cdots \mathrm{d}\mathbf{x}_P}{\int_V \cdots \int_V \mathrm{e}^{-\beta V^{\text{eff}}(\mathbf{x}_1,\mathbf{x}_2,\ldots,\mathbf{x}_P)} \mathrm{d}\mathbf{x}_1 \mathrm{d}\mathbf{x}_2 \cdots \mathrm{d}\mathbf{x}_P}$$

$$= \frac{1}{\beta} \frac{\int_V \cdots \int_V \frac{1}{2} \sum_{j=1}^N \sum_{i=1}^P \mathrm{e}^{-\beta V^{\text{eff}}(\mathbf{x}_1,\mathbf{x}_2,\ldots,\mathbf{x}_P)} \left[\frac{\partial}{\partial \mathbf{x}_i^j} \cdot \mathbf{x}_i^j \right] \mathrm{d}\mathbf{x}_1 \mathrm{d}\mathbf{x}_2 \cdots \mathrm{d}\mathbf{x}_P}{\int_V \cdots \int_V \mathrm{e}^{-\beta V^{\text{eff}}(\mathbf{x}_1,\mathbf{x}_2,\ldots,\mathbf{x}_P)} \mathrm{d}\mathbf{x}_1 \mathrm{d}\mathbf{x}_2 \cdots \mathrm{d}\mathbf{x}_P}$$

$$= \frac{1}{\beta} \frac{3NP}{2}.$$
(7.72)

With these, the equality in Eq. (7.60) is proven and one can use the virial estimator as given in Eq. (7.61) to calculate the finite-temperature internal energy of the poly-atomic system under investigation. In practice, a slight variation of Eq. (7.61) is often used. This variation is based on the following equation:

$$\left\langle \frac{1}{2P} \sum_{j=1}^N \sum_{i=1}^P \mathbf{x}_i^j \cdot \frac{\partial V(\mathbf{x}_i^1,\ldots,\mathbf{x}_i^N)}{\partial \mathbf{x}_i^j} \right\rangle$$

$$= \left\langle \frac{1}{2P} \sum_{j=1}^N \sum_{i=1}^P \mathbf{x}_c^j \cdot \frac{\partial V(\mathbf{x}_i^1,\ldots,\mathbf{x}_i^N)}{\partial \mathbf{x}_i^j} \right\rangle$$

$$+\left\langle \frac{1}{2P}\sum_{j=1}^{N}\sum_{i=1}^{P}(\mathbf{x}_i^j - \mathbf{x}_c^j)\cdot\frac{\partial V(\mathbf{x}_i^1,\ldots,\mathbf{x}_i^N)}{\partial\mathbf{x}_i^j}\right\rangle$$

$$=\left\langle \frac{1}{2}\sum_{j=1}^{N}\mathbf{x}_c^j\cdot\mathbf{F}_c^j\right\rangle + \left\langle \frac{1}{2P}\sum_{j=1}^{N}\sum_{i=1}^{P}(\mathbf{x}_i^j - \mathbf{x}_c^j)\cdot\frac{\partial V(\mathbf{x}_i^1,\ldots,\mathbf{x}_i^N)}{\partial\mathbf{x}_i^j}\right\rangle$$

$$=\frac{3N}{2\beta} + \left\langle \frac{1}{2P}\sum_{j=1}^{N}\sum_{i=1}^{P}(\mathbf{x}_i^j - \mathbf{x}_c^j)\cdot\frac{\partial V(\mathbf{x}_i^1,\ldots,\mathbf{x}_i^N)}{\partial\mathbf{x}_i^j}\right\rangle, \qquad (7.73)$$

where $\mathbf{F}_c^j$ stands for the effective force imposed on the centroid of the jth atom and $\mathbf{x}_c^j$ stands for its centroid position, and the estimator in Eq. (7.61) further changes into

$$\langle E\rangle = \frac{3N}{2\beta} + \left\langle \frac{1}{2P}\sum_{j=1}^{N}\sum_{i=1}^{P}(\mathbf{x}_i^j - \mathbf{x}_c^j)\cdot\frac{\partial V(\mathbf{x}_i^1,\ldots,\mathbf{x}_i^N)}{\partial\mathbf{x}_i^j}\right\rangle$$

$$+\left\langle \sum_{i=1}^{P}\frac{1}{P}V\left(\mathbf{x}_i^1,\ldots,\mathbf{x}_i^N\right)\right\rangle. \qquad (7.74)$$

One can use either the estimator in Eq. (7.74) or the one in Eq. (7.57) to evaluate the internal energy in real poly-atomic systems.

We note that the internal energy calculated this way includes contribution from the QNEs. As mentioned above, an MD simulation, in which only the thermal nuclear effects are included, can also give us an expectation value of this quantity. Therefore, a comparison between results obtained from these two simulations in principle can give us the zero-point energy of a real poly-atomic system. To understand how this works in practice, we show a sketch for the evolution of the internal energy in the MD and PIMD (using different P) simulations as a function of temperature in Fig. 7.4. At 0 K, the internal energy equals the static geometry-optimized total energy in the MD simulation. With the increase in the temperature, it increases linearly due to the classical virial theorem. In the PIMD simulations, this internal energy evolves differently from the one obtained from the MD simulations, and this difference originates from the QNEs. Its value increases with the number of beads P till convergence. At higher Ts, a small value of P is good energy to describe this difference. At lower Ts, larger P is needed. At 0 K, since an infinite P is needed for the path integral sampling, the PIMD simulation loses its precision too. However, an extrapolation of the QNEs from finite T still presents a good estimator for the zero-point energy.

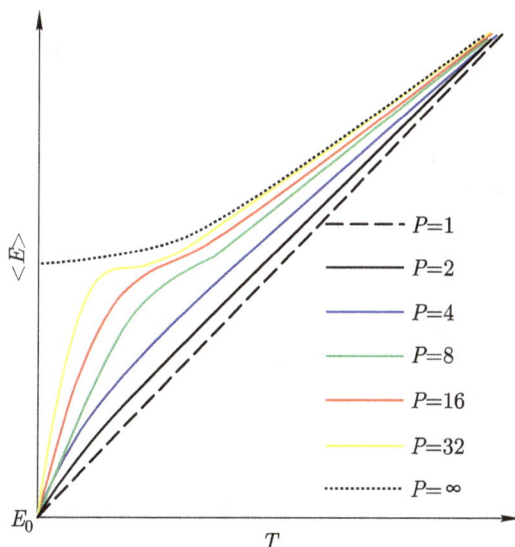

Figure 7.4 Illustration of how the internal energy's expectation value changes with temperature in the MD and PIMD simulations. In the MD simulation, as T approaches $0\,K$, the internal energy approaches the static geometry-optimized total energy E_0. With the increase in T, it increases linearly because of thermal fluctuations. In the PIMD simulations, since the QNEs are included, at finite T, there is a difference in this value from its classical result. This difference increases with the number of beads P till convergence. At higher temperatures, a small P is already good energy to describe this difference. At lower temperatures, larger Ps are needed. At $0\,K$, due to the fact that an infinite P is needed for the path integral sampling, the PIMD method also loses its power. However, an extrapolation of the difference between the internal energy obtained from the converged PIMD simulations and MD simulations at finite T (indicated by the dotted line) still presents a good estimator for the zero-point energy.

7.2 Extensions Beyond the Statistical Studies

So far, all our discussions have been restricted to the statistics. The associated time-averaged quantities can be used to study the impact of the QNEs on the equilibrium statistical properties of a poly-atomic system under investigation at finite temperatures. Another aspect of the real quantum world, i.e. the dynamics, however, has never been touched. We note that descriptions of such nuclear dynamics, with relevant electronic structures computed accurately "on-the-fly", poses a "grand challenge" to both theoretical physics and chemistry. As a matter of fact, illustrations of many key physical/chemical properties in the real world, e.g. the transport properties, the chemical reaction rates, and the neutron or light scattering spectra, etc., require their descriptions. In a many-body (poly-atomic) system, we know that the key quantity in describing such dynamics is the so-called time-correlation function. Taking the infrared absorption spectrum as an example, it is directly related to the dipole–dipole time-correlation function of the system. The translational diffusion coefficient, on the other hand, can be understood using the velocity autocorrelation function, etc. Because of these, inclusion of the QNEs in descriptions of such time-correlation functions becomes an issue of considerable interest in present studies in both theoretical physics and chemistry [74, 79, 91, 352–357].

A natural choice for the calculation of such time-correlation functions is to solve the time-dependent Schrödinger equation of the nuclei in real poly-atomic systems. This implies propagating the nuclear quantum dynamics on the Born–Oppenheimer potential energy surfaces (BO-PES) which are precomputed with very accurate electronic structure theories. The multi-configuration time-dependent Hartree method (MCTDH) is such an example [358, 359]. In the past years, it has been very successful in describing some gas phase chemical reactions and the dynamical properties of small molecules [60, 61, 360–362]. However, one notes that the scaling of their computational cost with system size makes it unfeasible to many practical systems quickly as the nuclear degree of freedom increases. Alternative methods in describing such nuclear dynamics, where scaling behavior is much better so that simulations can be performed in systems of relevance in practical studies, must be used.

7.2.1 Different Semiclassical Dynamical Methods

Similar to the route we have chosen in studies of the statistical mechanics, we now resort to the path integral representation of the quantum mechanics. Within this picture, the most rigorous method in quantifying such a time-correlation function naturally involves using a complex-time path integral sampling technique [67, 74]. In this method, the thermal effects are treated as imaginary time in the complex-time space and the real-time axis takes care of the dynamics. Extension of the path integral equations from statistics to dynamics is straightforward [74]. However, different from the success of PIMD/PIMC techniques in addressing the statistical properties in simulations of real poly-atomic systems, practical simulations based on such extension of the path integral sampling methods to the complex-time space has continued to be problematic due to an extensive phase cancellation originating from the paths with weights that are non-positive in character [78, 79]. In spite of these difficulties, by defining a symmetrized time-correlation function that lends itself to PIMD/PIMC simulations, significant progress has still been made over the last 30 years within this complex-time path integral scheme [75–77]. As a prominent example, in a recent development of this method by Nakayama and Makri for studies of the subcritical liquid *para*-hydrogen, the authors have shown that accurate quantum mechanical results for the initial 0.2 ps segment of the symmetrized velocity autocorrelation function, as well as the incoherent dynamic structure factor at certain momentum transfer values at moderate temperatures and densities, can be obtained [352]. But for more general problems involving longer time dynamics and more complicated systems with higher density, which is clearly of more practical use, this method becomes less practical and one needs to resort to less accurate quantum dynamical methods.

Following the summary by Braams and Manolopoulos [91], here we categorize these less accurate quantum dynamical methods that have been applied to condensed matter systems, essentially into three classes, noting that each method has its own strengths and weaknesses. The first class of methods simulate the system using an imaginary-time propagator only. Then, an inverse Wick rotation is used to infer the thermodynamically averaged real-time-correlation function from the imaginary-time one based on the Baym–Mermin theorem [363]. This trick is similar to the analytical continuation of the self-energy from the imaginary frequency axis to the real frequency one for the calculation of the quasi-particle energies as introduced in Sec. 4.4.5. There are two practical schemes for the time-correlation function to be calculated, i.e. the numerical analytic continuation (NAC) [75–77, 364–367] and the quantum mode-coupling theory (QMCT) [368–373]. The advantage of these methods is that the short-time behavior of the time-correlation function can normally be described accurately and the imaginary-time treatment makes the correlation function easy to compute. However, we note that the numerical stability of the analytical continuation is much worse than its counterparts in the calculation of the self-energies. Some standard methods, such as the Padé technique, are of limited use in practice [374, 375], and these methods are not exact at the classical limit.

The second class of methods combine an exact treatment of the quantum Boltzmann operator with an approximate treatment of the real-time evolution based on classical dynamics. These methods include the linearized semiclassical-initial value representation (LSC-IVR) method (see e.g. Refs. [355, 356, 376–379], etc.), the Feynman–Kleinert linearized path-integral (FK-LPI) method (see e.g. Refs. [380–383], etc.), and the forward–backward semiclassical dynamics (FB–SD) (see e.g. Refs. [353, 354, 357], etc.). Their main advantage is that they are exact in three important limits, i.e., the short-time limit, the limit of a harmonic potential, and the classical limit, while the main disadvantage is that the classical trajectories do not, in general, conserve the quantum mechanical equilibrium distribution functions [91].

The third class of methods, which is currently of more practical use in condensed matter, includes the so-called CMD by Cao and Voth [80–84, 384] and the RPMD by Manolopoulos and his coworkers [85–93]. A key difference between these two methods and those in the earlier two classes is that mathematically there are only slight modifications in the standard PIMD method as we have introduced in Sec. 7.1, although conceptually these differences are fundamental, and they all originate from generalizations of the physical concepts implied in the PIMD simulations, with the CMD method that appears earlier. Therefore, in the following, we will give a detailed explanation of these two methods in a chronological order with a special focus on their similarities and differences with the normal statistical PIMD simulations, starting from a precursor of the CMD method, i.e. Gillan's generalization of Feynman's path centroid concept to its applications in the TST.

7.2.2   *Centroid Molecular Dynamics and Ring-Polymer Molecular Dynamics*

In the seminal book by Feynman and Hibbs [67], it has been shown that the concept of path centroid can be used to interpret the impact of quantum effects on the effective potential the particle under investigation feels. Later, this concept was extended by Feynman and Kleinert so that we have an effective centroid potential [385], which is of practical use in molecular simulations. As one example, Gillan has employed it in calculating the probability of finding a quantum particle at its transition state during a rare-event transition process, which could be used later to describe the transition rate between two stable states at the quantum mechanical level, see e.g. Refs. [386, 387]. We note that theories behind characterizing such transition rates of slow processes, such as the chemical reactions and the diffusion events, are the so-called TSTs. It is both statistical and dynamical in the sense that a statistical property (the probability of finding the system at its TS) is assumed to be proportional to a dynamical property (the TS). Due to this fundamental assumption, theories behind characterizing this probability of finding the system in its transition state lie at the heart of the TST.

Already at the static level, searching for this TS is non-trivial in a poly-atomic system, due to the high-dimensional feature of the BO PES associated with the nuclear degrees of freedom. Currently, there are several schemes in which such a hunting can be carried out, including the constrained optimization (CO) [19], nudged elastic band (NEB) [20–22], Dewar, Healy and Stewart (DHS) [23], Dimer [24–26], activation–relaxation technique (ART) [27, 28] and one-side growing string (OGS) [29] as well as their various combinations. Here, we suppose that a reasonable estimation of the TS and its associated hyperplane separating the reactant and product states is already carried out. Our discussions only concern further thermal and quantum nuclear effects on the static energies based on this knowledge.

At the classical level, suppose that we have a well-defined reaction coordinate s, which properly separates the reactant and product states. According to the TST, the transition rate is

$$k_{\text{TST}} = \frac{\langle v_\perp \rangle}{2\delta} \frac{Z^{\text{T}} \delta}{Z} = \frac{\langle v_\perp \rangle}{2} \frac{Z^{\text{T}}}{Z}, \qquad (7.75)$$

where δ is defined as the interval for the reaction coordinate s with which we think that the system is at the TS. With this definition, $Z^{\text{T}}\delta/Z$ stands for the probability of finding the system at the TS while $\langle v_\perp \rangle/(2\delta)$ stands for the escape rate from the TS. Putting these two factors together, one obtains a transition rate through the TS from the reactant to the product, which is sensitive to the choice of the "TS". One notes that this estimation of the transition rate with the TST in Eq. (7.75) sets the upper bound for the real transition rate [331], since moving the "TS" used here away from the real one will clearly increase Z^{T}/Z. Therefore, it is highly recommended that one searches for the TS and its associated hyperplane dividing the reactant and product states so that the estimated rate in Eq. (7.75) reaches a minimum. In doing so, one obtains the best evaluation of the transition rate. Such a variational method is called variational transition-state theory (VTST) [331, 388, 389].

Then one tries to include the QNEs. There are several extensions of this TST to its quantum version [390–395], with the simplest ones only replacing the classical statistical averaging with the quantum one. In these simplest methods, we note that the extension of the centroid's concept from the traditional statistical PIMD method is already used in calculating the free energy difference between the reactant and the TSs. One prominent example is the work by Gillan in the 1980s [386, 387]. In this example, the probability of finding the system in the TS compared to that in the reactant was calculated using the reversible work. In particular, a series of locations along the transition path were selected and a PIMD simulation is carried out by fixing the centroid of the path integral chain of the transition particle at each point. In doing so, the average over all the remaining quantum degrees of freedom at this point can be properly evaluated and one obtains an effective mean force on the centroid of the transition particle along the transition path. By integrating this effective force along the transition path using reversible work, one gets the free energy difference between the reactant and the TSs. The relative density of the centroid at this TS is then calculated using this free energy difference and by assuming that the transition rate is proportional to this relative density, one finally gets the transition rate.

Following the idea that effective forces on the centroids of the quantum particles can be calculated by fixing the centroid positions and doing statistics over all other quantum degrees of freedom, the CMD method was introduced by Cao and Voth in 1993 as an approximate method to compute the real-time quantum correlation function for the dynamical properties of a real poly-atomic system to be described [384]. The central point of this CMD method is the assumption that the real-time evolution of the centroid positions on their potential of mean force (PMF) surface can be used to generate approximate quantum dynamical properties of real poly-atomic systems. Their evolution respects the Newton equations:

$$\dot{\mathbf{x}}_c^j = \frac{\mathbf{p}_c^j}{m},$$
$$\dot{\mathbf{p}}_c^j = F^j(\mathbf{x}_c),$$

(7.76)

where j again runs through the atomic index, $\mathbf{x}_c$ represents a spatial configuration of the centroids and $\mathbf{x}_c^j$ means the centroid position of the jth atom. $F^j(\mathbf{x}_c)$ is the derivative of potential of mean force surface with respect to $\mathbf{x}_c^j$, i.e. the mean field centroid force at $\mathbf{x}_c$. It is mathematically defined as

$$F^j(\mathbf{x}_0) = \frac{\iint d\mathbf{x}_1 \cdots d\mathbf{x}_P \delta(\mathbf{x}_0 - \mathbf{x}_c) F_0^j(\mathbf{x}_c) e^{-\beta V^{\text{eff}}(\mathbf{x}_1,\ldots,\mathbf{x}_P)}}{\iint d\mathbf{x}_1 \cdots d\mathbf{x}_P \delta(\mathbf{x}_0 - \mathbf{x}_c) e^{-\beta V^{\text{eff}}(\mathbf{x}_1,\ldots,\mathbf{x}_P)}}.$$

(7.77)

Here, $V^{\text{eff}}(\mathbf{x}_1, \ldots, \mathbf{x}_P)$ is the effective potential defined in Eq. (7.18) and $F_0^j(\mathbf{x}_c)$ is the instantaneous Hellmann–Feynman force imposed on the centroid, given by

$$F_0^j(\mathbf{x}_c) = \frac{1}{P} \sum_{i=1}^{P} \frac{\partial V(\mathbf{x}_1, \ldots, \mathbf{x}_P)}{\partial \mathbf{x}_i^j}, \tag{7.78}$$

and $\mathbf{x}_0$ represents the instantaneous centroid configuration to which $\mathbf{x}_c$ should be restricted. Due to the use of the Newton equations in describing the centroid propagation, this method is intrinsically a semiclassical method, with the QNEs rigorously described only at the statistical level when the mean field centroid force is calculated. However, we note that the quantum correction to the effective potential sometimes already incorporates the dominant elements of QNEs in descriptions of the nuclear dynamics and it is currently used as a standard routine to investigate the impact of QNEs on the dynamical property of condensed matter systems.

One point implied in the procedure described above for the CMD method is that a fully statistical PIMD (or PIMC) simulation should be carried out at each centroid configuration before it propagates to the next centroid configuration. For complex poly-atomic systems, however, this is inapplicable due to the computational cost associated. As a simplified version of this method, the adiabatic approximation can be used [396]. For a better explanation of this idea, we go back to the normal-mode coordinate as explained in Sec. 7.1.4. Mathematically, the fundamental difference between a PIMD simulation in the primitive Cartesian coordinate and the normal-mode coordinate is that a coordinate transformation should be made at each PIMD step in order to convert the forces and the spatial configurations of the polymer, so that the forces can be calculated in the Cartesian space and the equation of motion can be propagated in the normal-mode one. The first normal mode describes the propagation of the centroid, while the other modes describe interbead vibrations. When the masses used in the PIMD simulation are chosen according to Eq. (7.49), one ensures that all the "artificial" interbead vibrations have the same frequency, whose value is determined by the constant c. Frequency associated with the centroid vibration is determined by the interatomic potential which is real and physical. Therefore, intuitively, one can set this constant c to a very small value so that the masses associated with the "artificial" interbead vibrations are small and they can adiabatically react to the motion of the centroid. Since the centroid mode moves much slower than the other ones, during a characteristic time for its vibration which is determined by the interatomic potential, the other interbead vibrational modes can already perform a very good statistical averaging over their degrees of freedom. In doing so, the potential each centroid feels upon characterizing the physical vibrations can be calculated "on-the-fly". As a cost of not doing a PIMD simulation at each centroid configuration, a much smaller time step should be used in order to address the fast interbead vibrations originating from the small masses associated with them.

We note that this "on-the-fly" simplified calculation of the mean field centroid force was first proposed in Ref. [83], where different time intervals were suggested for the propagation of the centroid and the much faster interbead motions. In between the centroid propagation time steps, a series of interbead propagations should be made, subject to the constraint in Eq. (7.77) so that a statistical averaging over the centroid force can be obtained before propagating the centroid. Therefore, different from a rigorous implementation of the CMD method, there is only one trajectory. This simplified version of the CMD method is called the adiabatic centroid molecular dynamics (ACMD) method [83]. In 2006, it is further simplified so that a single small time interval is used for the propagation of both the centroid and interbead vibrations [397] and the corresponding simulation is called partially adiabatic centroid molecular dynamics (PACMD) simulations [397]. We note that nowadays this PACMD method is the frequently used method in practical simulations of complex systems, and in practice, this distinction between PACMD, ACMD, and CMD is often obviated and one simply refers to PACMD as CMD [398]. Here, we follow such a tradition and refer to PACMD as CMD in later discussions.

Now, we look at the differences between the statistical PIMD and the dynamical CMD simulations. For this comparison to be as simple as possible, we use the normal-mode coordinate for the PIMD simulations. In statistical normal-mode PIMD simulations, the constant c in Eq. (7.49) is set as one and the time step is determined by the frequency of real interatomic vibrations. Instead, in a CMD simulation, since the interbead vibrations have much higher frequency due to their small artificial masses, this c takes a small value between zero and one, and a much smaller time step than that of the interatomic vibrations should be used to ensure the adiabatic approximation works. For the thermostating strategy, each mode should be coupled to an efficient thermostat in both cases. Therefore, mathematically these two simulations are very similar, although conceptually they are fundamentally different.

Then we compare the CMD method with the more recently proposed RPMD [85–93]. The differences are mainly located in three aspects. First, the RPMD method chooses the kinetic mass M_i^j in Eq. (7.26) as $M_i^j = m^j/P$, where m^j is the mass of the jth nucleus, if the dynamics is done at real temperature T. Such a setting ensures that the mass of each bead associated with its potential $V(\mathbf{x}_i^1, \ldots, \mathbf{x}_i^N)/P$ in the case of Eq. (7.26) gives the physical interatomic vibration frequency when interbead interactions are neglected. If the dynamics is performed at PT, then the potential part in Eq. (7.26) will be

$$\sum_{j=1}^{N} \sum_{i=1}^{P} \left[\frac{1}{2} P m_j \omega_P^2 (\mathbf{x}_i^j - \mathbf{x}_{i-1}^j)^2 + V(\mathbf{x}_i^1, \ldots, \mathbf{x}_i^N) \right], \qquad (7.79)$$

and the mass will be chosen as $M_i^j = m^j$.

The second difference between the CMD and RPMD methods is that in CMD, the centroid dynamics is used to calculate the time-correlation function, while in the RPMD method, the dynamics in each image is calculated separately and then the time-correlation function for the whole system is an average over all images. Because of this difference, in the RPMD method, the interbead vibrations are also accounted for when the vibrational spectrum of the quantum system is calculated, and these artificial vibrational frequencies are evenly distributed on the frequency axis, which often pollute the real physical vibrational frequencies associated with the interatomic motion. In the CMD method, on the other hand, the time-correlation function is calculated from the propagation of the centroid. In doing so, the vibration of the centroid will not be polluted by the interbead vibrations. However, at low temperatures, taking the OH stretching mode as an example, the centroid of the H atom often falls much closer to the oxygen atom compared with its physical value within each image. This induces the so-called "curvature problem" in CMD simulations, which artificially softens the covalent bond stretching frequencies [399].

The third difference between the CMD and RPMD methods is that the RPMD method needs to be carried out in a Hamiltonian manner when the time-correlation function is calculated. In other words, no thermostat should be added when the trajectory under construction will be used in the calculation of the time-correlation function. The temperature effect should be included during the thermal equilibrium process when the canonical distribution of the snapshots starting from which the micro-canonical simulations are carried out is generated. In the CMD method, on the other

hand, the canonical ensemble is used for the single trajectory to be generated. Because of this difference, although RPMD does not need a very small
time step, many trajectories are needed in order for the thermal averaging
on the time-correlation function to be sufficiently sampled.

In recent years, there are several studies aiming at comparing the performance of CMD and RPMD in some models and real poly-atomic systems [91, 397, 398]. In Ref. [91], Braams and Manolopoulos showed that
the Kubo-transformed autocorrelation functions obtained from the RPMD
simulations are accurate on the time scale up to the sixth order for the
position and the fourth for the velocity; that of the CMD method leads to
an accuracy of the fourth order and second order for these two quantities,
respectively. Hone *et al.*, on the other hand, showed results clearly in favor
of CMD [397], where simulations on *para*-hydrogen demonstrates that the
CMD method gives better agreement with experiments. Later, Perez *et al.*
pointed out that when such a comparison is made, the differences in the
setting of the simulations as mentioned above must be kept in mind [398].
Furthermore, in complex systems when the accuracy of the inter-atomic
potential is unclear, comparison with experimental results cannot be used
to judge which one is more accurate, since it is impossible to discern how
much of the discrepancy with experiment is due to the accuracy of quantum
dynamics and how much is due to the interatomic potential. Rather, an
alternative method for such a comparison should be used. In this paper, it
is suggested that one uses the same numerical treatment to infer the time-
correlation function from the real-time axis as obtained from the CMD and
RPMD methods to the imaginary-time axis. Then, these results can be
compared with the numerically exact results from imaginary-time PIMD
or PIMC simulations. In doing so, the performance of these two methods
on quantum dynamics is compared solely. We highly recommend such a
choice of criterion for future studies in this direction.

7.3   Free Energy with Anharmonic QNEs

In the previous chapter, we have introduced the thermodynamic integration method. Using this method, the anharmonic contribution from the
nuclear thermal fluctuations to the free energy can be calculated in real
poly-atomic systems, as long as a well-defined reference state exists. The
QNEs, however, stay on the level of the harmonic approximation. In reality, we know that these QNEs also have anharmonic contributions to their

vibrations/rotations and consequently the free energy. To include such effects, one needs to extend the thermodynamic integration method as introduced in Chapter 6 so that the QNEs on the free energy beyond the harmonic approximation are also accounted for. In practice, this can be done through an extension of the thermodynamic integration method in the framework of PIMD/PIMC, as will be introduced below.

The starting point for this discussion is the quantum partition function for the real poly-atomic system, defined as

$$Z^Q = \lim_{P\to\infty} Z_P = \lim_{P\to\infty} \left[\prod_{j=1}^{N} \left(\frac{m_j P}{2\beta\pi\hbar^2} \right)^{\frac{P}{2}} \right]$$

$$\cdot \int_V \int_V \cdots \int_V e^{-\beta V^{\mathrm{eff}}(\mathbf{x}_1,\mathbf{x}_2,\ldots,\mathbf{x}_P)} d\mathbf{x}_1 d\mathbf{x}_2 \cdots d\mathbf{x}_P, \qquad (7.80)$$

where

$$V^{\mathrm{eff}}(\mathbf{x}_1, \mathbf{x}_2, \ldots, \mathbf{x}_P) = \sum_{i=1}^{P} \sum_{j=1}^{N} \left[\frac{1}{2} m_j \omega_P^2 (\mathbf{x}_i^j - \mathbf{x}_{i-1}^j)^2 \right] + \sum_{i=1}^{P} \frac{1}{P} V(\mathbf{x}_i^1, \ldots, \mathbf{x}_i^N).$$

$$(7.81)$$

It is clear from our earlier discussion that the free energy associated with this partition function is the free energy of the quantum poly-atomic system, given by

$$F^Q = -\frac{1}{\beta} \ln Z^Q. \qquad (7.82)$$

Now, we look at the effective potential in Eq. (7.82), we note that we can replace the second term in it by the effective potential on the centroid as

$$V^{\mathrm{eff}}(\mathbf{x}_1, \mathbf{x}_2, \ldots, \mathbf{x}_P) = \sum_{i=1}^{P} \sum_{j=1}^{N} \left[\frac{1}{2} m_j \omega_P^2 (\mathbf{x}_i^j - \mathbf{x}_{i-1}^j)^2 \right] + \sum_{i=1}^{P} \frac{1}{P} V(\mathbf{x}_c^1, \ldots, \mathbf{x}_c^N).$$

$$(7.83)$$

Here, $\mathbf{x}_c^j$ means the centroid position of the jth nucleus, which does not depend on the bead index i. By inputting this equation into Eq. (7.80), we can see that the partition function Z becomes

$$Z^{\mathrm{C}} = \lim_{P \to \infty} \left[\prod_{j=1}^{N} \left(\frac{m_j P}{2\beta \pi \hbar^2} \right)^{\frac{P}{2}} \right] \int_V \cdots \int_V \mathrm{e}^{-\beta \sum_{j=1}^{N} V(\mathbf{x}_c^1, \ldots, \mathbf{x}_c^N)}$$

$$\cdot \, \mathrm{e}^{-\beta \sum_{j=1}^{N} \sum_{i=1}^{P} \frac{1}{2} m_j \omega_P^2 \left(\mathbf{x}_i^j - \mathbf{x}_{i-1}^j \right)^2} \mathrm{d}\mathbf{x}_1 \cdots \mathrm{d}\mathbf{x}_P. \tag{7.84}$$

At this time, if we resort to the normal-mode coordinate as defined in Eqs. (7.47) and (7.48), Eq. (7.84) can be rewritten as

$$Z^{\mathrm{C}} = \lim_{P \to \infty} \left[\prod_{j=1}^{N} \left(\frac{m_j P}{2\beta \pi \hbar^2} \right)^{\frac{P}{2}} \right] \int_V \mathrm{d}\mathbf{u}_0 \mathrm{e}^{-\beta V(\mathbf{u}_0^1, \ldots, \mathbf{u}_0^N)}$$

$$\cdot \prod_{i=-n}^{-1} \prod_{i=1}^{n+1} \int_V \mathrm{e}^{-\beta \sum_{j=1}^{N} \frac{1}{2} m_j \lambda_i \omega_P^2 (\mathbf{u}_i^j)^2} \mathrm{d}\mathbf{u}_i$$

$$= \lim_{P \to \infty} \left[\prod_{j=1}^{N} \left(\frac{m_j P}{2\beta \pi \hbar^2} \right)^{\frac{P}{2}} \right] \int_V \mathrm{d}\mathbf{u}_0 \mathrm{e}^{-\beta V(\mathbf{u}_0^1, \ldots, \mathbf{u}_0^N)}$$

$$\cdot \prod_{i=-n}^{-1} \prod_{i=1}^{n+1} \int_V \mathrm{e}^{-\frac{1}{2} \sum_{j=1}^{N} \frac{m_j P}{\beta \hbar^2} (2 \sin \frac{i\pi}{P})^2 (\mathbf{u}_i^j)^2} \mathrm{d}\mathbf{u}_i. \tag{7.85}$$

Here, the n is related to P by $P = 2n+2$ and the normal modes are aligned as $\mathbf{u}_{-n}^j, \ldots, \mathbf{u}_{-1}^j, \mathbf{u}_0^j, \mathbf{u}_1^j, \ldots, \mathbf{u}_{n+1}^j$, as introduced in Sec. 7.1.4. Making use of the property that $\prod_{i=-n}^{-1} \prod_{i=1}^{n+1} 2 \sin \frac{i\pi}{P} = P$, the Z^{C} in Eq. (7.85) can be further rewritten into

$$Z^{\mathrm{C}} = \lim_{P \to \infty} \left[\prod_{j=1}^{N} \left(\frac{m_j P}{2\beta \pi \hbar^2} \right)^{\frac{P}{2}} \right] \left\{ \prod_{j=1}^{N} \left[\prod_{i=-n}^{-1} \prod_{i=1}^{n+1} \left(\frac{2\pi \beta \hbar^2}{m_j P} \right)^{-\frac{1}{2}} \right. \right.$$

$$\left. \left. \times \left(2 \sin \frac{i\pi}{P} \right)^{-1} \right] \right\} \cdot \int_V \mathrm{d}\mathbf{u}_0 \mathrm{e}^{-\beta V(\mathbf{u}_0^1, \ldots, \mathbf{u}_0^N)}$$

$$= \lim_{P \to \infty} \left[\prod_{j=1}^{N} \left(\frac{m_j P}{2\beta \pi \hbar^2} \right)^{\frac{P}{2}} \right] \left\{ \prod_{j=1}^{N} \left[\left(\frac{2\pi \beta \hbar^2}{m_j P} \right)^{-\frac{P-1}{2}} \frac{1}{P} \right] \right\}$$

$$\times \int_V \mathrm{d}\mathbf{u}_0 e^{-\beta V(\mathbf{u}_0^1, \ldots, \mathbf{u}_0^N)}$$

$$= \left[\prod_{j=1}^{N} \left(\frac{m_j}{2\beta\pi\hbar^2} \right)^{\frac{1}{2}} \right] \int_V \mathrm{d}\mathbf{x}_c e^{-\beta V(\mathbf{x}_c^1, \ldots, \mathbf{x}_c^N)}, \tag{7.86}$$

which is simply the classical partition function. In other words, the partition function defined by Eq. (7.80) evolves into a classical partition function if one sets the effective potential $V^{\mathrm{eff}}(\mathbf{x}_1, \mathbf{x}_2, \ldots, \mathbf{x}_P)$ as

$$\sum_{i=1}^{P}\sum_{j=1}^{N} \left[\frac{1}{2} m_j \omega_P^2 (\mathbf{x}_i^j - \mathbf{x}_{i-1}^j)^2 \right] + \sum_{i=1}^{P} \frac{1}{P} V(\mathbf{x}_c^1, \ldots, \mathbf{x}_c^N). \tag{7.87}$$

From Sec. 6.4, we know that the free energies of two systems (F_1 and F_0) with potentials (U_1 and U_0) can be linked by a thermodynamic integral. In the above introduction, we also understand that the free energies of the "artificial" polymer corresponds to the free energies of the quantum and classical systems, respectively, if one takes the effective potential as Eqs. (7.81) and (7.83). Based on this analysis, one can easily introduce an artificial effective potential between the classical and quantum systems as

$$V^{\mathrm{eff}}(\mathbf{x}_1, \mathbf{x}_2, \ldots, \mathbf{x}_P; \lambda) = \sum_{i=1}^{P}\sum_{j=1}^{N} \frac{1}{2} m_j \omega_P^2 \left(\mathbf{x}_i^j - \mathbf{x}_{i-1}^j \right)^2$$

$$+ \sum_{i=1}^{P} \frac{1}{P} \left[\lambda V(\mathbf{x}_i^1, \ldots, \mathbf{x}_i^N) + (1 - \lambda) V(\mathbf{x}_c^1, \ldots, \mathbf{x}_c^N) \right]. \tag{7.88}$$

Using this effective potential, one can calculate the free energy of the "artificial system" between the classical and quantum ones through

$$F(\lambda) = -\frac{1}{\beta} \ln \left[Z(\lambda) \right], \tag{7.89}$$

with $F(1)$ giving the quantum free energy and $F(0)$ giving the classical one. The difference between them, in terms of thermodynamic integration, can be calculated using

$$\Delta F = F(1) - F(0) = \int_0^1 \mathrm{d}\lambda F'(\lambda), \tag{7.90}$$

where

$$F^{'}(\lambda) = \left\langle \frac{1}{P} \sum_{i=1}^{P} \left[V(\mathbf{x}_i^1, \ldots, \mathbf{x}_i^N) - V(\mathbf{x}_c^1, \ldots, \mathbf{x}_c^N) \right] \right\rangle_{V^{\mathrm{eff}}(\lambda)} . \qquad (7.91)$$

Similar to Sec. 6.4, the symbol $\langle \cdots \rangle_{V^{\mathrm{eff}}(\lambda)}$ means that ensemble is generated using the effective potential in Eq. (7.88). From Eqs. (7.90) and (7.91), the free energy difference between the classical and quantum systems can be rigorously evaluated. If the free energy of the classical system is known, the free energy of the quantum system will be obtainable. We note that this introduction follows the algorithm by Morales and Singer in Ref. [400]. The only thing one needs to take care in is the numerical stability, especially in the strong quantum case when $V(\mathbf{x}_i^1, \ldots, \mathbf{x}_i^N)$ and $V(\mathbf{x}_c^1, \ldots, \mathbf{x}_c^N)$ differ significantly, as pointed out in Ref. [333]. In such cases, nonlinear interpolation of the effective potential can be used. For details of this extension, please see Ref. [333].

7.4 Examples

For a better understanding of the principles underlying the path integral molecular simulations, similar to Chapter 5, we also use some examples to show how they work in practice.

7.4.1 Impact of QNEs on Structures of the Water–Hydroxyl Overlayers on Transition Metal Surfaces

The first example concerns the problem on how the impact of the QNEs is like that on the structure of the water–metal interface. This problem was investigated by Li *et al.* in Ref. [286]. Here, we use some of their results to show how the results of PIMD simulations are analyzed. The system chosen is composed of a transition metal substrate and a hexagonal water–hydroxyl overlayer.

Concerning the importance of such interfaces, it was already well known in the study of surface physics/chemistry that under ambient conditions, most surfaces are covered in a film of water [401]. These wet surfaces are of pervasive and fundamental importance in processes like corrosion, friction, and ice nucleation. On many such surfaces, it was also well known that the first contact layer of water does not comprise pure water, but instead a mixture of water and hydroxyl molecules [401–411]. Physically, these overlayers form because they provide the optimal balance of the hydrogen bonding within the overlayer and the bonding of these overlayers to the surface, and now they have been observed on several oxide, semiconductor, and metal surfaces.

For the characterization of such overlayer structures, from the experimental perspective, it is fair to say that these water–hydroxyl wetting layers are now most well characterized on close-packed metal surfaces under ultrahigh vacuum (UHV) conditions [401]. In these experiments, it is widely accepted that the molecules in the overlayer are "pinned" in registry with the substrate, with the oxygen atoms sitting right above individual metal atoms in the hydrogen bonded network (see Fig. 7.5). Because of this feature, we can say that the distances between the adjacent molecules are determined mainly by the substrate, being relatively large on a metal with a large lattice constant (e.g. $\sim$2.83 Å on average on Pt(111)) and small on a metal with a relatively small lattice constant (e.g. $\sim$2.50 Å on average on Ni(111)). From earlier studies of water in other environments, e.g. certain phases of bulk ice, it was known that the behavior of the shared proton in intermolecular hydrogen bonds varies dramatically over such a large range of O–O distances. For example, under ambient pressures, bulk ice is a conventional molecular crystal, with O–O separations of $\sim$2.8 Å. At very high pressures ($\geqslant$ 70 GPa), however, the O–O separations can decrease to $\sim$2.3 Å. In the meantime, ice loses its integrity as a molecular crystal and the protons become delocalized between the O nuclei (see, e.g., Refs. [339, 412, 413]). Now, if we make a direct comparison between the behavior of proton in ice under pressure and that of the water–hydroxyl overlayer on metal surfaces, it is reasonable to expect that in the latter system, pronounced substrate dependence of QNEs might exist.

To describe the influence of the QNEs on the structure of such overlayers as well as its substrate dependence, as mentioned, one can take a series of systems and perform both *ab initio* MD and PIMD simulations. A comparison between the MD and PIMD results illustrates in a clear manner how such an influence of the QNEs will be, and analysis on the differences between the impact of QNEs on different substrates shows us the substrate dependence. Based on this consideration, we choose three substrates, i.e. Pt(111), Ru(0001), and Ni(111), and perform *ab initio* MD and PIMD simulations at 160 K. These three substrates, in descending order

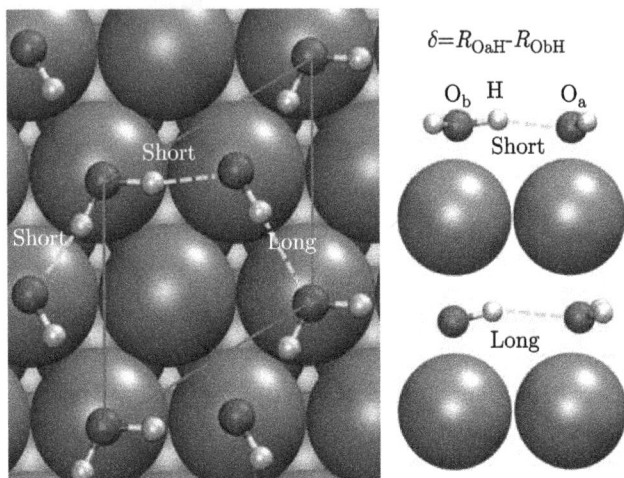

Figure 7.5   Static geometry optimized structure of the $\sqrt{3} \times \sqrt{3}$-$R30°$ overlayer (with classical nuclei) that forms on the transition metal surfaces. Side views on right show the cases when the proton is donated from water to hydroxyl (upper, labeled "short") and from hydroxyl to water (lower, labeled "long"). The short and long hydrogen bond lengths (denoted by the dashed lines) are ~1.7 and ~2.1 Å on Pt, ~1.6 and ~1.9 Å on Ru, and ~1.4 and ~1.6 Å on Ni, respectively. The coordinate for proton transfer δ is defined as $R_{O_aH} - R_{O_bH}$, where R_{O_aH} and R_{O_bH} are the instantaneous O–H distances between O_a and H and O_b and H, respectively. For a proton equidistant from its two neighbors, $\delta = 0$ and upon transfer from one O to another δ changes its sign.

of magnitude of the lattice constant, give an average O–O distance of ~2.8, ~2.7, and ~2.5 Å. Ni was chosen here because of its relatively small lattice constant, although we acknowledge that water–hydroxyl films have not yet been characterized on it [401]. The simulation package chosen is the famous Cambridge Sequential Total Energy Package (CASTEP) [414]. The Perdew–Burke–Ernzerhof (PBE) exchange–correlation (XC) functional is used for the descriptions of the electronic interactions within the DFT [129], together with a $\sqrt{3} \times \sqrt{3}$-$R30°$ water–hydroxyl overlayer (see e.g. Refs. [401–407, 415, 416]).

Our discussions start from the properties of these overlayers at the classical level. These overlayers are composed of hexagonal hydrogen bonded networks of water and hydroxyl bonded above metal atoms of the substrate in a $\sqrt{3} \times \sqrt{3}$-$R30°$ periodicity. Both types of the molecules lie almost parallel to the surface, forming a perfect extended 2D network. Because OH is a better acceptor of hydrogen bonds than a donor, there is an asymmetry in the overlayer with each molecule involved in two short and one long hydrogen bond at the classical static ground state (Fig. 7.5). At finite temperature, *ab initio* MD simulations with classical nuclei show that this asymmetry is still kept, although thermal fluctuations cause the peaks associated with the long and short hydrogen bonds to overlap, particularly on Ni (which has the smallest lattice constant). This asymmetry is illustrated in Fig. 7.6 where we show the probability distributions of O–H and O–O distances on Pt, Ru, and Ni. In addition to this asymmetric feature, the probability distribution of O–H distances also shows that the overlayer comprises individual H_2O and OH molecules of hydrogen bonded to each other. This is reflected by the sharp peak at ~ 1.0 Å, characteristic of the covalent bonds of water and hydroxyl, and broader peaks at ~ 1.7–2.1 Å, ~ 1.6–1.9 Å, and ~ 1.5 Å, characteristic of the hydrogen bonds on Pt, Ru, and Ni, respectively. The probability distributions of the O–H bond length between these peaks characteristic of the covalent and hydrogen bonds are negligible.

Then we turn on the QNEs and see what happens in the PIMD simulations. A key result is that there is no longer a clear division between short covalent and longer hydrogen bonds. This is explicitly shown in Fig. 7.6. On Pt, the population of covalent O–H bonds is reduced by one-third and replaced with a clearly non-zero probability distribution over the entire range of 1–1.5 Å (Fig. 7.6(a)). Likewise, the proportion of the short O–O distances is reduced from two-thirds to one-third, and the center of the peak associated with the short O–O distances moves from ~ 2.7 Å to ~ 2.5 Å (Fig. 7.6(b)). These changes are associated with one-third of the shared protons being delocalized between the two oxygen atoms to which they are bonded. In turn, this delocalization proton further "drags" the oxygen atoms sharing it closer and in doing so creates an "H_3O_2" complex. We note that in this complex, the shared proton belongs to neither of the two oxygen atoms. A typical snapshot from the *ab initio* PIMD simulation is shown in Fig. 7.6(g) with the H_3O_2 complex located along one particular O–O axis. This snapshot also shows how when the two oxygen atoms on either side of the shared proton are drawn close, the distances to their other oxygen neighbors increase. It is this effect that leads to a larger proportion of the long O–O distances than that was observed in the classical simulation (Fig. 7.6(b)).

On Ru, similarly, delocalization of the proton was observed and again the structure contained H_3O_2 complexes (Fig. 7.6(h)). The smaller lattice constant of Ru also means that only a small variation in the proportion of the short O–O separation (~ 2.5 Å) is required to enable proton

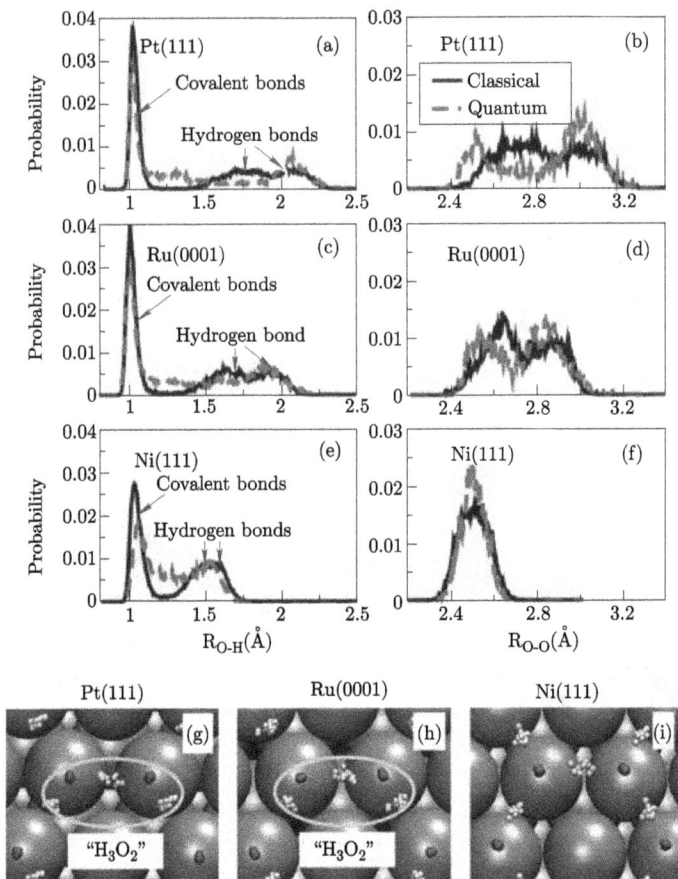

Figure 7.6   Statistical structural information from the *ab initio* MD and PIMD simulations of the water–metal interfaces, using some selected structural properties. More specific, probability distributions of the O–H ((a), (c), (e)) and O–O distances ((b), (d), (f)) on Pt(111), Ru(0001), and Ni(111) are chosen. Results obtained from *ab initio* MD simulations with classical nuclei were labeled "classical" and shown by solid lines. Those from the *ab initio* PIMD simulations with quantum nuclei at the statistical level were labeled "quantum" and shown by dashed lines. A key difference between the MD and PIMD results is that in the PIMD simulations, a non-negligible distribution of the O–H distance between the covalent and hydrogen bond peaks was observed. This feature is absent in the MD simulations with classical nuclei and it originates from some spatial configurations of the system during the simulation in which one proton is equally shared by two oxygen atoms. In panels (g)–(i), we show some snapshots for typical spatial configurations of the overlayer on Pt, Ru, and Ni obtained from the PIMD simulations (using 16 beads). On Pt and Ru, at any given snapshot, one proton is equally shared by two of the oxygen atoms yielding an intermediate "H_3O_2" complex. On Ni at any given snapshot, several protons can simultaneously be shared between the oxygens.

delocalization (Fig. 7.6(d)). Upon moving to Ni, the influence of the QNEs on these structural properties becomes even larger. This is shown by the larger magnitude for the distribution of the O–H distances between the peaks characterizing the covalent and hydrogen bonds (Fig. 7.6(e)). Because of Ni's smaller lattice constant, the quantum delocalization of the proton within the overlayer becomes possible without any major rearrangement of the oxygen nuclear "skeleton". A snapshot from the PIMD simulation on Ni, in which several protons are delocalized simultaneously and the distinction between covalent and hydrogen bonds is completely lost, is shown in Fig. 7.6(i).

The obviously different probability distributions observed in the MD and PIMD simulations mean that the QNEs significantly change the structures of the water–hydroxyl overlayer on the transition metal surfaces studied. For a more rigorous characterization of difference from a statistical perspective, we further calculated the free energy profiles for the protons along the intermolecular axes. This free energy profile is calculated using $\Delta F(\delta) = -k_B T \ln P(\delta)$, where $P(\delta)$ is the probability distribution of δ and δ is the proton transfer reaction coordinate as defined in Fig. 7.5. k_B is the Boltzmann constant. For an unbiased analysis, we take all inequivalent hydrogen bonds in the system into account. In other words, the free energy profile calculated here is an average over all hydrogen bonds in the overlayer. The results are shown in Fig. 7.7. In the MD simulations with classical nuclei, the free energy profiles are characterized by two partially overlapping valleys. On Pt (Ru), they are located at $\delta \sim 0.7$ (0.6) and $\delta \sim 1.1$ (0.9) Å. On Ni, these two valleys almost completely overlap at ~ 0.5 Å, since, as we have said, thermal broadening obscures the distinction between short and long hydrogen bonds on this surface. Concerning proton transfer, it is a rare event, as reflected by the presence of large classical free energy barriers on all substrates, at $\delta = 0$.

Then we move to the free energy profiles obtained from the PIMD simulation, in which the QNEs are included in the theoretical descriptions. It is clear in Fig. 7.7 that they differ significantly from the MD ones. On the Pt and Ru substrates, the minima for the long hydrogen bond remain. But we note that those associated with the short hydrogen bonds completely disappear due to the formation of the intermediate H_3O_2 complexes as mentioned before. On Ni, the single valley feature was kept. However, it was softened and its position shifted from $\delta \sim 0.5$ Å to $\delta \sim 0.4$ Å. We note that the key difference between the quantum and classical free energy profiles

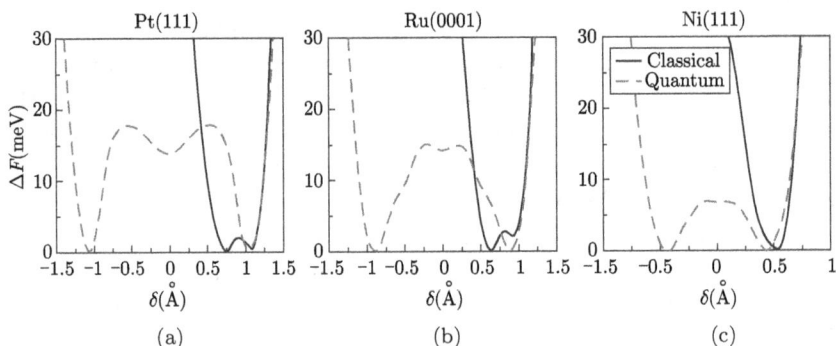

Figure 7.7   Free energy profile (denoted as ΔF) for the protons along the intermolecular axes within the water–hydroxyl overlayers on Pt (a), Ru (b), and Ni (c) from the *ab initio* MD and PIMD simulations at 160 K. The MD results were labeled classical and shown by solid lines and the PIMD results were labeled quantum and shown by dashed lines.

is that in the quantum simulations, the proton transfer energy barriers are significantly smaller than the classical ones. Upon going from Pt through Ru to Ni, the height of the barrier and the area beneath it decreases, indicating that proton transfer probability increases as the lattice constant is reduced, and a plateau appears on all three quantum free energy profiles.

To understand how this plateau appears, we correlate the location of the proton along the intermolecular axes (δ) with the corresponding O–O distances (R_{O-O}) and plot the probability distribution as a function of these two variables in Fig. 7.8. In the MD simulations (Figs. 7.8(a), 7.8(d), and 7.8(g)), these functions are characterized by negligible distributions at $\delta = 0$, consistent with the fact that the proton transfer is a rare event and the protons hop from one side of the hydrogen bond to the other. The O–O distribution has two peaks for the short and long hydrogen bonds, respectively, on Pt and Ru, but they merge on Ni. When the QNEs are taken into account, finite distributions at $\delta = 0$ appear on all three substrates. These distributions correspond to the delocalized protons, as shown by the snapshots in Fig. 7.6. To understand the behavior of this "delocalized" proton from a more rigorous perspective, one can focus on the most active proton, i.e. the proton which at any given snapshot in the PIMD simulations has the smallest magnitude of δ. On Pt and Ru, this is the proton located along the hydrogen bond with the smallest O–O distance. On Ni, due to the fact that the average O–O distance is only $\sim$2.5 Å, the most active proton need not necessarily be the one with the shortest O–O distance.

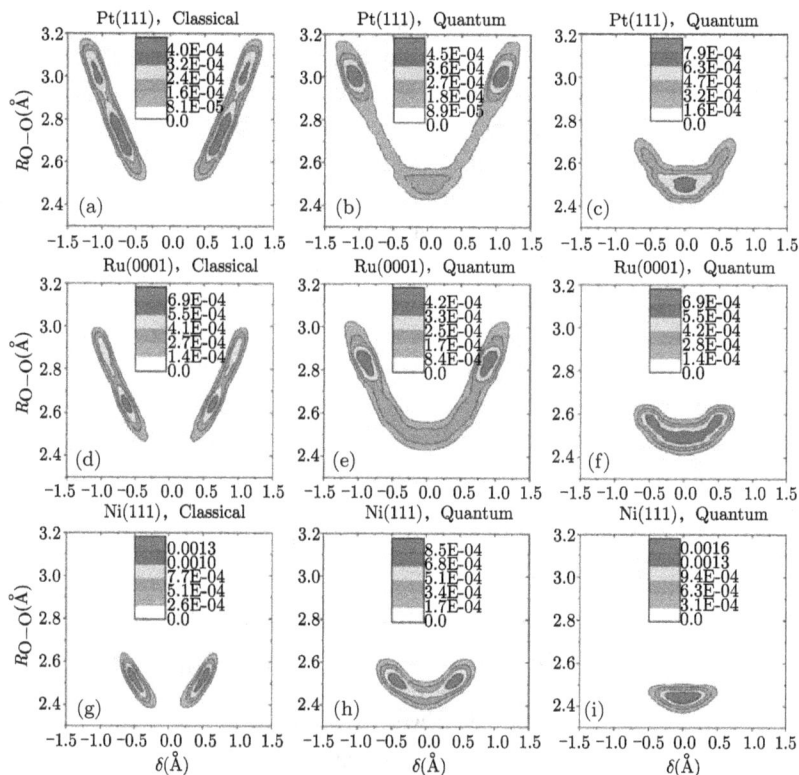

Figure 7.8   Probability distribution in the MD and PIMD simulations as a function of δ and R_{O-O}. Similar to the earlier figures, the MD simulations are labeled as classical and the PIMD simulations are labeled as quantum. The left and middle columns show results obtained from all hydrogen bonds in the overlayer. On the right column, only data from the most active hydrogen bond are chosen in the PIMD simulations. The most active proton is defined as the one with the smallest δ. All MD and PIMD distribution functions have been symmetrized with respect to δ.

The results are shown in Figs. 7.8(c), 7.8(f), and 7.8(i). The key feature is that different from the panels on the left and middle columns in Fig. 7.8, where the mean peak is located at δ with large magnitude of the absolute value, the distribution peaks on the right column clearly are located around $\delta = 0$. Therefore, the corresponding free energy barrier for the transfer of the most active proton is zero and the classical proton transfer energy barrier is wiped out by the QNEs.

Another mean feature of the distribution functions in Fig. 7.8 is that a "horseshoe" shape exists. On Pt and Ru, this indicates that the covalently bonded proton requires the oxygen atoms to move close first. When this O–O distance is smaller than a certain value, the classical proton transfer energy barrier for the most active proton will become so small that it can be easily wiped out by its zero-point energy. In this case, the quantum nature of the proton results in an "adiabatic" response to the movement of the oxygen atoms and the proton quickly becomes delocalized along this short hydrogen bond. In this case, the H_3O_2 complex as shown in the earlier discussions appears, and it persists till the thermal fluctuations of the oxygen force them to move apart. When this O–O distance is larger than a certain value, the quantum zero-point energy fails to wipe out the classical proton transfer energy barrier and consequently it falls to either side and becomes covalently bonded to one of the oxygen atoms. Therefore, the mechanism for proton transfer on Pt and Ru is the so-called "adiabatic proton transfer" [417], as predicted for the diffusion of the excess proton in water and ice at certain pressures [321, 412].

For more details of these simulations, please see Ref. [286].

7.4.2   Impact of Quantum Nuclear Effects on the Strength of Hydrogen Bonds

The second example we want to show here, in which the quantum nature of the nuclei is explicitly addressed, concerns a fundamental problem in physics and chemistry, i.e. what will be the impact of QNEs on the strength of hydrogen bonds.

We all know that hydrogen bonds are weak intermolecular interactions which hold much of soft matter together, as well as the condensed phases of water, network liquids, and many ferroelectric crystals. The small mass of hydrogen, as shown already in the above example, means that they are inherently quantum mechanical in nature, and effects such as zero-point motion and tunneling must be taken into account in descriptions of the properties related to it. As a prominent example, from the statistical point of view, it is well known that by replacing H by D, the hydrogen bond strength changes. However, as direct as it looks, a simple picture, in which the impact of QNEs on the strength of hydrogen bonds and consequently the structure of the hydrogen bonded systems, can be rationalized, has been absent for a long time.

As a matter of fact, this problem concerning the influence of QNEs on the strength of hydrogen bonds is a fundamental problem in physics and chemistry. Already in the 1950s, it was observed experimentally that in some hydrogen bonded molecular crystals, by replacing the hydrogen with deuterium, the heavy atom (e.g. O–O) distances change [418, 419]. This phenomenon is known as the Ubbelohde effect. The conventional Ubbelohde effect causes an elongation of the O–O distance upon replacing H with D, indicating that the QNEs strengthen the hydrogen bond, although a negative Ubbelohde effect has also been observed in several systems [418, 419]. From the molecular simulation perspective, starting from the early 1980s, when PIMD and PIMC simulations became a conventional routine to investigate the influence of QNEs on real poly-atomic systems, computer simulations on this issue had been carried out in a wide range of sample systems. A general conclusion is that the result is strongly system-dependent. In liquid hydrogen fluoride (HF), for example, *ab initio* MD and PIMD simulations using DFT for the description of the electronic structures have shown that when the QNEs are accounted for, the first peak in the F–F radial distribution function (RDF) sharpens and shifts to a shorter F–F distance [420]. The implication of this increase in the structuring of the RDF in the liquid is that the hydrogen bond is strengthened upon including the QNEs. In contrast, similar simulations for liquid water show that the O–O radial distribution function is less peaked when simulations with quantum nuclei are compared with those with classical nuclei [338], suggesting a decrease of the overall hydrogen bond strength. We note, however, that although this conclusion is probably correct, it is the opposite of what was observed in an earlier *ab initio* study [345].

Besides these discussions concerning hydrogen bonded crystals and liquids, the influence of the QNEs on the hydrogen bonds has also been widely discussed in studies of gas phase clusters [349, 421, 422]. Specifically, in water clusters up to hexamer, it is predicted that the QNEs weaken the hydrogen bonds, whereas in simulations of the HF clusters, both strengthening and weakening is predicted depending on the size of the cluster [349, 421, 423]. For clusters smaller than tetramer, a weakening of intermolecular hydrogen bond is predicted upon including the QNEs. For clusters larger than tetramer, a strengthening of the hydrogen bond is expected. In tetramer, the influence is negligible. Clearly, it would be very useful to rationalize these various results within a single conceptual framework and identify the underlying factors that dictate the influence of the QNEs on hydrogen bond strength for a broad class of materials. In a recent study [322], Li *et al.* gave a simple picture to rationalize these different results using analysis based on *ab initio* MD and PIMD simulations. Here, we use some of their key results to show how it is done in practice.

First of all, a broad range of hydrogen bonded systems are chosen, including HF clusters (dimer to hexamer), H_2O clusters (dimer, pentamer, and octamer), charged, protonated, and hdyroxylated water and ammonia clusters ($H_9O_5^-$, $H_9O_4^+$, $H_7O_4^-$, and $N_2H_5^-$), organic dimers (formic acid and formamide), and solids (HF, HCl, and squaric acid $C_4H_2O_4$). For each system, both conventional *ab initio* MD simulations, in which the nuclei are treated as classical point-like particles, and more state-of-the-art *ab initio* PIMD simulations, in which the QNEs are accounted for, were performed. With these two complementary sets of simulations, one can identify in a very clear manner the precise influence of the QNEs on the statistical properties of interest at finite temperatures.

Before we start, let us first make the following points clear. First, the quantities we focus on when characterizing H-bonds are: (i) the heavy-atom (X–X, where X is either O, Cl, C, N, or F) distances, which characterize the intermolecular separations, (ii) the H-bond angles (X–H–···–X), which are associated with H-bond bending (libration) modes, and (iii) the X–H covalent bond lengths, characteristic of the covalent bond stretching in the H-bond donor molecules. In later discussion, it will become clear that these quantities provide an indication of H-bond strength. However, as the main measure of H-bond strength, we still use a standard estimate based on the computed red-shift (softening) in the X–H stretching frequency of the H-bond donor molecule. We note that there is no perfect measure for H-bond strength [424], however the red-shift of the stretching frequency is a widely used measure [see e.g. Refs. [425, 426]]. This measure is particularly useful here because it allows us to discriminate between different types of H-bond in the same complex and can be used for both neutral and charged systems. In Fig. 7.9(b), it is shown that this estimator correlates well with the computed binding energy per H-bond in the neutral systems we study. This binding energy is defined as the difference between the total energy of the system and the sum over its unrelaxed components, as in Ref. [424]. When the red-shift of the stretching frequency (measured as the ratio of the X–H stretching frequency in the H-bonded cluster to that in the free monomer) gets larger, the H-bond becomes stronger.

With the definition of the above-mentioned quantities in mind, we first look at the results for the impact of the QNEs on the strength of hydrogen bonds. Upon comparing these results for the various hydrogen bonded systems, an interesting correlation can be established between the H-bond strength and the change in intermolecular separations. This correlation is shown in Fig. 7.9(a) where we see that as the H-bond gets stronger, the heavy-atom separations in the PIMD simulations with quantum nuclei go from being longer than those in the MD simulations with classical nuclei (positive Δ(X–X)) to being shorter (negative Δ(X–X)). Thus, the QNEs result in longer hydrogen bonds in weak hydrogen-bonded systems and shorter hydrogen bonds in relatively strong hydrogen-bonded systems. We note that the hydrogen-bond strength increases upon going from small to large clusters and from water to HF. The trend reported in Fig. 7.9 is a key finding and in the following, we explain why it emerges and discuss the implications it has for H-bonded materials in general.

To understand the reason for this correlation between the impact of the QNEs on the strength of hydrogen bond and the strength of hydrogen bond itself, it is useful to look at the HF clusters. These provide the ideal series because upon increasing the cluster size, the hydrogen-bond strength increases, and the influence of the QNEs switches from a tendency to lengthen to a tendency to shorten the intermolecular separations (as seen in Ref. [423]). Our analysis is summarized in Fig. 7.10, where we plot the distance and angle distributions from MD and PIMD simulations for these three HF clusters separately. The left column shows the final results, where one can see that in the dimer, the averaged F–F distance is increased by including the QNEs; in the tetramer, there is no difference between the averaged quantum and classical F–F distances; in the pentamer, the F–F distance is clearly shortened by including the QNEs. The key to understanding this variation of the heavy-atom distances is in recognizing that there are also related differences between MD and PIMD in the covalent F–H bond lengths (center) and H-bond angles (right). Because of anharmonic quantum fluctuations, these two geometric properties also show systematic changes. First of all, the F–H bonds are longer in the quantum compared to the classical simulations, and this elongation becomes more pronounced as the H-bonds get stronger. Second, the hydrogen bonds are more bent in the quantum than in the classical simulations, and this bending generally becomes less pronounced as the hydrogen bonds get stronger. In order to understand the influence of these variations in structure, analysis of various dimer configurations was performed. This analysis reveals that the covalent bond stretching increases the intermolecular interaction whereas hydrogen bond bending decreases it. Taking the HF dimer as an example, a 0.04 Å

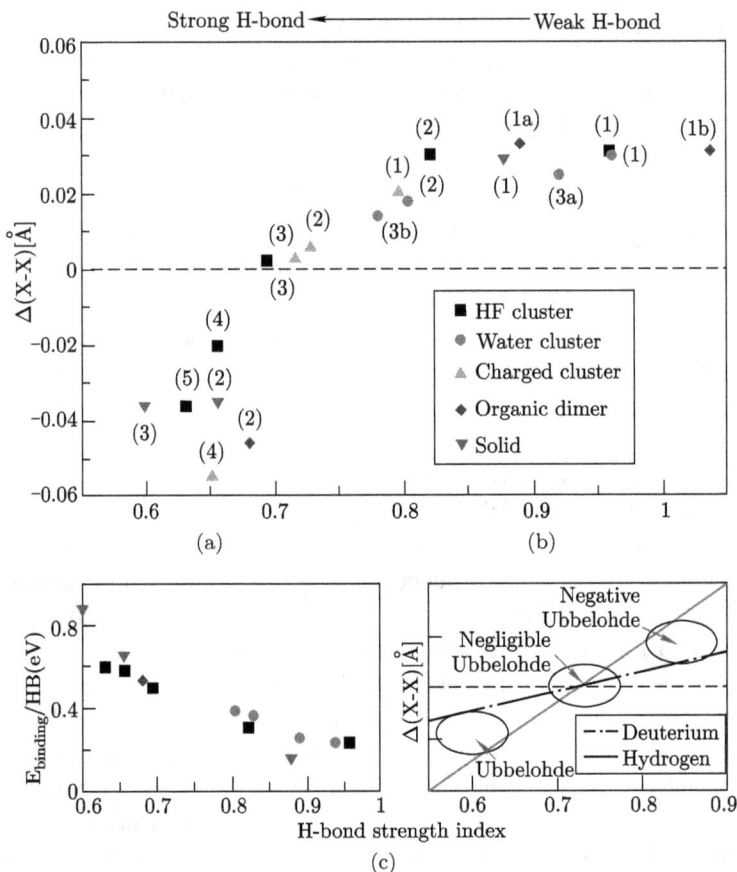

Figure 7.9 Correlation between the impact of the QNEs and the hydrogen bond strength. In panel (a), the differences between the shortest heavy-atom distances obtained from the PIMD and MD simulations $(X–X)^{PIMD}_{average}–(X–X)^{MD}_{average}$, denoted by $\Delta(X–X)$, is chosen as the y axis. It characterizes the impact of the QNEs on the strength of the hydrogen bonds. This influence is drawn as a function of the hydrogen-bond strength. As mentioned in the prose, this hydrogen bond strength is defined as the ratio of the X–H stretching frequency in the hydrogen-bonded system to that in the free monomer. In panel (b), the correlation between this hydrogen-bond strength index and the binding energy per hydrogen bond in the neutral systems is given. In panel (c), simplified schematic illustration of the expected isotope (Ubbelohde) effect on the differences in heavy-atom distances. We suggest that three regimes of positive, negligible, and negative Ubbelohde effect depending on the hydrogen-bond strength exist. For the HF clusters, labels (1)–(5) denote the hydrogen bonds in the dimer to the hexamer. For the water clusters, labels (1), (2), (3a), and (3b) refer to the hydrogen bonds in the dimer, pentamer, and the long (short) hydrogen bond in the octamer. For the charged clusters, labels 1–4 refer to $H_9O_5^-$, $H_9O_4^+$, $H_7O_4^-$, and $N_2H_5^-$, respectively.

Figure 7.9 (*continued*) For the organic dimers, labels (1a), (1b), and (2) refer to the red-shifted and blue-shifted hydrogen bond in the formamide and the red-shifted hydrogen bond in the formic acid. For the solids, labels (1)–(3) refer to the hydrogen bonds in HCl, HF, and squaric acid. The same labels are applied in Fig. 7.11. For the water cluster in panel (b), the octamer is not included since there are two kinds of hydrogen bonds. Results for the trimer and tetramer are added to further test the correlation.

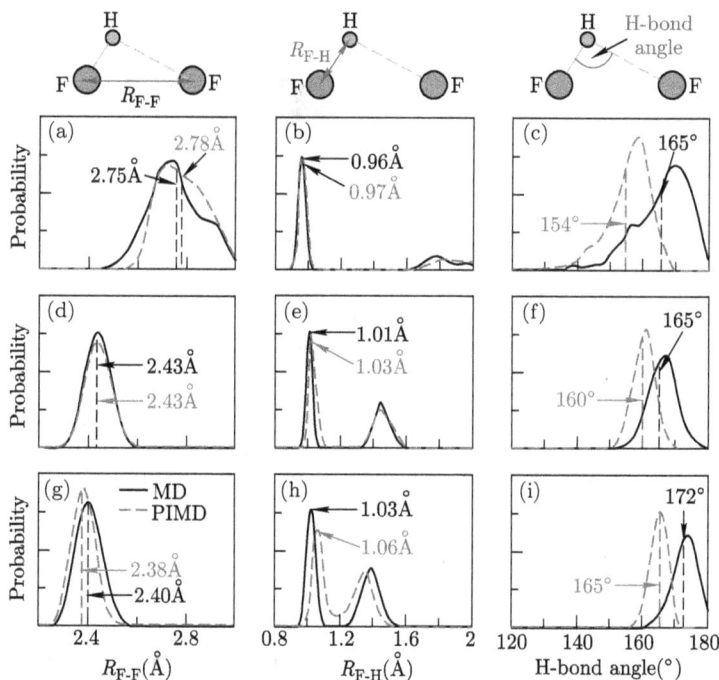

Figure 7.10 HF clusters as examples for detailed analysis of the QNEs. Distributions of the F–F distances (left), the F–H bond lengths (center), and the intermolecular bending (F–H··· F angle, right) from the MD (solid lines) and PIMD (dashed grey lines) for a selection of systems: the HF dimer (top), the HF tetramer (middle), and the HF pentamer (bottom). The MD and PIMD averages are shown in black and grey vertical dashes, respectively.

increase in the F–H bond length of the donor leads to a 40 meV increase in interaction energy within the dimer, whereas in contrast, a 21° reduction in H-bond angle leads to a 16 meV decrease in interaction energy. This analysis provides a qualitative understanding of the trend observed. In short, the F–F distance increases in the dimer as a result of a large decrease in hydrogen bond angle, but only a small increase in the covalent F–H bond length.

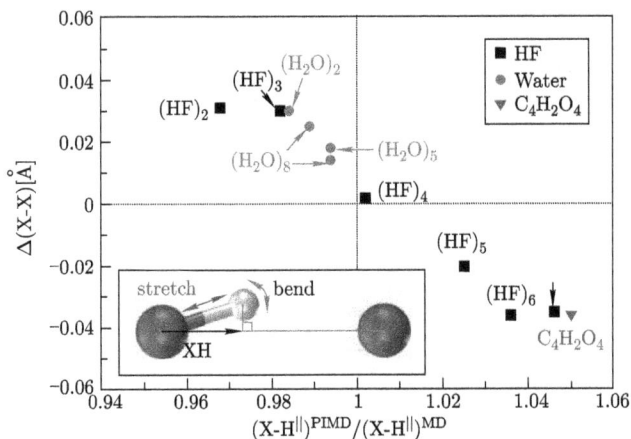

Figure 7.11   A quantification for the competition between the quantum fluctuations on the stretching and bending modes. Differences in average shortest heavy-atom distances between PIMD and MD simulations (Δ(X–X), vertical axis) vs. the ratio of the projection of the donor X–H covalent bond along the intermolecular axis from PIMD and MD simulations (horizontal axis). For the meaning of the labels, please refer to the caption of Fig. 7.9. x larger (smaller) than 1 indicates a dominant contribution from the stretching (bending) mode when the QNEs are included. Negative values of Δ(X–X) indicate that quantum nuclear effects decrease the intermolecular separation. An almost linear correlation between the two variables can be observed: When the contribution from stretching becomes more dominant, the QNEs turn from weakening to strengthening the H-bonds. The inset illustrates the geometry used for projecting the donor covalent X–H bond onto the intermolecular axis. The curved arrow represents the intermolecular bending and the straight arrow represents the intra-molecular stretching.

While in the tetramer, the F–H stretching is sufficiently pronounced to compensate for the increase in hydrogen-bond bending, leaving the overall F–F distance unchanged, in the pentamer, the F–F distance decreases because the F–H covalent bond stretching dominates over the H-bond bending.

For a rigorous examination of this picture and a quantitative description of this competition for all systems studied, one can further calculate the projection (X–H$^\parallel$) of the donor molecule's covalent bond along the intermolecular axis (see inset of Fig. 7.11). Since X–H$^\parallel$ increases upon intra-molecular stretching but decreases upon intermolecular bending, this quantity itself allows the balance between stretching and bending to be evaluated, to a certain extent. The influence of the QNEs is quantified by the ratio of the PIMD and MD projections, i.e. $x = $ (X–H$^\parallel$)^{PIMD}/(X–H$^\parallel$)^{MD}. When this value is clearly greater than one, it indicates that when the QNEs are included, the main influence is on the stretching of the covalent bond, and when this value is clearly smaller than 1, it indicates that when the QNEs are included, the main influence is on the bending of the hydrogen bond. When one plots this ratio against the variations in intermolecular separations, $y = \Delta(X-X)$ (which we used to quantify the impact of the QNEs). A striking (almost linear) correlation is observed (Fig. 7.11). For all systems where hydrogen-bond bending dominates (x clearly smaller than 1), the heavy-atom distances are longer in PIMD than in MD ($y > 0$). In cases where covalent bond stretching is dominant (x clearly larger than 1), the heavy-atom distances are shorter in PIMD than in MD ($y < 0$). With the increase of x, quantum fluctuations on the stretching mode become more dominant and the QNEs turn from weakening the hydrogen bonds to strengthening them. Thus, the overall influence of the QNEs on the hydrogen bonding interaction quantitatively comes down to this delicate interplay between covalent bond stretching and intermolecular bond bending. One notes that this explanation arrived at here for the general case is consistent with what Manolopoulos and coworkers have elegantly shown for liquid water in Ref. [427].

Given the ubiquity of the hydrogen bonds in the physical, chemical, and biological sciences, there are a number of implications in this finding. Considering that liquid HF comprises long polymer chains and rings whereas liquid water is widely considered to be made up of small clusters, these results shed light on why the QNEs strengthen the structure of liquid HF but weaken that of liquid water [338, 420]. More generally, one can use the trend observed in Fig. 7.9 as a simple rule of thumb to estimate the impact of the QNEs on hydrogen-bonded systems without performing expensive PIMD simulations. All that is required is an estimate of hydrogen-bond strength, which can be obtained from the red-shift in the covalent stretching frequency or from other commonly used measures of hydrogen-bond strength such as hydrogen-bond length. Thus, the trend may be particularly useful to biological systems such as α-helixes and β-sheets for which many crystal structures have been determined and where cooperative effects lead to particularly strong H bonds [428].

In addition to the implications mentioned, this trend also allows one to rationalize the Ubbelohde effect over a broad range of H-bond regimes (Fig.7.9(c)). Specifically speaking, traditional Ubbelohde ferroelectrics such as potassium dihydrogen phosphate fall in the relatively strong H-bond regime where a positive Ubbelohde effect (i.e., an increase of the X–X distance upon replacing H with D) is observed in the experiment and also in recent PIMD studies [419, 429], and in this context, the squaric acid, the solid HF, and the larger HF clusters are expected to exhibit a traditional Ubbelohde effect upon replacing H with D. In contrast, the smaller hydrogen-bonded clusters studied here and solid HCl are expected to exhibit a negative Ubbelohde effect (a decrease of the X–X distance upon replacing H with D). Hydrogen-bonded materials of intermediate strength such as large water clusters and ice at ambient pressure are predicted to exhibit a negligible Ubbelohde effect because in this regime, the QNEs have little influence on the intermolecular separations. Indeed, this observation is consistent with experimental and theoretical observations for the ferroelectric hydrogen-bonded crystals, ice, and gas phase dimers [418, 419, 422].

A further prediction stemming from this work is that ice under pressure will exhibit the traditional Ubbelohde effect. However, one cautions that at very high pressure, ice possesses such strong hydrogen bonds, with shared symmetric protons [412], that the picture sketched in Fig. 7.11 is not likely to apply. Indeed, this note of caution applies to all ultra-strong hydrogen bonds, where the proton is shared symmetrically by the two heavy atoms already in the classical perspective. In this case, the distinction between a relatively short covalent bond and a relatively long hydrogen bond is lost and bond stretching along the X–X axis does not lead to any strengthening of the intermolecular interactions. The gas phase Zundel complex, $H_5O_2^+$, is an example of one such ultra-strong hydrogen bond and the calculations in Ref. [322] show an approximate increase of 0.016 Å in the O–O distance, which is consistent with the previous studies [326, 430].

Another class of very strong H-bonded systems are the so-called "low-barrier" H-bonds, e.g. $H_3O_2^-$, $N_2H_7^+$, and $N_2H_5^-$. In these systems, there remains a clear distinction between covalent and hydrogen bonds, and the picture we have presented still holds. This fact can be seen from our data for $N_2H_5^-$ in Fig. 7.9. We caution, however, that in these very strong H-bonded systems, errors associated with the underlying XC functional can have a qualitative impact on the results and that the accuracy of the underlying PES is of critical importance. For example, in $H_3O_2^-$ and $N_2H_7^+$, using the PBE XC functional yields a shared symmetric proton already in the classical MD simulations. But earlier studies with the more accurate second-order Møller–Plesset perturbation theory and also the Becke–Lee–Yang–Parr XC functional [326, 430, 431] show that protons actually feel a double-well potential and in this case, quantum nuclear effects strengthen the hydrogen bond, consistent with the model presented here. In addition to this, since both inter- and intramolecular vibrations are relevant to the QNEs, this work also highlights the need for flexible anharmonic monomers in force field simulations of the quantum nuclear effects. Specifically, if this feature is absent, only hydrogen-bond bending will be present in the simulation and consequently the intermolecular interaction will be "artificially" weakened. For more details concerning such discussions and the numerical details of the calculations, please see Ref. [322].

7.4.3 Quantum Simulation of the Low-Temperature Metallic Liquid Hydrogen

The third example we show here, in which the *ab initio* PIMD simulation is used to study the fundamental properties of condensed matter, concerns the existence of a low-temperature quantum metallic liquid, which exists in high-pressure hydrogen.

Concerning the importance for the existence of this low-temperature metallic liquid phase, one can track back to a very famous conjecture about hydrogen under pressure. This conjecture was first proposed by Wigner and Huntington in 1935, and states that solid molecular hydrogen would dissociate and form an atomic metallic phase at high pressures [432]. Ever since this prediction, the phase diagram of hydrogen has been the focus of intense experimental and theoretical studies in condensed matter and high-pressure physics [278, 281, 433–436].

Due to the advancement of many experimental techniques, notably diamond anvil cell approaches, it is possible nowadays to explore hydrogen at pressures up to about 360 GPa [433, 437–439], and one notes that new types of diamond anvil cell may be able to access even higher pressures [440]. These experiments, together with numerous theoretical studies, have revealed a remarkably rich and interesting phase diagram comprising regions of stability for a molecular solid, a molecular liquid and an atomic liquid, and within the solid region, four distinct phases have been detected [438, 439, 441]. In high-temperature shock-wave experiments, metallic liquid hydrogen has also been observed [442, 443]. It is accepted to be a major component of gas giant planets, such as Jupiter and Saturn [443]. Despite the tremendous and rapid progress, important gaps in our understanding of the phase diagram of high-pressure hydrogen still remain, with arguably the least well-understood issue being the solid to liquid melting transition at very high pressures. Indeed, the melting curve is only established experimentally and theoretically up to around 200 GPa [278, 436]. From 65 GPa to about 200 GPa, the slope of the melting curve is negative (that is, the melting point drops with increasing pressure), which suggests that at yet higher pressures, a low-temperature liquid state of hydrogen might exist or, as suggested by Ashcroft [444], perhaps even a metallic liquid state at 0 K. Further interest in hydrogen at pressures well above 200 GPa stems from other remarkable suggestions, such as superfluidity [435] and superconductivity at room temperature [445, 446], all of which imply that hydrogen at extreme pressures could be one of the most interesting and exotic materials among all condensed matter types.

In Ref. [323], Chen *et al.* used computer simulation techniques to probe the low-temperature phase diagram of hydrogen in the ultra-high 500–1200 GPa regime to try and find this potential low-temperature liquid state of hydrogen. Concerning the proton motion in this condensed phase, *ab initio* PIMD as introduced earlier in this chapter has been used. To compute the melting curve, the solid and liquid phases in coexistence were simulated [278, 447, 448]. This coexistence approach, otherwise called the two-phase simulation method, minimizes hysteresis effects arising from superheating or supercooling during the phase transition. With this combination of approaches, they have found a low-temperature metallic atomic liquid phase at pressures of 900 GPa and above, down to the lowest temperature they can simulate reliably, 50 K. The existence of this low-temperature metallic atomic liquid is associated with a negative slope of the melting curve between atomic liquid and solid phases at pressures between 500 and 800 GPa. This low-temperature metallic atomic liquid is strongly quantum in nature, as treating the nuclei as classical particles using the *ab initio* MD method significantly raises the melting curve of the atomic solid to ~300 K over the whole pressure range. The classical treatment of the nuclei does not reproduce a notable negative slope of the melting curve and consequently does not predict a low-temperature liquid phase.

For a clear explanation of these results, we go through the detailed procedures of their study here. One problem which is essential in a theoretical description of the hydrogen phase diagram is that in its solid phase, many local minima on the potential energy surface exist. Therefore, in order for the *ab initio* MD and PIMD simulations to make sense, extensive computational searches for low-enthalpy solid structures of hydrogen must be performed. From earlier studies using DFT methods [15, 449–452], a metallic phase of $I41/amd$ space group symmetry has been widely reported to be stable from about 500 to 1200 GPa, when quasi-harmonic proton zero-point motion was included. Accordingly, they have used this phase as the starting point for their finite-temperature exploration of the phase diagram and melting curve [323]. With the coexistence method, they have performed a series of two-phase solid–liquid simulations at different temperatures (from 50 K to 300 K), which are then used to bracket the melting temperature from above and below. We begin by considering the 500–800 GPa pressure regime and show an example of the data they have obtained from the coexistence simulations at 700 GPa in Fig. 7.12. At this pressure, one can see that for $T \geqslant 125$ K, the system transforms into a liquid state, whereas for $T \leqslant 100$ K, it ends up as a solid. To characterize these states, they have used a pair-distribution function and averaged out its angular dependence. The result is a function of interatomic separations, which is denoted by $g(r)$ throughout this section. In a liquid, this is the so-called radial distribution function. As can be seen in Fig. 7.12(d), upon moving from 100 to 125 K, the system clearly possesses less structure, indicating that a transition from solid to liquid occurs. These phases were also characterized by the variations in the mean-square displacement (MSD) of the nuclei of the particles over time. As PIMD rigorously provides only thermally averaged information, they have used the partially adiabatic centroid MD (PACMD) approach within the path integral scheme to obtain real-time quantum dynamical information [397]. Again, as shown in Fig. 7.12(e), the distinction between the solid phase at 100 K and liquid phase at 125 K is clear.

The same coexistence procedure was used to locate the melting point at 500 GPa and 800 GPa, leading to the melting curve shown in Fig. 7.13. The up (down) triangles indicate the highest (lowest) temperatures at which the solid (liquid) phases are stable, bracketing the melting temperatures within a 25 K window. From this, we see that the melting temperature is between 150 K and 175 K at 500 GPa, and that it drops rapidly with increasing pressure, yielding a melting temperature that is only between 75 K and 100 K at 800 GPa. Thus, the melting curve has a substantial negative slope ($dP/dT < 0$) in this pressure range. Across this entire pressure range, the molten liquid state is atomic, and the solid phase, which grows, is the original atomic $I41/amd$ phase that was used as the starting structure. Given that molecular phases have been observed at pressures lower than 360 GPa in both experimental and theoretical studies [437, 439], they have suggested that a molecular-to-atomic solid–solid phase transition should

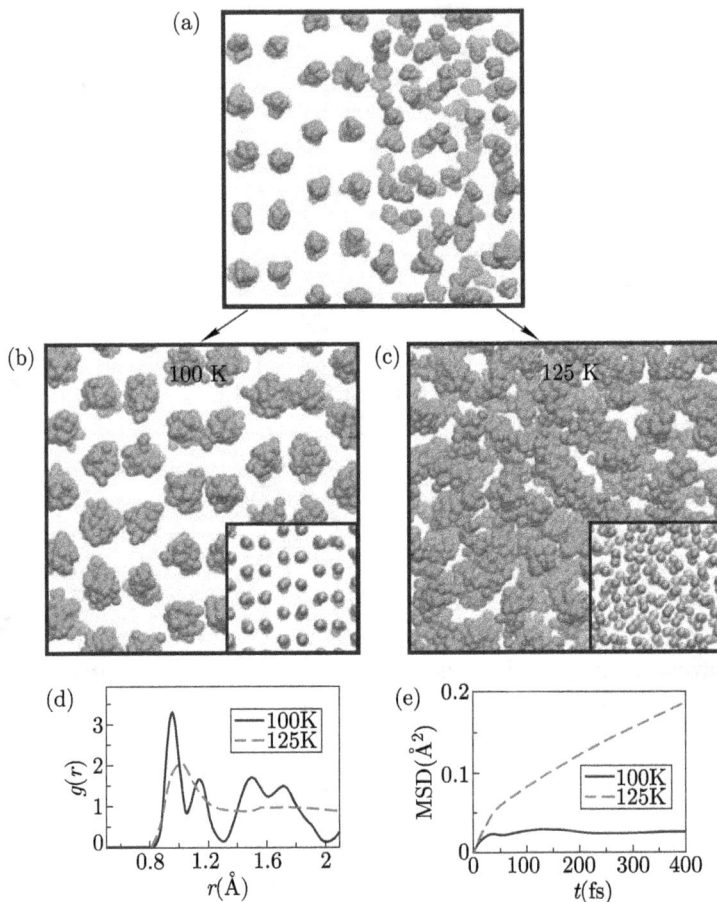

Figure 7.12 *Ab initio* PIMD simulations of solid–liquid coexistence and melting. Snapshots of the PIMD simulations at 700 GPa showing (a) the starting structure, (b) the final state at 100 K, and (c) the final state at 125 K. Around 32 beads (pictured in the larger squares) were used to represent the imaginary-time path integral for each atom. The balls in the insets of (b) and (c) correspond to the centroid of each atom. (d) The angularly averaged pair-distribution function $g(r)$ for the same two simulations at 100 K and 125 K. At 100 K, the solid state persists (solid line) as indicated by the relatively sharp peaks. At 125 K (dashed line), these peaks are much broader and the $g(r)$ is characteristic of a liquid. This is further supported by the data in panel (e), where the MSDs as a function of time from separate adiabatic CMD simulations within the path integral framework are shown. The MSD for the 100 K solid-phase saturates rapidly, whereas for the liquid phase at 125 K, it rises approximately linearly with time, resulting in a finite-diffusion coefficient.

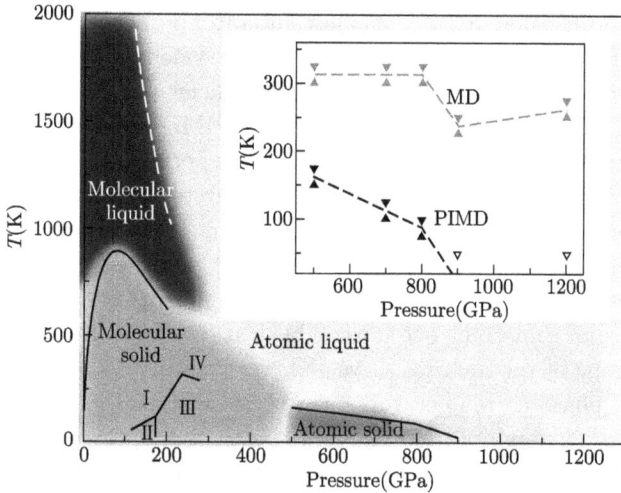

Figure 7.13   Phase diagram of hydrogen and the low-temperature metallic liquid phase. Regions of stability for the molecular solid, molecular liquid, atomic solid, and atomic liquid are indicated by the various shadings. The dashed line separating the molecular and atomic liquid phases is taken from quantum MC calculations [282]. The solid line separating the molecular solid and molecular liquid phases is taken from *ab initio* MD simulations [284], whose negative slope has been confirmed by experiment [436]. The thick black line is the melting curve obtained in this study from the *ab initio* PIMD coexistence simulations. The solid lines separating phases I, II, III, and IV are from Refs. [441, 453]. The inset shows how the high-pressure melting curve (dashed lines) are established here. The black and grey triangles (inset) correspond to the PIMD and MD results, respectively. The solid up triangles give the highest temperatures for solidification and the solid down triangles show the lowest temperatures for liquefaction. At 900 GPa and 1200 GPa, the so-called degeneracy temperature is ∼40 K, below which the exchange of nuclei will be important. Accordingly, 50 K was the lowest temperature examined in our PIMD simulations. At this temperature, each simulation yields a liquid state, and so the two open triangles at 900 GPa and 1200 GPa indicate upper bounds for the melting temperature.

occur between 360 GPa and 500 GPa (the lowest pressure they have considered in their simulations of the melting).

The negative slope of the melting curve up to 800 GPa suggests that at even higher pressures, a lower-temperature liquid phase might exist. Motivated by this, they also carry out simulations at 900 GPa and 1200 GPa. However, in this pressure range, one needs to consider nuclear exchange effects, which are neglected in the PIMD simulations, but could potentially become significant. Indeed, analysis of their simulations reveals that at these pressures, the dispersion of the beads in the path integral ring polymer becomes comparable to the smallest interatomic separations when the temperature is below $\sim$40 K. This is the so-called quantum degeneracy temperature below which the exchange of nuclei will be important, and consequently simulations with a (standard) PIMD approach are expected to be inaccurate. With this in mind, they have performed all simulations in this very high-pressure regime at $T \geqslant 50$ K. Interestingly, they find that at 50 K, at both 900 GPa and 1200 GPa, the systems are already in the liquid state, revealing that the melting temperature at these pressures is below 50 K. Whether the liquid phase is the 0 K ground state of hydrogen at these pressures is not something one can establish at this stage. However, the large negative slope of the melting curve at lower pressures and the observation of a liquid phase at temperatures as low as 50 K provide strong support for Ashcroft's low-temperature liquid metallic state of hydrogen [444], and it implies that any room temperature superconductor in this regime would have to be a liquid.

In order to understand the role the QNEs play in inducing the properties discussed, it is informative to compare the results of the *ab initio* PIMD simulations with those obtained from the *ab initio* MD approach in which the nuclei are approximated by classical point-like particles. To this end, they have performed a second complete set of coexistence simulations with *ab initio* MD across the entire 500–1200 GPa range. The *ab initio* MD melting curve is shown by the grey data in the inset of Fig. 7.13, where it can be seen that the melting temperatures obtained from the MD simulations are much higher than those from the fully quantum PIMD simulations. The *ab initio* MD melting temperature is well above 200 K across the pressure range 500–1200 GPa, and the slope of the melting curve is small. A melting curve with a negative slope was also found above 90 GPa in the *ab initio* MD simulations of hydrogen by Bonev *et al.* [278] and above 10 GPa in lithium [454]. In Ref. [323], *ab initio* MD simulations with classical nuclei exhibit considerably higher melting temperatures than the *ab initio* PIMD ones at pressures above 500 GPa, which shows that the quantum description of the protons strongly depresses the melting point. The entropy arising from the greater delocalization of the protons in the quantum description has a crucial role in stabilizing the low-temperature liquid.

Before we end our discussion, it is worthwhile to note that serious analysis of the accuracy of the simulations should always be carried out in molecular simulations in general. Taking the low-temperature metallic liquid phase we discussed above as an example, the main conclusion is that the melting line of solid hydrogen has a negative slope, and that the quantum fluctuations of the nuclei lead to a low temperature (<50 K) metallic liquid phase at pressures higher than 900 GPa. We use the remaining part of this section to discuss the accuracy of the simulations from which these conclusions are drawn. This analysis includes: (i) the accuracy of the electronic structures, and (ii) the convergence of the *ab initio* MD and PIMD simulations with respect to the number of beads used in representing the finite-temperature imaginary-time path integral of the nuclei, the simulation cell size and simulation time, as well as the significance of nuclear exchange effects, a factor which is not accounted for in the PIMD method. Besides these, the superconducting properties of the solid atomic phase will also be discussed.

We start with the accuracy of the Brillouin-zone integrations and plane wave basis set cut-off energies used in our MD and PIMD calculations. In the main manuscript of Ref. [323], a Monkhorst–Pack k-point mesh of spacing $2\pi \times 0.05\,\text{Å}^{-1}$ was used for the Brillouin-zone integration and a 500 eV cut-off was used for the expansion of the electronic wave functions. Figure 7.14 shows the variation of the relative static lattice enthalpies of various relevant structures over the pressure range 500–1200 GPa. These results are in very good agreement with those reported in an even earlier study [451]. The molecular *Cmca* phase is found to be the most stable one at 500 GPa, and the phase transition from *Cmca* to the atomic *I41/amd* phase occurs at about 500 GPa. The *I41/amd* phase has the lowest static lattice enthalpy from about 500 GPa to over 1200 GPa.

Then one investigates how well the above k-point mesh spacing and cut-off energy perform when thermal and quantum fluctuations of the nuclei are included in the calculations. For this purpose, six snapshots (three from simulations for solids at low temperatures and three from liquid phases) were chosen at random from the PIMD simulations at 700 GPa. The centroid of each atom is used, and they have performed single point calculations for the total energy of these structures using a higher energy cut-off (600 eV) and a denser k-point mesh ($8 \times 8 \times 8$, which corresponds to a grid of spacing $2\pi \times 0.025\,\text{Å}^{-1}$ in the Monkhorst–Pack k-point mesh). The differences between the results obtained with these settings and those used in the MD and PIMD simulations (500 eV and a $4 \times 4 \times 4$ k-point mesh) are smaller than 1 meV, see Fig. 7.15(a). These errors are negligible compared with the

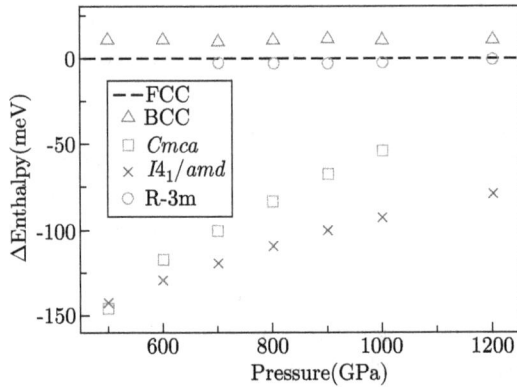

Figure 7.14   Static lattice ground state enthalpies of different crystal structures relative to FCC in solid hydrogen as a function of pressure.

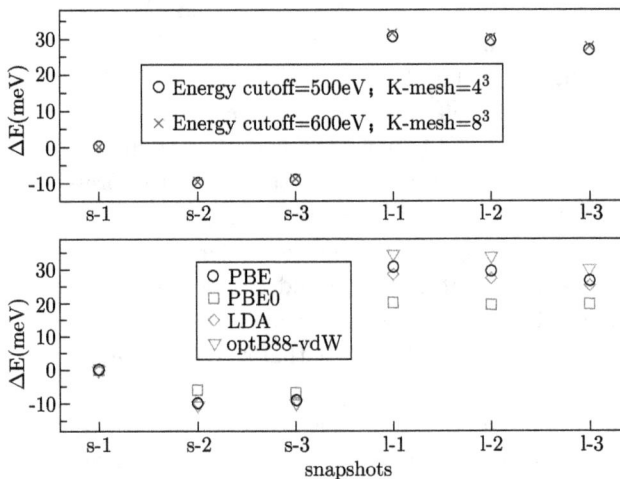

Figure 7.15   Single point total energies of snapshots from the thermalized state of the two-phase PIMD simulations at 700 GPa. The centroid position is used for simplicity. s-1, s-2, and s-3 correspond to snapshots at low temperature (100 K) with hydrogen in the solid $I41/amd$ phase. l-1, l-2, and l-3 correspond to snapshots of the liquid phase at high temperature (150 K).

energy differences of several tens of meV between the internal energy of the liquid and solid phases.

One notes that the PBE XC functional was used in MD and PIMD simulations reported. This functional suffers from self-interaction errors which can be significant in systems containing hydrogen. One can investigate the potential role of the self-interaction errors by comparing the total energies obtained with PBE and PBE0. PBE0 is a hybrid functional containing 25% Hartree–Fock exact exchange [455]. Therefore, the self-interaction error arising from the PBE0 functional is expected to be smaller than that of the PBE functional. In addition to this, comparisons with other functionals such as LDA and optB88-vdW [456, 457] within the van der Waals density functional (vdW-DF) scheme [458, 459] should also give some insights. Based on this consideration, in Fig. 7.15(b), we compare the relative energies of the six snapshots using LDA, PBE, optB88-vdW and PBE0. LDA and optB88-vdW give very similar results to the PBE ones. PBE0 gives lower total energies for the liquid phase than for the solid, and consequently, it favors melting of the solid. It is therefore likely that using a more accurate density functional than PBE would lead to stabilization of the liquid phase at even lower temperatures.

In PIMD simulations, the number of beads used to sample the imaginary-time path integral is a very important parameter in the description of the quantum nuclear effects. A series of tests at 700 GPa were therefore performed and the melting temperature was calculated using 1 (MD), 4, 8, 16, 24, 32, 48, and 64 beads to check if the melting temperature converges with respect to the number of beads. The results are shown in Fig. 7.16. We find that 32 beads are required to ensure that the melting temperature is converged within a window of 25 K. In Ref. [323], they have therefore used 32 beads for the main calculations reported in the main manuscript.

The results of the two-phase simulations also depend on the size and shape of the simulation cell. In Fig. 7.17, one checks the dependence of the results on the cell size in the MD simulations. Using a cell containing 200 atoms gives results identical to those from 432 to 576 atoms. In the PIMD simulation reported in Ref. [323], they have also compared results using 200 and 300 atoms and found that these simulations gave essentially identical results. Therefore, they believe that using a cell containing 200 atoms in the simulations is accurate (at least) for a qualitative description of the phenomena reported.

In the two-phase PIMD simulations, solidification and melting happen on a time scale of 1 ps. To ensure that the systems have equilibrated, Chen *et al.* have run all simulations for 10 ps and calculated the angularly averaged pair-distribution function $g(r)$, as explained above, using different time intervals. They already found very good convergence with respect to the simulation time at 3 ps (Fig. 7.18).

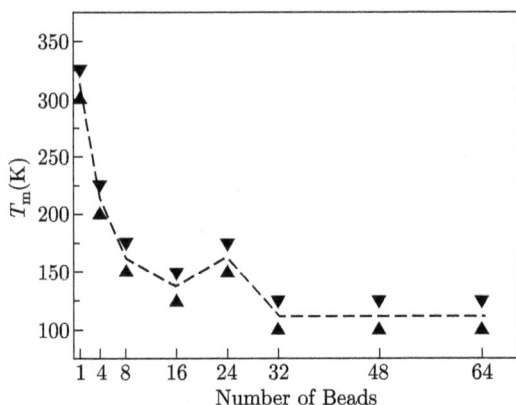

Figure 7.16   Melting temperatures calculated at 700 GPa using different numbers of beads. A bead number of 1 means an MD simulation. The upper and lower limits of the melting temperatures from the two-phase simulations are indicated by down and up triangles, respectively. The dashed line indicates the middle of the upper and lower limits.

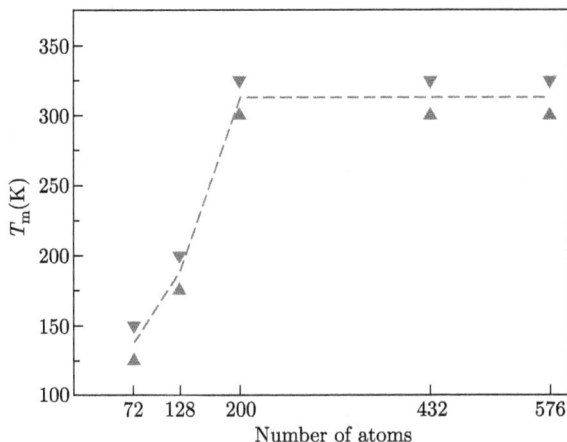

Figure 7.17   Melting temperatures calculated at 700 GPa using different numbers of atoms in the *ab initio* MD simulations. Upper and lower limits of melting temperatures from two-phase simulations are plotted with down and up triangles. The dashed line indicates the middle of the upper and lower limits.

In standard PIMD simulations, the exchange of nuclei is neglected. To estimate if the neglect of nuclear exchange effects has a significant effect on the accuracy of the simulations, Chen *et al.* have also examined the distributions of distances between beads in the same nucleus and between beads in neighboring nuclei. Results at 1200 GPa (900 GPa) and 50 K are

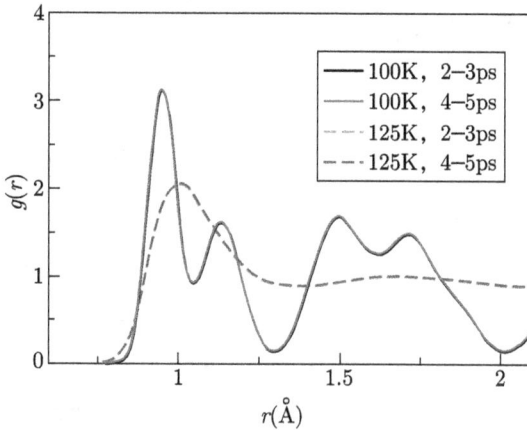

Figure 7.18   The angularly averaged pair-distribution function $g(r)$, as explained above, calculated using different intervals during a two-phase PIMD simulation of hydrogen at 700 GPa and different temperatures. Black (from 2 ps to 3 ps) and red solid (from 4 ps to 5 ps) lines give $g(r)$ from simulation at 100 K when the system solidifies. Green and blue dashed lines are results at 125 K when the hydrogen melts.

reported in Fig. 7.19, where the distribution of distances between bead 1 and bead $N/2+1$ of the ring polymer in the 32-bead simulation ($N = 32$) was compared with the distribution of H–H distances for the same bead. The distances between bead 1 and bead $N/2+1$ of the ring polymer give the solid curves and the H–H distances of the same bead give the dashed curves. The absence of any significant overlap between the peaks of these two curves in this highest pressure and lowest temperature simulation suggests that exchange effects are unlikely to be an issue in the simulations reported. As the pressure decreases from 1200 GPa to 900 GPa, the overlap of the two curves becomes even smaller.

The last property of interest in the simulations is the superconductivity feature of the hydrogen at this region of the phase diagram. It has been widely reported that metallic hydrogen formed at these pressures could be a high T_c superconductor [445, 446, 460]. Therefore, Chen et al. have also calculated T_c for the $I41/amd$ solid phase using the Allen–Dynes equation [461] and the QUANTUM-Espresso code [462]. They found that over the entire pressure range examined (500–1200 GPa), T_c is predicted to be around room temperature or above, which is consistent with previous predictions for $I41/amd$ at these pressures in Ref. [446]. At 500 GPa, for example, the conservative estimate of T_c is 358 K. To understand the physical

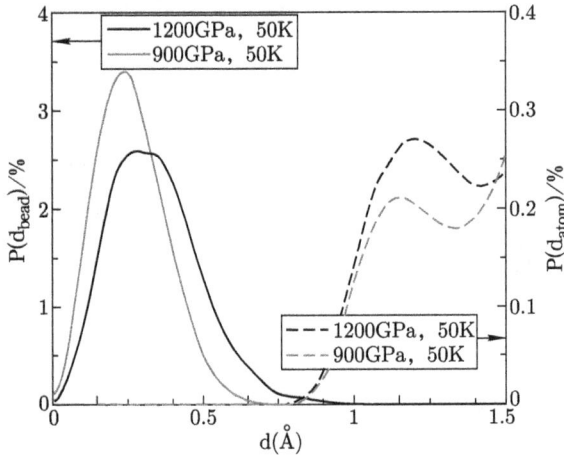

Figure 7.19 Probability distribution of the distances between the first and $N/2+1$(th)beads in the same atom (solid lines scale on left) and probability distribution of the distances between the first bead in two neighboring atoms of different molecules (dashed lines scale on right). Distributions are reported from a 32-bead two-phase PIMD simulation at 1200 GPa and 50 K, the highest pressure and lowest temperature case investigated, and for comparison, a simulation at 900 GPa and 50 K.

origin of this high T_c phase, we show details of its electronic and vibrational properties at 500 GPa in Fig. 7.20. This reveals a high electronic density of states (DOS) at the Fermi level (Fig. 7.20(a)), strong electron–phonon coupling (Fig. 7.20(d)), and consequently a high value of λ (2.15 as shown in Fig. 7.21) which leads to a high value of T_c within Bardeen–Cooper–Schrieffer (BCS) theory [463].

These results were obtained using a dense q-point mesh ($8 \times 8 \times 8$). Their convergence with the energy cut-off for the PAW pseudopotential is shown in Fig. 7.21. A cut-off energy of 80 Ryd gives good convergence for both T_c and the electron–phonon interaction parameter λ. In Fig. 7.21(b), we also plot T_c versus μ^* for the LDA and PBE functionals. We found that the LDA and PBE results for T_c are similar, and both of them give T_c values which are much higher than the melting temperature of the solid phase. In Ref. [323], we have used $\mu^* = 0.1$ to obtain the value of $T_c = 358$ K reported. This value of μ^* is close to the value of 0.085 obtained from the Bennemann–Garland formula [464] and larger than the value of 0.089 used in Ref. [446]. From Fig. 7.21(b), it is clear that T_c decreases with increasing μ^*. Considering the fact that they have chosen a large value of μ^* and that their value of T_c is still much higher than the melting temperature of the solid phase, it is

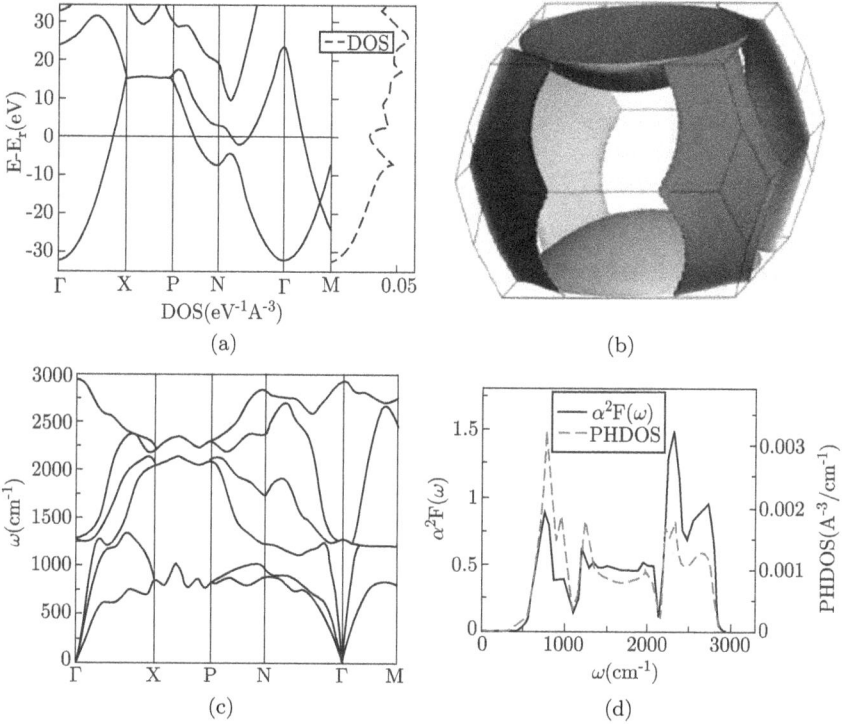

Figure 7.20   Electron and phonon properties of the $I41/amd$ structure of solid hydrogen at 500 GPa, with a volume of 2.28 Å^3 for its primitive cell: (a) electronic band structure and DOS, (b) Fermi surface in the Brillouin-zone, (c) phonon dispersion curves, and (d) phonon DOS (dashed line) and $\alpha^2 F(\omega)$ (solid line).

reasonable to suppose that the atomic solid phase under the melting line is superconducting. As the crystal melts well below room temperature, their results also rule out a room temperature superconducting phase of solid hydrogen at the pressures considered here and concluded that any room temperature superconductor in this regime would have to be a liquid. For more details concerning this study, please see Ref. [323].

7.5   Summary

In summary, we have discussed some extensions of the molecular simulation methods as introduced in the earlier chapters to descriptions of the QNEs in this chapter. The language we have used is the path integral representation

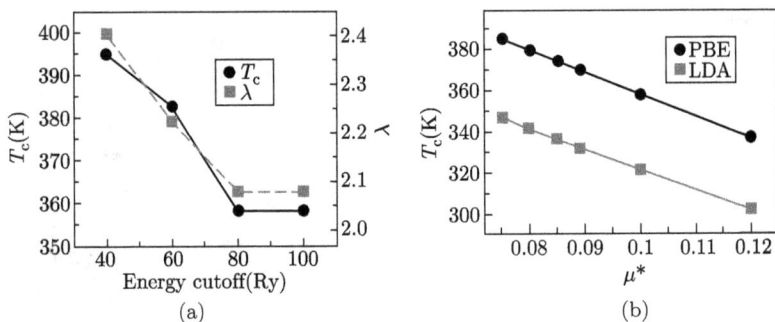

Figure 7.21   Superconductivity of the $I41/amd$ phase at 500 GPa: (a) superconducting critical temperatures T_c (circles) and the electron–phonon interaction parameter λ (squares) as a function of the plane wave cut-off energy using the PBE functional, and (b) superconducting critical temperatures T_c as a function of the effective Coulomb interaction parameter μ^* using the PBE (circle) and LDA (square) functionals.

of the quantum mechanics. Based on this language, the general theory behind the statistical PIMD and PIMC methods as well as their extensions to the dynamical regime were explained. A combination between the thermodynamic integration and PIMD methods was also presented, and some examples for the practical simulations of these computational methods were shown. These introductions, together with the computational methods for the calculation of the electronic structures and simulations of the molecular dynamics as presented in the earlier chapters, aim to set up a framework of concepts concerning molecular simulations of molecules and condensed matters. We sincerely hope this framework can help graduate students working on computer simulations of molecules and condensed matter find the proper method for tackling the problems of interest.

Appendix A: Useful Mathematical Relations

In this appendix, we provide a summary of useful mathematical formulae that are used throughout the text.

A.1 Spherical Harmonics

Definition: In the Condon Shortley phase convention, the spherical harmonics are defined as

$$Y_{l,m}(\theta, \phi) \equiv (-1)^m \sqrt{\frac{2l+1}{4\pi} \frac{(l-m)!}{(l+m)!}} P_l^m(\cos\theta) e^{im\phi}, \qquad \text{(A.1)}$$

where $P_l^m(x)$ is the corresponding Legendre polynomial (see Ref. [465]).

Recurrence relations

$$Y_{0,0}(\theta, \phi) = \sqrt{\frac{1}{4\pi}}, \qquad \text{(A.2a)}$$

$$Y_{1,0}(\theta, \phi) = \sqrt{\frac{3}{4\pi}} \cos(\theta), \qquad \text{(A.2b)}$$

$$Y_{1,1}(\theta, \phi) = -\sqrt{\frac{3}{8\pi}} \sin(\theta) e^{i\phi}, \qquad \text{(A.2c)}$$

$$Y_{1,-1}(\theta, \phi) = -Y_{1,1}(\theta, \phi), \qquad \text{(A.2d)}$$

$$Y_{l,l}(\theta, \phi) = -\sqrt{\frac{2l+1}{2l}} \sin(\theta) e^{i\phi} Y_{l-1,l-1}, \qquad \text{(A.2e)}$$

$$Y_{l,m}(\theta,\phi) = \sqrt{\frac{(2l-1)(2l+1)}{(l-m)(l+m)}}\cos(\theta)Y_{l-1,m}(\theta,\phi)$$

$$-\sqrt{\frac{(l-1+m)(l-1-m)(2l+1)}{(2l-3)(l-m)(l+m)}}Y_{l-2,m}(\theta,\phi). \qquad \text{(A.2f)}$$

Conjugation

$$Y_{l,-m}(\theta,\phi) = (-1)^m Y^*_{l,m}(\theta,\phi). \qquad \text{(A.3)}$$

Inversion

$$Y_{l,m}(\hat{r}) = (-1)^l Y_{l,m}(-\hat{r}). \qquad \text{(A.4)}$$

A.2   Plane Waves

Rayleigh expansion

$$e^{i\mathbf{G}\cdot\mathbf{r}} = 4\pi \sum_{\lambda=0}^{\infty} \sum_{\mu=-\lambda}^{+\lambda} i^\lambda j_\lambda(Gr) Y^*_{\lambda,\mu}(T^{-1}\hat{G}) Y_{\lambda,\mu}(T^{-1}\hat{r})$$

$$= 4\pi \sum_{\lambda=0}^{\infty} \sum_{\mu=-\lambda}^{+\lambda} i^\lambda j_\lambda(Gr) Y_{\lambda,\mu}(T^{-1}\hat{G}) Y^*_{\lambda,\mu}(T^{-1}\hat{r}). \qquad \text{(A.5)}$$

A.3   Fourier Transform

Definition: We use the following convention for the time-frequency Fourier Transform

$$F(\omega) = \int_{-\infty}^{\infty} F(t)e^{i\omega t}dt,$$

$$F(t) = \frac{1}{2\pi}\int_{-\infty}^{\infty} F(\omega)e^{-i\omega t}d\omega. \qquad \text{(A.6)}$$

Imaginary axes: The Fourier transform between imaginary axes work like its counterpart on the real axes, except that additional factors of $\pm i$

have to be included

$$F(i\omega) = -i \int_{-\infty}^{\infty} F(i\tau)e^{-i\omega\tau}d\tau,$$

$$F(i\tau) = \frac{i}{2\pi} \int_{-\infty}^{\infty} F(i\omega)e^{i\omega\tau}d\omega. \tag{A.7}$$

A.4 Spherical Coordinates

Derivatives

$$\partial x \pm i\partial y = \sin\theta e^{\pm i\phi}\frac{\partial}{\partial r} + \frac{e^{\pm i\phi}}{r}\left(\cos\theta\frac{\partial}{\partial\theta} \pm \frac{i}{\sin\theta}\frac{\partial}{\partial\phi}\right),$$

$$\partial z = \cos\theta\frac{\partial}{\partial r} - \frac{1}{r}\sin\theta\frac{\partial}{\partial\theta}. \tag{A.8}$$

A.5 The Step(Heaviside) Function

Definition:

$$\Theta(\mathbf{r}) = \begin{cases} 1, & \mathbf{r} \in \text{interstitial}, \\ 0, & \mathbf{r} \notin \text{interstitial}. \end{cases} \tag{A.9}$$

Since the step function $\Theta(\mathbf{r})$ has the periodicity of the lattice, we may expand it in a Fourier series as

$$\Theta(\mathbf{r}) = \sum_{\mathbf{G}} \tilde{\Theta}_{\mathbf{G}} e^{i\mathbf{G}\cdot\mathbf{r}}, \tag{A.10}$$

where $\tilde{\Theta}_{\mathbf{G}}$ can be calculated analytically, giving

$$\tilde{\Theta}_{\mathbf{G}} = \begin{cases} 1 - \sum_a \frac{4\pi r_a^3}{3\Omega}, & \mathbf{G} = 0, \\ -\frac{4\pi}{\Omega G} \sum_a j_1(Gr_a)\, r_a^2 e^{i\mathbf{G}\cdot\mathbf{r}_a}, & \mathbf{G} \neq 0. \end{cases} \tag{A.11}$$

Appendix B: Expansion of a Non-Local Function

In calculations of the non-local self-energy, many other non-local operators, such as the bare Coulomb potential, the polarizability, and the screened Coulomb potential, etc. need to be calculated in a matrix form for the basis set chosen. In this appendix, we discuss how these non-local operators are expanded.

We assume $f(\mathbf{r}_1, \mathbf{r}_2, \tau)$ is the general form of this non-local operator, where τ can be either a time or a frequency coordinate, or nothing. This non-local operator possesses a lattice translational symmetry with respect to $\mathbf{R}$ (e.g. v, W, P, ε, etc.)

$$f(\mathbf{r}_1 + \mathbf{R}, \mathbf{r}_2 + \mathbf{R}, \tau) = f(\mathbf{r}_1, \mathbf{r}_2, \tau). \tag{B.1}$$

To calculate a function of this type, we use the expansion in a complete set of Bloch functions $\{\chi_i^{\mathbf{q}}(\mathbf{r})\}$ in the following way:

$$\begin{cases} f(\mathbf{r}_1, \mathbf{r}_2, \tau) = \sum_{\mathbf{q}}^{BZ} \sum_{\mathbf{q}'}^{BZ} \sum_{i,j} \chi_i^{\mathbf{q}}(\mathbf{r}_1) f_{i,j}(\mathbf{q}, \mathbf{q}', \tau) (\chi_j^{\mathbf{q}'}(\mathbf{r}_1))^*, \\ f_{i,j}(\mathbf{q}, \mathbf{q}', \tau) = \int_V \int_V (\chi_i^{\mathbf{q}}(\mathbf{r}_1))^* f(\mathbf{r}_1, \mathbf{r}_2, \tau) \chi_j^{\mathbf{q}'}(\mathbf{r}_2) d\mathbf{r}_2 d\mathbf{r}_1. \end{cases} \tag{B.2}$$

Since $\chi_i^{\mathbf{q}}(\mathbf{r}_1)$ is a Bloch function $(\chi_i^{\mathbf{q}}(\mathbf{r} - \mathbf{R}) = e^{-i\mathbf{q}\cdot\mathbf{R}}\chi_i^{\mathbf{q}}(\mathbf{r}))$ normalized to unity in the crystal with volume V, the matrix element $f_{i,j}(\mathbf{q}, \mathbf{q}', \tau)$ can be evaluated by

$$f_{i,j}(\mathbf{q}, \mathbf{q}', \tau) = \int_V \int_V (\chi_i^{\mathbf{q}}(\mathbf{r}_1))^* f(\mathbf{r}_1, \mathbf{r}_2, \tau) \chi_j^{\mathbf{q}'}(\mathbf{r}_2) d\mathbf{r}_2 d\mathbf{r}_1$$

$$= \sum_{\mathbf{R}, \mathbf{R}'} \int_\Omega \int_\Omega (\chi_i^{\mathbf{q}}(\mathbf{r}_1 - \mathbf{R}))^* f(\mathbf{r}_1 - \mathbf{R}, \mathbf{r}_2 - \mathbf{R} - \mathbf{R}', \tau)$$

$$\times \chi_j^{q'}(\mathbf{r}_2 - \mathbf{R} - \mathbf{R}')d\mathbf{r}_2 d\mathbf{r}_1$$

$$= \sum_{\mathbf{R}} \int_{\Omega} \sum_{\mathbf{R}'} \int_{\Omega} e^{i\mathbf{q}\cdot\mathbf{R}} (\chi_i^{\mathbf{q}}(\mathbf{r}_1))^* f(\mathbf{r}_1, \mathbf{r}_2 - \mathbf{R}', \tau) e^{-i\mathbf{q}'\cdot\mathbf{R}} e^{-i\mathbf{q}'\cdot\mathbf{R}'}$$

$$\times \chi_j^{q'}(\mathbf{r}_2)d\mathbf{r}_2 d\mathbf{r}_1$$

$$= \sum_{\mathbf{R}} e^{i(\mathbf{q}-\mathbf{q}')\cdot\mathbf{R}} \int_{\Omega} \sum_{\mathbf{R}'} \int_{\Omega} (\chi_i^{\mathbf{q}}(\mathbf{r}_1))^* f(\mathbf{r}_1, \mathbf{r}_2 - \mathbf{R}', \tau)$$

$$\times e^{-i\mathbf{q}'\cdot\mathbf{R}'} \chi_j^{q'}(\mathbf{r}_2)d\mathbf{r}_2 d\mathbf{r}_1$$

$$= N_c \delta_{\mathbf{q},\mathbf{q}'} \int_{\Omega} \int_{\Omega} (\chi_i^{\mathbf{q}}(\mathbf{r}_1))^* \sum_{\mathbf{R}'} f(\mathbf{r}_1, \mathbf{r}_2 - \mathbf{R}', \tau)$$

$$\times e^{-i\mathbf{q}\cdot\mathbf{R}'} \chi_j^{\mathbf{q}}(\mathbf{r}_2)d\mathbf{r}_2 d\mathbf{r}_1, \tag{B.3}$$

where we have made use of this relation for the Bravais lattice

$$\sum_{\mathbf{R}} e^{-i(\mathbf{q}-\mathbf{q}')\cdot\mathbf{R}} = N_c \delta_{\mathbf{q},\mathbf{q}'}. \tag{B.4}$$

N_c is the number of cells in the crystal.

With these treatments, the expansion of Eq. (B.2) is written as

$$\begin{cases} f(\mathbf{r}_1, \mathbf{r}_2, \tau) = \sum_{\mathbf{q}}^{BZ} \sum_{i,j} \chi_i^{\mathbf{q}}(\mathbf{r}_1) f_{i,j}(\mathbf{q}, \tau)(\chi_j^{\mathbf{q}}(\mathbf{r}_2))^*, \\ f_{i,j}(\mathbf{q}, \tau) = \int_V \int_V (\chi_i^{\mathbf{q}}(\mathbf{r}_1))^* f(\mathbf{r}_1, \mathbf{r}_2, \tau) \chi_j^{\mathbf{q}}(\mathbf{r}_2)d\mathbf{r}_2 d\mathbf{r}_1, \end{cases} \tag{B.5}$$

where the integration must be done on the whole volume of the crystal, or

$$\begin{cases} f(\mathbf{r}_1, \mathbf{r}_2, \tau) = \sum_{\mathbf{q}}^{BZ} \sum_{i,j} \chi_i^{\mathbf{q}}(\mathbf{r}_1) f_{i,j}(\mathbf{q}, \tau)(\chi_j^{\mathbf{q}}(\mathbf{r}_2))^*, \\ f_{i,j}(\mathbf{q}, \tau) = N_c \int_{\Omega} \int_{\Omega} (\chi_i^{\mathbf{q}}(\mathbf{r}_1))^* \sum_{\mathbf{R}} f(\mathbf{r}_1, \mathbf{r}_2 - \mathbf{R}, \tau) e^{-i\mathbf{q}\cdot\mathbf{R}} \chi_j^{\mathbf{q}}(\mathbf{r}_2)d\mathbf{r}_2 d\mathbf{r}_1. \end{cases} \tag{B.6}$$

The integration must be performed only on the Wigner–Seitz cell.

If we have a product of operators, say

$$h(\mathbf{r}_1, \mathbf{r}_2, \tau) = \int_V f(\mathbf{r}_1, \mathbf{r}_3, \tau) g(\mathbf{r}_3, \mathbf{r}_2, \tau)d\mathbf{r}_3, \tag{B.7}$$

then, according to Eq. (B.5), the expansion of h in the set of functions $\{\chi_i^{\mathbf{q}}(\mathbf{r})\}$ is

$$h_{i,j}(\mathbf{q}, \tau) = \int_V \int_V [\chi_i^{\mathbf{q}}(\mathbf{r}_1)]^* h(\mathbf{r}_1, \mathbf{r}_2, \tau) \chi_j^{\mathbf{q}}(\mathbf{r}_2) d\mathbf{r}_2 d\mathbf{r}_1$$

$$= \int_V \int_V [\chi_i^{\mathbf{q}}(\mathbf{r}_1)]^* \left(\int_V f(\mathbf{r}_1, \mathbf{r}_3, \tau) g(\mathbf{r}_3, \mathbf{r}_2, \tau) d\mathbf{r}_3 \right) \chi_j^{\mathbf{q}}(\mathbf{r}_2) d\mathbf{r}_2 d\mathbf{r}_1.$$

$$\text{(B.8)}$$

We can now use the second line of Eq. (B.5) for f and g and the orthogonality of the basis to get

$$h_{i,j}(\mathbf{q}, \tau) = \int_V \int_V [\chi_i^{\mathbf{q}}(\mathbf{r}_1)]^* \left(\int_V f(\mathbf{r}_1, \mathbf{r}_3, \tau) g(\mathbf{r}_3, \mathbf{r}_2, \tau) d\mathbf{r}_3 \right) \chi_j^{\mathbf{q}}(\mathbf{r}_2) d\mathbf{r}_2 d\mathbf{r}_1$$

$$= \int_V \int_V [\chi_i^{\mathbf{q}}(\mathbf{r}_1)]^* \left(\int_V \sum_{\mathbf{q}_1}^{\mathrm{BZ}} \sum_{l,m} \chi_l^{\mathbf{q}_1}(\mathbf{r}_1) f_{l,m}(\mathbf{q}_1, \tau) [\chi_m^{\mathbf{q}_1}(\mathbf{r}_3)]^* \right.$$

$$\times \left. \sum_{\mathbf{q}_2}^{\mathrm{BZ}} \sum_{n,p} \chi_n^{\mathbf{q}_2}(\mathbf{r}_3) g_{n,p}(\mathbf{q}_2, \tau) [\chi_p^{\mathbf{q}_2}(\mathbf{r}_2)]^* d\mathbf{r}_3 \right) \chi_j^{\mathbf{q}}(\mathbf{r}_2) d\mathbf{r}_2 d\mathbf{r}_1$$

$$= \sum_{\mathbf{q}_1}^{\mathrm{BZ}} \sum_{\mathbf{q}_2}^{\mathrm{BZ}} \sum_{l,m} \sum_{n,p} \left(\int_V [\chi_i^{\mathbf{q}}(\mathbf{r}_1)]^* \chi_l^{\mathbf{q}_1}(\mathbf{r}_1) d\mathbf{r}_1 \right) f_{l,m}(\mathbf{q}_1, \tau)$$

$$\times \left(\int_V [\chi_m^{\mathbf{q}_1}(\mathbf{r}_3)]^* \chi_n^{\mathbf{q}_2}(\mathbf{r}_3) d\mathbf{r}_3 \right) g_{n,p}(\mathbf{q}_2, \tau)$$

$$\times \left(\int_V [\chi_p^{\mathbf{q}_2}(\mathbf{r}_2)]^* \chi_j^{\mathbf{q}}(\mathbf{r}_2) d\mathbf{r}_2 \right)$$

$$= \sum_{\mathbf{q}_1}^{\mathrm{BZ}} \sum_{\mathbf{q}_2}^{\mathrm{BZ}} \sum_{l,m} \sum_{n,p} \delta(\mathbf{q}, \mathbf{q}_1) \delta_{i,l} f_{l,m}(\mathbf{q}_1, \tau)$$

$$\times \delta(\mathbf{q}_1, \mathbf{q}_2) \delta_{m,n} g_{n,p}(\mathbf{q}_2, \tau) \delta(\mathbf{q}_2, \mathbf{q}) \delta_{pj}. \tag{B.9}$$

With this, one arrives at the expected expression

$$h_{i,j}(\mathbf{q}, \tau) = \sum_l f_{i,l}(\mathbf{q}, \tau) g_{l,j}(\mathbf{q}, \tau). \tag{B.10}$$

Appendix C: The Brillouin-Zone Integration

In Sec. 4.4.4, we introduced the ideas of the $\mathbf{q}$-dependent linear tetrahedron method for the calculation of the polarization matrix. Compared with the one introduced by Rath and Freeman in Ref. [223], the frequency dependence and the variation of the operator to be integrated within each tetrahedron are included. For a clear illustration, we begin with the formula of the traditional linear tetrahedron method and the idea of isoparametric transformation. Then we extend these ideas to the $\mathbf{q}$-dependent case and illustrate the different configurations of the possible integration region. The frequency dependence is discussed in the end.

C.1   The Linear Tetrahedron Method

The task of this Brillouin-zone integration is to calculate the average expectation value of an operator satisfying the form

$$\langle X \rangle = \frac{1}{V_G} \sum_n \int_{V_G} X_n(\mathbf{k}) f(\epsilon_n(\mathbf{k})) d\mathbf{k}, \qquad (C.1)$$

where

$$X_n(\mathbf{k}) = \langle \varphi_n(\mathbf{k}) | X | \varphi_n(\mathbf{k}) \rangle. \qquad (C.2)$$

This $X_n(\mathbf{k})$ is the expectation value of this operator on the state $(n, \mathbf{k})$. V_G is the volume of the reciprocal unit cell. $f(\epsilon)$ is the Fermi function. An exact evaluation of Eq. (C.1) requires calculating the expectation value of this operator over all its occupied states, including an infinite number of $\mathbf{k}$ points in the Brillouin zone. In practice, this average expectation value is determined from a set of sample points in the Brillouin zone; each has

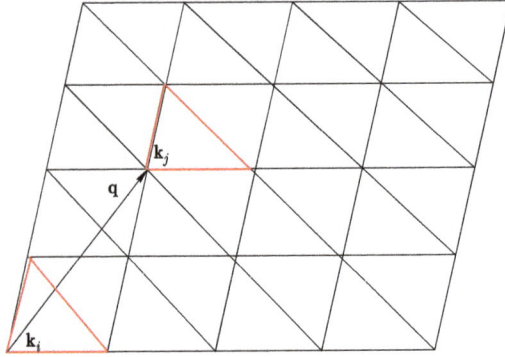

Figure C.1   The 2D sketch of the BZ in the tetrahedron method. In this case, the space is divided into a list of triangles. One triangle is related to another by a vector **q**.

a certain weight addressing the integration of Eq. (C.1) over the region around it.

In the tetrahedron method, this is obtained by dividing the Brillouin zone into a set of tetrahedra using a grid (as shown in Fig. C.1 for the 2D case). The values of $X_n(\mathbf{k})$ are calculated on the discrete set of vectors $\mathbf{k}_i$ at the vertices of all these tetrahedra, namely the grid points. A function $\bar{X}_n(\mathbf{k})$ obtained by linearly interpolating the function $X_n(\mathbf{k})$ within the tetrahedra using its expectation values on the vertices can be written as a superposition of functions $w_i(\mathbf{k})$, such that

$$\bar{X}_n(\mathbf{k}) = \sum_i X_n(\mathbf{k}_i) w_i(\mathbf{k}), \qquad (C.3)$$

where $w_i(\mathbf{k}_j) = \delta_{ij}$ and it is linear within the corresponding tetrahedron and zero outside of it. Now, replacing $X_n(\mathbf{k})$ in Eq. (C.1) by its linear approximation, one has

$$\langle X \rangle \cong \frac{1}{V_G} \sum_n \int_{V_G} \bar{X}_n(\mathbf{k}) f(\epsilon_n(\mathbf{k})) d\mathbf{k}$$

$$= \frac{1}{V_G} \sum_n \int_{V_G} \sum_i X_n(\mathbf{k}_i) w_i(\mathbf{k}) f(\epsilon_n(\mathbf{k})) d\mathbf{k}$$

$$= \sum_n \sum_i X_n(\mathbf{k}_i) \frac{1}{V_G} \int_{V_G} w_i(\mathbf{k}) f(\epsilon_n(\mathbf{k})) d\mathbf{k}.$$

Defining

$$w_{n,i} = \frac{1}{V_G} \int_{V_G} w_i(\mathbf{k}) f(\epsilon_n(\mathbf{k})) d\mathbf{k}, \qquad (C.4)$$

one can write the average expectation value of X in Eq. (C.1) as a weighted sum over the discrete set of $\mathbf{k}$ points

$$\langle X \rangle = \sum_{i,n} X_n(\mathbf{k}_i) w_{n,i}. \qquad (C.5)$$

Since $w_i(\mathbf{k})$ is zero for all $\{\mathbf{k}_j\}$ except $\mathbf{k}_i$, we can rewrite the weights as

$$w_{n,i} = \frac{1}{V_G} \sum_{T_i} \iiint_{V_T} w_i(\mathbf{k}) f(\epsilon_n(\mathbf{k})) d^3k = \sum_{T_i} w_{n,i}^{1T}, \qquad (C.6)$$

where T_i means that the sum runs only over those tetrahedra containing $\mathbf{k}_i$ as one of its vertices, and one has defined

$$w_{n,i}^{1T} = \frac{1}{V_G} \iiint_{V_T} w_i(\mathbf{k}) f(\epsilon_n(\mathbf{k})) d^3k.$$

With this, it is clear that the integration in Eq. (C.1) can be approximated by a sum of the form in Eq. (C.5) where $w_{n,i}$ can be calculated by summing its contribution from each tetrahedron containing this $\mathbf{k}_i$ as a vertex. The next job is to define the function $w_i(\mathbf{k})$ in order to calculate these $w_{n,i}$. For this, one needs the isoparametric transformation.

C.1.1   The Isoparametric Transfromation

In Eq. (C.3), the function behavior is approximated inside each tetrahedron by a linear interpolation between the function values at the vertices. Let $\mathcal{F}$ be such a function, and x, y and z be the coordinates, then

$$\mathcal{F} = Ax + By + Cz + D, \qquad (C.7)$$

where the constants A, B, C and D are to be determined. Substituting $x = x_i$, $y = y_i$ and $z = z_i$ where $i = 0, 1, 2, 3$ label the vertices, the values of $\mathcal{F}_i$ at the vertices (which are known) can be written as

$$\mathcal{F}_i = Ax_i + By_i + Cz_i + D. \tag{C.8}$$

Clearly, Eq. (C.8) for $i = 0$ can be used to eliminate the constant D. Then we have

$$\mathcal{F} - \mathcal{F}_0 = A(x - x_0) + B(y - y_0) + C(z - z_0). \tag{C.9}$$

The constants A, B and C are determined by solving the system of equations:

$$\begin{aligned}
\mathcal{F}_1 - \mathcal{F}_0 &= A(x_1 - x_0) + B(y_1 - y_0) + C(z_1 - z_0), \\
\mathcal{F}_2 - \mathcal{F}_0 &= A(x_2 - x_0) + B(y_2 - y_0) + C(z_2 - z_0), \\
\mathcal{F}_3 - \mathcal{F}_0 &= A(x_3 - x_0) + B(y_3 - y_0) + C(z_3 - z_0),
\end{aligned} \tag{C.10}$$

with solution

$$\begin{pmatrix} A \\ B \\ C \end{pmatrix} = \begin{pmatrix} x_1 - x_0 & y_1 - y_0 & z_1 - z_0 \\ x_2 - x_0 & y_2 - y_0 & z_2 - z_0 \\ x_3 - x_0 & y_3 - y_0 & z_3 - z_0 \end{pmatrix}^{-1} \begin{pmatrix} \mathcal{F}_1 - \mathcal{F}_0 \\ \mathcal{F}_2 - \mathcal{F}_0 \\ \mathcal{F}_3 - \mathcal{F}_0 \end{pmatrix}. \tag{C.11}$$

If one defines a coordinate (ξ, η, ζ) inside this tetrahedron, with each vertex 0, 1, 2, 3 having coordinates (0,0,0), (1,0,0), (0,1,0), (0,0,1), respectively, the function $\mathcal{F}$ can be linearly interpolated as

$$\mathcal{F} - \mathcal{F}_0 = \xi(\mathcal{F}_1 - \mathcal{F}_0) + \eta(\mathcal{F}_2 - \mathcal{F}_0) + \zeta(\mathcal{F}_3 - \mathcal{F}_0). \tag{C.12}$$

Putting Eq. (C.10) into the above equation, we have

$$\begin{aligned}
\mathcal{F} - \mathcal{F}_0 &= \begin{pmatrix} \xi & \eta & \zeta \end{pmatrix} \begin{pmatrix} \mathcal{F}_1 - \mathcal{F}_0 \\ \mathcal{F}_2 - \mathcal{F}_0 \\ \mathcal{F}_3 - \mathcal{F}_0 \end{pmatrix} \\
&= \begin{pmatrix} \xi & \eta & \zeta \end{pmatrix} \begin{pmatrix} x_1 - x_0 & y_1 - y_0 & z_1 - z_0 \\ x_2 - x_0 & y_2 - y_0 & z_2 - z_0 \\ x_3 - x_0 & y_3 - y_0 & z_3 - z_0 \end{pmatrix} \begin{pmatrix} A \\ B \\ C \end{pmatrix}.
\end{aligned} \tag{C.13}$$

On the other hand, Eq. (C.9) can be written as

$$\mathcal{F} - \mathcal{F}_0 = \begin{pmatrix} x - x_0 & y - y_0 & z - z_0 \end{pmatrix} \begin{pmatrix} A \\ B \\ C \end{pmatrix}. \tag{C.14}$$

Comparing Eq. (C.14) with Eq. (C.13), we have

$$x - x_0 = \xi(x_1 - x_0) + \eta(x_2 - x_0) + \zeta(x_3 - x_0),$$
$$y - y_0 = \xi(y_1 - y_0) + \eta(y_2 - y_0) + \zeta(y_3 - y_0), \tag{C.15}$$
$$z - z_0 = \xi(z_1 - z_0) + \eta(z_2 - z_0) + \zeta(z_3 - z_0).$$

Combining Eq. (C.15) with Eq. (C.12), we see that the *same* expression holds for the function $\mathcal{F}$ as well as for the coordinates x, y, and z. This coordinate transition from outside the tetrahedron to inside the tetrahedron is called as an isoparametric transformation. The functions $w_i(\mathbf{k})$ used in Eq. (C.3) can be simply written as

$$w_0(\xi, \eta, \zeta) = 1 - \xi - \eta - \zeta,$$
$$w_1(\xi, \eta, \zeta) = \xi,$$
$$w_2(\xi, \eta, \zeta) = \eta, \tag{C.16}$$
$$w_3(\xi, \eta, \zeta) = \zeta$$

in terms of this internal coordinates. The energy eigenvalue of the state $(n, \mathbf{k})$ with the coordinate (ξ, η, ζ) inside this tetrahedron is linearly interpolated as

$$\epsilon_n(\xi, \eta, \zeta) = (\epsilon_{n,1} - \epsilon_{n,0})\xi + (\epsilon_{n,2} - \epsilon_{n,0})\eta + (\epsilon_{n,2} - \epsilon_{n,0})\zeta + \epsilon_{n,0}, \tag{C.17}$$

where $\epsilon_{n,i}$ is the energy eigenvalue on the vertex i.

C.1.2   Integrals in One Tetrahedron

The integral of any function $\mathcal{F}$ inside one tetrahedron, after applying the isoparametric transformation, is given by

$$\iiint\limits_{V_T} \mathcal{F}(x,y,z)f(\epsilon_n(x,y,z))\mathrm{d}x\mathrm{d}y\mathrm{d}z$$

$$= \int_0^1 \int_0^{1-\zeta} \int_0^{1-\zeta-\eta} \left[\xi(\mathcal{F}_1 - \mathcal{F}_0) + \eta(\mathcal{F}_2 - \mathcal{F}_0)\right.$$

$$\left. + \zeta(\mathcal{F}_3 - \mathcal{F}_0) + \mathcal{F}_0\right] \left|\frac{\partial(xyz)}{\partial(\xi\eta\zeta)}\right| f(\epsilon_n(\xi,\eta,\zeta))\mathrm{d}\xi\mathrm{d}\eta\mathrm{d}\zeta,$$

$$\text{(C.18)}$$

where V_T is the volume of the tetrahedron and $\left|\dfrac{\partial(xyz)}{\partial(\xi\eta\zeta)}\right|$ is the Jacobian determinant given by

$$\left|\frac{\partial(xyz)}{\partial(\xi\eta\zeta)}\right| = \begin{vmatrix} \frac{\partial x}{\partial\xi} & \frac{\partial x}{\partial\eta} & \frac{\partial x}{\partial\zeta} \\ \frac{\partial y}{\partial\xi} & \frac{\partial y}{\partial\eta} & \frac{\partial y}{\partial\zeta} \\ \frac{\partial z}{\partial\xi} & \frac{\partial z}{\partial\eta} & \frac{\partial z}{\partial\zeta} \end{vmatrix} = \begin{vmatrix} x_1 - x_0 & y_1 - y_0 & z_1 - z_0 \\ x_2 - x_0 & y_2 - y_0 & z_2 - z_0 \\ x_3 - x_0 & y_3 - y_0 & z_3 - z_0 \end{vmatrix}. \qquad \text{(C.19)}$$

This is just the volume of a parallelepiped whose sides are given by those of the tetrahedron, clearly

$$\left|\frac{\partial(xyz)}{\partial(\xi\eta\zeta)}\right| = 6V_T. \qquad \text{(C.20)}$$

Then Eq. (C.18) is just

$$\iiint\limits_{V_T} \mathcal{F}(x,y,z)\mathrm{d}x\mathrm{d}y\mathrm{d}z$$

$$= 6V_T \int_0^1 \int_0^{1-\zeta} \int_0^{1-\zeta-\eta} \left[\xi(\mathcal{F}_1 - \mathcal{F}_0) + \eta(\mathcal{F}_2 - \mathcal{F}_0) + \zeta(\mathcal{F}_3 - \mathcal{F}_0) + \mathcal{F}_0\right]$$

$$\cdot f(\epsilon_n(\xi,\eta,\zeta))\mathrm{d}\xi\mathrm{d}\eta\mathrm{d}\zeta. \qquad \text{(C.21)}$$

C.1.3   The Integration Weights

Let us take one of the tetrahedra with its four vertices denoted as 0, 1, 2, and 3. Using the $w_i(\mathbf{k})$ and $\epsilon_n(\mathbf{k})$ defined in Eq. (C.16) and Eq. (C.17), one can calculate the integration weights on these vertices. If the four energies are below the Fermi energy, the occupation is identically one and we have

$$
w_{n,i}^{1T} = \frac{6V_T}{V_G} \int_0^1 \int_0^{1-\zeta} \int_0^{1-\zeta-\eta} \zeta \, d\xi d\eta d\zeta = \frac{6V_T}{V_G} \int_0^1 \int_0^{1-\zeta} \zeta(1-\zeta-\eta) d\eta d\zeta
$$

$$
= \frac{6V_T}{V_G} \int_0^1 \frac{1}{2}\zeta(1-\zeta)^2 d\zeta = \frac{3V_T}{V_G}\left(\frac{1}{2} - \frac{2}{3} + \frac{1}{4}\right) = \frac{V_T}{4V_G}.
$$

$$(C.22)$$

Let us now take the case where only $\epsilon_{n,0} < \epsilon_F$ and, for the sake of simplicity, $\epsilon_{n,3} > \epsilon_{n,2} > \epsilon_{n,1} > \epsilon_{n,0}$, then the integration limits are changed, and one gets

$$
w_{n,3}^{1T} = \frac{6V_T}{V_G} \int_0^{\frac{\epsilon_F - \epsilon_{n,0}}{\epsilon_{n,3} - \epsilon_{n,0}}} \int_0^{\frac{\epsilon_F - \epsilon_{n,0} - \zeta(\epsilon_{n,3} - \epsilon_{n,0})}{\epsilon_{n,2} - \epsilon_{n,0}}} \int_0^{\frac{\epsilon_F - \epsilon_{n,0} - \zeta(\epsilon_{n,3} - \epsilon_{n,0}) - \eta(\epsilon_{n,2} - \epsilon_{n,0})}{\epsilon_{n,1} - \epsilon_{n,0}}}
$$

$$
\times \zeta d\xi d\eta d\zeta
$$

$$
= \frac{V_T}{4V_G} \frac{(\epsilon_F - \epsilon_{n,0})^4}{(\epsilon_{n,1} - \epsilon_{n,0})(\epsilon_{n,2} - \epsilon_{n,0})(\epsilon_{n,3} - \epsilon_{n,0})^2}. \qquad (C.23)
$$

A similar calculation for the rest of the vertices leads to

$$
w_{n,2}^{1T} = \frac{V_T}{4V_G} \frac{(\epsilon_F - \epsilon_{n,0})^4}{(\epsilon_{n,1} - \epsilon_{n,0})(\epsilon_{n,2} - \epsilon_{n,0})^2(\epsilon_{n,3} - \epsilon_{n,0})},
$$

$$
w_{n,1}^{1T} = \frac{V_T}{4V_G} \frac{(\epsilon_F - \epsilon_{n,0})^4}{(\epsilon_{n,1} - \epsilon_{n,0})^2(\epsilon_{n,2} - \epsilon_{n,0})(\epsilon_{n,3} - \epsilon_{n,0})},
$$

$$
w_{n,0}^{1T} = \frac{V_T}{V_G} \frac{(\epsilon_F - \epsilon_{n,0})^3}{(\epsilon_{n,1} - \epsilon_{n,0})(\epsilon_{n,2} - \epsilon_{n,0})(\epsilon_{n,3} - \epsilon_{n,0})} - w_{n,1}^{1T} - w_{n,2}^{1T} - w_{n,3}^{1T}.
$$

$$(C.24)$$

The last line in Eq. (C.24) can be calculated using $w_0(\mathbf{k}) = 1 - \xi - \eta - \zeta = w_t - w_1(\mathbf{k}) - w_2(\mathbf{k}) - w_3(\mathbf{k})$, where w_t means the total weight over this tetrahedron. Expressions for the remaining cases can be found in Ref. [224]. Since these vertices are also sample points in the grid mesh, the integration weight on each grid point can be calculated from Eq. (C.6).

C.2 Tetrahedron Method for q-Dependent Brillouin-Zone Integration

If one wants to calculate the mean value of a $\mathbf{q}$-dependent operator, the situation becomes more complicated. In this section, we discuss the case when the expectation value of this operator satisfies

$$\langle X(\mathbf{q}) \rangle = \frac{1}{V_G} \sum_{n,n'} \int_{V_G} X_{nn'}(\mathbf{k}, \mathbf{q}) f[\epsilon_n(\mathbf{k})] \left(1 - f[\epsilon_{n'}(\mathbf{k} - \mathbf{q})]\right) d^3k,$$

(C.25)

where

$$X_{n,n'}(\mathbf{k}, \mathbf{q}) = \langle \varphi_n(\mathbf{k}) | X(\mathbf{q}) | \varphi_{n'}(\mathbf{k} - \mathbf{q}) \rangle.$$

(C.26)

To evaluate this operator, one needs to know $X_{nn'}(\mathbf{k}, \mathbf{q})$ on each $\mathbf{k}$ point in the Brillouin zone in principle. In practice, again, this is obtained by calculating the expectation value of this operator on a set of sample points weighted by a certain factor. In addition to $\epsilon_n(\mathbf{k}_i)$ and $\varphi_n(\mathbf{k}_i)$ on the set of sample points $\{\mathbf{k}_i\}$, one also needs to know $\epsilon_{n'}(\mathbf{k}_i - \mathbf{q})$ and $\varphi_{n'}(\mathbf{k}_i - \mathbf{q})$ on another set of sample points $\{\mathbf{k}_i - \mathbf{q}\}$.

In Ref. [212], an even division of the Brillouin zone along each axis is made. Then, one takes the $\mathbf{q}$ vector from this mesh. With this treatment, the meshes of $\mathbf{k}_i$ and $\mathbf{k}_i - \mathbf{q}$ overlap totally with each other. One just needs to know the eigenwavefunctions and the energy eigenvalues in one mesh. A 2D sketch for the $\mathbf{k}$-mesh is shown in Fig. C.1.

Using this grid, the Brillouin zone is divided into a set of tetrahedra. The expectation values of the function $X_{n,n'}(\mathbf{k}, \mathbf{q})$ are calculated on the vertices of these tetrahedra, namely, the grid points, giving $X_{n,n'}(\mathbf{k}_i, \mathbf{q})$.

Following the same procedure as in the previous section, we interpolate the function $X_{n,n'}(\mathbf{k}, \mathbf{q})$ linearly within each tetrahedron using

$$\bar{X}_{n,n'}(\mathbf{k}, \mathbf{q}) = \sum_i X_{n,n'}(\mathbf{k}_i, \mathbf{q}) w_i(\mathbf{k}, \mathbf{q}),$$

(C.27)

where $w_i(\mathbf{k}_j, \mathbf{q}) = \delta_{i,j}$, and it is a linear function. Since the integration is over the vector $\mathbf{k}$ and this $w_i(\mathbf{k}_j, \mathbf{q})$ is only a function of the coordinates of $\mathbf{k}$ for a fixed $\mathbf{q}$, it is easy to see that we can get rid of the $\mathbf{q}$ dependence. Eq. (C.27) becomes

$$\bar{X}_{n,n'}(\mathbf{k}, \mathbf{q}) = \sum_i X_{n,n'}(\mathbf{k}_i, \mathbf{q}) w_i(\mathbf{k}).$$

(C.28)

For the expectation value, we get

$$\langle X(\mathbf{q}) \rangle = \sum_{i,n,n'} X_{n,n'}(\mathbf{k}_i, \mathbf{q}) w_{n,n',i}(\mathbf{q}),$$

(C.29)

with

$$w_{n,n',i}(\mathbf{q}) = \frac{1}{V_G} \int_{V_G} w_i(\mathbf{k}) f[\epsilon_n(\mathbf{k})] \left(1 - f[\epsilon_{n'}(\mathbf{k} - \mathbf{q})]\right) \mathrm{d}^3 k. \qquad (\text{C.30})$$

To calculate the weights, following the steps as in the previous section, one has

$$w_{n,n',i}(\mathbf{q}) = \sum_{T_i} w_{n,n',i}^{1T}(\mathbf{q}), \qquad (\text{C.31})$$

where

$$w_{n,n',i}^{1T}(\mathbf{q}) = \frac{1}{V_G} \iiint_{V_T} w_i(\mathbf{k}) f[\epsilon_n(\mathbf{k})] \left(1 - f[\epsilon_{n'}(\mathbf{k} - \mathbf{q})]\right) \mathrm{d}^3 k. \qquad (\text{C.32})$$

T_i runs over all the tetrahedra in which the sample point $\mathbf{k}_i$ serves as a vertex.

C.2.1 *Isoparametric Transformation*

Now, we perform the isoparametic transformation to calculate the integration of Eq. (C.32) in one tetrahedron. If we denote the vertices of this tetrahedron as 0, 1, 2, and 3, respectively, we have

$$
\begin{aligned}
w_0(\mathbf{k}) &= w_0(\xi, \eta, \zeta) = 1 - \xi - \eta - \zeta, \\
w_1(\mathbf{k}) &= w_1(\xi, \eta, \zeta) = \xi, \\
w_2(\mathbf{k}) &= w_2(\xi, \eta, \zeta) = \eta, \\
w_3(\mathbf{k}) &= w_3(\xi, \eta, \zeta) = \zeta, \\
\epsilon_n(\mathbf{k}) &= \epsilon_n(\xi, \eta, \zeta) = \xi(\epsilon_{n,1} - \epsilon_{n,0}) + \eta(\epsilon_{n,2} - \epsilon_{n,0}) \\
&\quad + \zeta(\epsilon_{n,3} - \epsilon_{n,0}) + \epsilon_{n,0}, \\
\epsilon_{n'}(\mathbf{k} - \mathbf{q}) &= \epsilon_{n'}(\xi, \eta, \zeta) = \xi(\epsilon_{n',1} - \epsilon_{n',0}) \\
&\quad + \eta(\epsilon_{n',2} - \epsilon_{n',0}) + \zeta(\epsilon_{n',3} - \epsilon_{n',0}) + \epsilon_{n',0},
\end{aligned}
\qquad (\text{C.33})
$$

where we have used the shorthand notation $\epsilon_{n,i}$ and $\epsilon_{n',i}$ to represent the energy eigenvalues of the state $(n, \mathbf{k})$ and $(n', \mathbf{k} - \mathbf{q})$ on the vertices of this tetrahedron.

Then, the general formula for the contribution of one tetrahedron to the weight is

$$
\begin{aligned}
w_{n,n',i}^{1T}&(\mathbf{q}) \\
&= \frac{6V_T}{V_G} \int_0^1 \int_0^{1-\zeta} \int_0^{1-\zeta-\eta} w_i(\xi,\eta,\zeta)\Theta[\epsilon_F - \xi(\epsilon_{n,1}-\epsilon_{n,0}) \\
&\quad - \eta(\epsilon_{n,2}-\epsilon_{n,0}) - \zeta(\epsilon_{n,3}-\epsilon_{n,0}) - \epsilon_{n,0}] \\
&\quad \times \Theta[\xi(\epsilon_{n',1}-\epsilon_{n',0}) + \eta(\epsilon_{n',2}-\epsilon_{n',0}) + \zeta(\epsilon_{n',3}-\epsilon_{n',0}) \\
&\quad + \epsilon_{n',0} - \epsilon_F]d\xi d\eta d\zeta,
\end{aligned} \tag{C.34}
$$

where Θ is the step function to address the Fermi function in Eq. (C.32).

C.2.2 The Integration Region

From Eq. (C.34), we see that the Θ functions determine the integration region within this tetrahedron. For insulators and semiconductors, this region is either the full tetrahedron or zero. For metals, the situation becomes more complicated. If the integration region does not include the full tetrahedron, that is, the $\epsilon_{n,i}$s are smaller or bigger than ϵ_F, the Fermi surface represented by the first Θ function in Eq. (C.34) will intersect with this tetrahedron, leading to only part of it satisfying the condition, the first Θ function equals one. Another case is when the $\epsilon_{n',i}$ are smaller or bigger than ϵ_F — the Fermi surface represented by the second Θ function in Eq. (C.34) will intersect with this tetrahedron, leading to only part of it satisfying the condition, and the second Θ function equals one. If neither of these cases happen, the integration region is either the full tetrahedron or zero. Otherwise, the integration region is determined by the intersection of these Fermi surfaces with this tetrahedron (Fig. 4.7 shows one example when both of them intersect with this tetrahedron).

There are in total nine different configurations for this region. They are shown in Fig. C.2 except for the simplest case of a tetrahedron. All of them can be subdivided into smaller tetrahedra. Then, we perform one further isoparametric transformation inside each of these small tetrahedra. The weight on each of its vertices is

$$
w_0 = \frac{6V_T}{V_G}\frac{V_{ST}}{V_T}\int_0^1\int_0^{1-z}\int_0^{1-y-z}(1-x-y-z)dxdydz = \frac{V_{ST}}{4V_G}, \tag{C.35}
$$

$$
w_1 = w_2 = w_3 = w_0,
$$

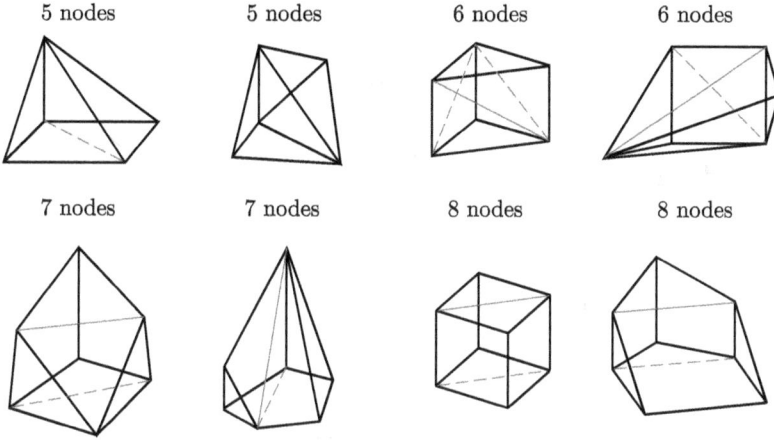

| 5 nodes | 5 nodes | 6 nodes | 6 nodes |

| 7 nodes | 7 nodes | 8 nodes | 8 nodes |

Figure C.2 The configurations for the region to be integrated. How these regions are decomposed into the principle units of the small tetrahedra is shown by the red lines in the graph.

where V_{ST} is the volume of the small tetrahedron, and w_i ($i = 0, 1, 2, 3$) represents the weight on each vertex. We further distribute these weights linearly into the vertices of the big tetrahedron. Assuming the coordinates of one vertex of this small tetrahedron is (ξ_1, η_1, ζ_1) in the big tetrahedron before the second parametric transformation, the integration weight on this point will be distributed with the ration $1 - \xi_1 - \eta_1 - \zeta_1$, ξ_1, η_1, and ζ_1 to the vertices 0, 1, 2, and 3 of the big tetrahedron.

C.2.3 *Polarizability*

As has already been mentioned in Sec. 4.4.4, for the polarizability, we cannot assume both the energies and the integrand to be simultaneously linear in the coordinates of the tetrahedron. In this case, we have to include the energy-dependent factor of Eq. (4.50) into the analytical integration. In Sec. 4.4.5, we have discussed the frequency integrations in the *GW* calculations, where we pointed out that we calculate all the frequency-dependent properties on the imaginary frequency axis. The polarizability is such a property. In this section, we will discuss the integration weight of the polarizability on both the real and imaginary frequency axes. The latter is the one used in the *GW* calculation. The former can be used to calculate the macroscopic dielectric constant.

C.2.3.1 *Polarizability on the Real Frequency Axis*

On the real frequency axis, the polarization matrix is

$$P_{i,j}(\mathbf{q}, \omega) = \frac{N_c}{\hbar} \sum_{\mathbf{k}}^{BZ} \sum_{n}^{occ} \sum_{n'}^{unocc} M_{n,n'}^i(\mathbf{k}, \mathbf{q}) [M_{n,n'}^i(\mathbf{k}, \mathbf{q})]^*$$

$$\cdot \left\{ \frac{1}{\omega - \epsilon_{n',\mathbf{k}-\mathbf{q}} + \epsilon_{n,\mathbf{k}} + i\eta} - \frac{1}{\omega - \epsilon_{n,\mathbf{k}} + \epsilon_{n',\mathbf{k}-\mathbf{q}} - i\eta} \right\}. \tag{C.36}$$

We define the weight as

$$w_{n,n',i}(\mathbf{q}, \omega) = \sum_{T_i} w_{n,n',i}^{1T}(\mathbf{q}, \omega), \tag{C.37}$$

where

$$w_{n,n',i}^{1T}(\mathbf{q}, \omega) = \frac{1}{V_G} \iiint_{V_T} w_i(\mathbf{k}) f[\epsilon_n(\mathbf{k})](1 - f[\epsilon_{n'}(\mathbf{k}-\mathbf{q})])$$

$$\times \left\{ \frac{1}{\omega - \epsilon_{n'}(\mathbf{k}-\mathbf{q}) + \epsilon_n(\mathbf{k}) + i\eta} - \frac{1}{\omega - \epsilon_n(\mathbf{k}) + \epsilon_{n'}(\mathbf{k}-\mathbf{q}) - i\eta} \right\} d^3\mathbf{k}. \tag{C.38}$$

Following the procedures in Sec. C.2.2, the weight on each vertex of the small tetrahedron is calculated by

$$w_0 = \frac{6V_{ST}}{V_G} \int_0^1 \int_0^{1-z} \int_0^{1-y-z} \frac{2(1-x-y-z)}{\omega^2 - (x\Delta_{1,0} + y\Delta_{2,0} + z\Delta_{3,0} + \Delta_0)^2} dxdydz,$$

$$w_1 = \frac{6V_{ST}}{V_G} \int_0^1 \int_0^{1-z} \int_0^{1-y-z} \frac{2x}{\omega^2 - (x\Delta_{1,0} + y\Delta_{2,0} + z\Delta_{3,0} + \Delta_0)^2} dxdydz,$$

$$w_2 = \frac{6V_{ST}}{V_G} \int_0^1 \int_0^{1-z} \int_0^{1-y-z} \frac{2y}{\omega^2 - (x\Delta_{1,0} + y\Delta_{2,0} + z\Delta_{3,0} + \Delta_0)^2} dxdydz,$$

$$w_3 = \frac{6V_{ST}}{V_G} \int_0^1 \int_0^{1-z} \int_0^{1-y-z} \frac{2z}{\omega^2 - (x\Delta_{1,0} + y\Delta_{2,0} + z\Delta_{3,0} + \Delta_0)^2} dxdydz. \tag{C.39}$$

Here, $\Delta_i = \epsilon_{n',i} - \epsilon_{n,i}$ and $\Delta_{i,j} = \Delta_i - \Delta_j$.

It is more complicated to solve the first equation in Eq. (C.39) analytically compared to the other three due to the presence of three variables in the numerator. So, we solve the other three respectively and then calculate the total integration weight over this tetrahedron with

$$
w_\text{t} = \frac{6V_\text{ST}}{V_\text{G}} \int_0^1 \int_0^{1-z} \int_0^{1-y-z} \frac{2}{\omega^2 - (x\Delta_{1,0} + y\Delta_{2,0} + z\Delta_{3,0} + \Delta_0)^2} dx\, dy\, dz.
$$

(C.40)

The correponding w_0 is then calculated from $w_0 = w_\text{t} - w_1 - w_2 - w_3$.

Even with this treatment, solving this analytical integration is very complicated. We use Mathematica to treat it. There exists a general solution. To restrict the size of this appendix, we just list that of w_t here, which is the simplest case due to the absence of variables in the numerator in Eq. (C.39):

$$
f(\omega) = (\omega - \Delta_3)^3 \Delta_{1,0}^2 \Delta_{2,0} \Delta_{2,1}^2 \ln[|\omega - \Delta_3|]
$$

$$
-(\omega - \Delta_2)^3 \Delta_{1,0}^2 \Delta_{3,0} \Delta_{3,1}^2 \ln[|\omega - \Delta_2|],
$$

$$
f(\omega) = f(\omega) + [\Delta_{1,0}\Delta_{2,1}(\omega - \Delta_3)
$$

$$
-(\omega - \Delta_0)\Delta_{2,1}\Delta_{3,1} + (\omega - \Delta_2)\Delta_{1,0}\Delta_{3,1}]
$$

$$
\times \Delta_{2,0}\Delta_{3,0}\Delta_{3,2}(\omega - \Delta_1)^2 \ln[|\omega - \Delta_1|],
$$

$$
f(\omega) = f(\omega) + (\omega - \Delta_0)^3 \Delta_{2,1}^2 \Delta_{3,2}\Delta_{3,1}^2 \ln[|\omega - \Delta_0|]
$$

$$
-(\omega - \Delta_1)^2 \Delta_{1,0}\Delta_{2,0} \times \Delta_{2,1}\Delta_{3,0}\Delta_{3,1}\Delta_{3,2},
$$

$$
f(\omega) = \frac{f(\omega)}{6\Delta_{1,0}^2 \Delta_{2,0}\Delta_{2,1}^2 \Delta_{3,0}\Delta_{3,1}^2 \Delta_{3,2}},
$$

$$
w_\text{t} = \frac{6V_\text{ST}}{V_\text{G}} [f(\omega) + f(-\omega)]
$$

(C.41)

(this equation is written following the programming rules because it is too long). In this equation, it is required that $\omega \neq \Delta_i$ and $\Delta_{i,j} \neq 0$. $\Delta_{i,j} \neq 0$ is required because of this analytical solution. $\omega \neq \Delta_i$ is required because the denominators in Eq. (C.39) and Eq. (C.40) cannot be zero. When these conditions are not fulfilled, we use Mathematica to get the analytical solution of that specific case, respectively.

C.2.3.2   *Polarizability on the Imaginary Frequency Axis*

The polarization matrix of our calculation on the imaginary frequency axis is

$$P_{i,j}(\mathbf{q},\omega)$$

$$= \frac{N_c}{\hbar} \sum_{\mathbf{k}}^{BZ} \sum_{n}^{occ} \sum_{n'}^{unocc} M_{n,n'}^{i}(\mathbf{k},\mathbf{q})[M_{n,n'}^{i}(\mathbf{k},\mathbf{q})]^{*} \frac{-2(\epsilon_{n',\mathbf{k-q}} - \epsilon_{n,\mathbf{k}})}{\omega^2 + (\epsilon_{n',\mathbf{k-q}} - \epsilon_{n,\mathbf{k}})^2}.$$

$$(\text{C.42})$$

In this case, the procedure is essentially the same as the above, except for the fact that the weight on the vertices of each small tetrahedron is calculated with

$$w_0 = \frac{6V_{\text{ST}}}{V_G} \int_0^1 \int_0^{1-z} \int_0^{1-y-z} \frac{2(1-x-y-z)}{\omega^2 + (x\Delta_{1,0} + y\Delta_{2,0} + z\Delta_{3,0} + \Delta_0)^2} dxdydz,$$

$$w_1 = \frac{6V_{\text{ST}}}{V_G} \int_0^1 \int_0^{1-z} \int_0^{1-y-z} \frac{2x}{\omega^2 + (x\Delta_{1,0} + y\Delta_{2,0} + z\Delta_{3,0} + \Delta_0)^2} dxdydz,$$

$$w_2 = \frac{6V_{\text{ST}}}{V_G} \int_0^1 \int_0^{1-z} \int_0^{1-y-z} \frac{2y}{\omega^2 + (x\Delta_{1,0} + y\Delta_{2,0} + z\Delta_{3,0} + \Delta_0)^2} dxdydz,$$

$$w_3 = \frac{6V_{\text{ST}}}{V_G} \int_0^1 \int_0^{1-z} \int_0^{1-y-z} \frac{2z}{\omega^2 + (x\Delta_{1,0} + y\Delta_{2,0} + z\Delta_{3,0} + \Delta_0)^2} dxdydz.$$

$$(\text{C.43})$$

Again, we introduce w_t as

$$w_t = \frac{6V_{ST}}{V_G} \int_0^1 \int_0^{1-z} \int_0^{1-y-z} \frac{2}{\omega^2 + (x\Delta_{1,0} + y\Delta_{2,0} + z\Delta_{3,0} + \Delta_0)^2} dxdydz$$

$$(\text{C.44})$$

to avoid solving the first equation of Eqs. (C.43) directly. Its general solution is

$$f(\omega) = 2(\omega - \Delta_0^2)\Delta_{0,1}\Delta_{0,2}\Delta_{0,3}\Delta_{1,2}\Delta_{1,3}\Delta_{2,3},$$

$$f(\omega) = f(\omega) + 2\omega[3\Delta_0^4 - \omega^2(\Delta_1\Delta_2 + \Delta_2\Delta_3 + \Delta_3\Delta_1)$$

$$-3\Delta_0^2(\omega^2 + \Delta_2\Delta_3 + \Delta_1\Delta_2 + \Delta_1\Delta_3)$$

$$+2\Delta_0(\omega^2\Delta_2 + \omega^2\Delta_3 + \Delta_1\omega^2 + 3\Delta_1\Delta_2\Delta_3)]$$

$$\times \Delta_{1,2}\Delta_{1,3}\Delta_{2,3}\arctan[\Delta_0/\omega],$$

$$f(\omega) = f(\omega) + 2\omega(\omega^2 - 3\Delta_1^2)\Delta_{0,2}^2\Delta_{0,3}^2\Delta_{2,3}\arctan[\Delta_1/\omega],$$

$$f(\omega) = f(\omega) - 2\omega(\omega^2 - 3\Delta_2^2)\Delta_{0,1}^2\Delta_{0,3}^2\Delta_{1,3}\arctan[\Delta_2/\omega],$$

$$f(\omega) = f(\omega) + 2\omega(\omega^2 - 3\Delta_3^2)\Delta_{0,1}^2\Delta_{0,2}^2\Delta_{1,2}\arctan[\Delta_3/\omega],$$

$$f(\omega) = f(\omega) + [\Delta_0^4(\Delta_1 + \Delta_2 + \Delta_3) - 3\omega^2\Delta_1\Delta_2\Delta_3$$

$$-2\Delta_0^3(3\omega^2 + \Delta_2\Delta_3 + \Delta_1\Delta_2 + \Delta_1\Delta_3)$$

$$+3\Delta_0^2(\omega^2\Delta_1 + \omega^2\Delta_2 + \omega^2\Delta_3 + \Delta_1\Delta_2\Delta_3)]$$

$$\times\Delta_{1,2}\Delta_{1,3}\Delta_{2,3}\ln[\omega^2 + \Delta_0^2],$$

$$f(\omega) = f(\omega) + \Delta_1(3\omega^2 - \Delta_1^2)\Delta_{0,2}^2\Delta_{0,3}^2\Delta_{2,3}\ln[\omega^2 + \Delta_1^2],$$

$$f(\omega) = f(\omega) + \Delta_2(\Delta_2^2 - 3\omega^2)\Delta_{0,1}^2\Delta_{0,3}^2\Delta_{1,3}\ln[\omega^2 + \Delta_2^2],$$

$$f(\omega) = f(\omega) - \Delta_3(\Delta_3^2 - 3\omega^2)\Delta_{0,1}^2\Delta_{0,2}^2\Delta_{1,2}\ln[\omega^2 + \Delta_3^2],$$

$$w_t = \frac{6V_{ST}}{V_G}\frac{f(\omega)}{6\Delta_{0,1}^2\Delta_{0,2}^2\Delta_{0,3}^2\Delta_{1,2}\Delta_{1,3}\Delta_{2,3}}. \tag{C.45}$$

Again, in this equation, it is required that $\Delta_{i,j} \neq 0$. When this condition is not fulfilled, we use Mathematica to get the analytical solution of that case again specifically, the same way we do in the above section.

Appendix D: The Frequency Integration

In this section, we show how to perform the integration on the frequency axis in Eq. (4.57).

This integration is peaked around $\omega' = \omega$ when $\epsilon_{n\mathbf{k}+\mathbf{q}}$ is small. To handle this problem, one can add and subtract the term

$$\frac{1}{\pi}\int_0^\infty \frac{(\epsilon_{n',\mathbf{k}+\mathbf{q}} - i\omega)W_{i,j}^c(\mathbf{q},i\omega)}{(i\omega - \epsilon_{n',\mathbf{k}+\mathbf{q}})^2 + \omega'^2}d\omega' = \frac{1}{2}\mathrm{sgn}(\epsilon_{n',\mathbf{k}+\mathbf{q}})W_{i,j}^c(\mathbf{q},i\omega). \qquad \text{(D.1)}$$

Then we have

$$\mathbb{I} = \frac{1}{\pi}\int_0^\infty \frac{(\epsilon_{n',\mathbf{k}+\mathbf{q}} - i\omega)\left[W_{i,j}^c(\mathbf{q},i\omega') - W_{i,j}^c(\mathbf{q},i\omega)\right]}{(i\omega - \epsilon_{n',\mathbf{k}+\mathbf{q}})^2 + \omega'^2}$$

$$\times\, d\omega' + \frac{1}{2}\mathrm{sgn}(\epsilon_{n',\mathbf{k}+\mathbf{q}})W_{i,j}^c(\mathbf{q},i\omega). \qquad \text{(D.2)}$$

The integrand is now smooth and a Gaussian quadrature may be used. To solve the semi-infinite integral of Eq. (D.2) which has the form

$$\mathbb{I} = \int_0^\infty f(\omega)d\omega, \qquad \text{(D.3)}$$

we split it into (following Ref. [225])

$$\mathbb{I} = \mathbb{I}_1 + \mathbb{I}_2, \qquad \text{(D.4a)}$$

$$\mathbb{I}_1 = \int_0^{\omega_0} f(\omega)d\omega, \qquad \text{(D.4b)}$$

$$\mathbb{I}_2 = \int_{\omega_0}^\infty f(\omega)d\omega. \qquad \text{(D.4c)}$$

243

For $\mathbb{I}_1$, we make the change of variables $u = 2\omega/\omega_0 - 1$ and thus, $d\omega = \dfrac{\omega_0}{2}du$. Then we have

$$\mathbb{I}_1 = \int_0^{\omega_0} f(\omega)d\omega = \frac{\omega_0}{2}\int_{-1}^{1} f[(u+1)\omega_0/2]du, \qquad (D.5)$$

which can be solved by standard Gauss–Legendre quadrature. For $\mathbb{I}_2$, we make the change of variables $u = 2\omega_0/\omega - 1$ and thus, $d\omega = -\dfrac{2\omega_0}{(u+1)^2}du$. Then we have

$$\mathbb{I}_2 = \int_{\omega_0}^{\infty} f(\omega)d\omega = 2\omega_0\int_{-1}^{1} f\left[\frac{2\omega_0}{u+1}\right](u+1)^{-2}du, \qquad (D.6)$$

which can also be solved by standard Gauss–Legendre quadrature.

References

1. R. M. Martin, *Electronic Structure, Basic Theory and Practical Methods* (Univ. Princeton, Cambridge, 2004).
2. M. Born and J. R. Oppenheimer, *Ann. Phys.* **84**, 457 (1927).
3. P. Ehrenfest, *Z. Phys.* **45**, 455 (1927).
4. H. D. Meyer and W. H. Miller, *J. Chem. Phys.* **70**, 3214 (1979).
5. J. C. Tully, *Faraday Discussions* **110**, 407 (1998).
6. X. S. Li, J. C. Tully, H. B. Schlegel, and M. J. Frisch, *J. Chem. Phys.* **123**, 084106 (2005).
7. Y. C. Wang, J. Lv, L. Zhu, and Y. M. Ma, *Phys. Rev. B* **82**, 094116 (2010).
8. Y. C. Wang, J. Lv, L. Zhu, and Y. M. Ma, *Comput. Phys. Commun.* **183**, 2063 (2012).
9. J. Lv, Y. C. Wang, L. Zhu, and Y. M. Ma, *J. Chem. Phys.* **137**, 084104 (2012).
10. Y. C. Wang, M. S. Miao, J. Lv, L. Zhu, K. T. Yin, H. Y. Liu, and Y. M. Ma, *J. Chem. Phys.* **137**, 224108 (2012).
11. X. Y. Luo, J. H. Yang, H. Y. Liu, X. J. Wu, Y. C. Wang, Y. M. Ma, S.-H. Wei, X. G. Gong, and H. J. Xiang, *J. Am. Chem. Soc.* **133**, 16285 (2011).
12. X. X. Zhang, Y. C. Wang, J. Lv, C. Y. Zhu, Q. Li, M. Zhang, Q. Li, and Y. M. Ma, *J. Chem. Phys.* **138**, 114101 (2013).
13. C. J. Pickard and R. J. Needs, *J. Phys.: Condens. Matter* **23**, 053201 (2011).
14. C. J. Pickard and R. J. Needs, *Physica Status Solidi* (b) **246**, 536 (2009).
15. C. J. Pickard and R. J. Needs, *Nat. Phys.* **3**, 473 (2007).
16. D. J. Jacobs, A. J. Rader, L. A. Kuhn, and M. F. Thorpe, *Proteins–Struct. Funct. Genet.* **44**, 150 (2001).
17. E. S. Huang, R. Samudrala, and J. W. Ponder, *J. Mol. Biol.* **290**, 267 (1999).
18. R. J. Trabanino, S. E. Hall, N. Vaidehi, W. B. Floriano, V. W. T. Kam, and W. A. Goddard, *Biophys. J.* **86**, 1904 (2004).
19. Y. Abashkin and N. Russo, *J. Chem. Phys.* **100**, 4477 (1994).
20. G. Mills, H. Jónsson, and G. K. Schenter, *Surf. Sci.* **324**, 305 (1995).
21. G. Henkelman and H. Jónsson, *J. Chem. Phys.* **113**, 9978 (2000).
22. S. A. Trygubenko and D. J. Wales, *J. Chem. Phys.* **120**, 2082 (2004).

23. M. J. S. Dewar, E. F. Healy, and J. J. P. Stewart, *J. Chem. Soc. Faraday Trans. II.* **80**, 227 (1984).
24. G. Henkelman and H. Jónsson, *J. Chem. Phys.* **111**, 7010 (1999).
25. B. Peters and A. Heyden, *J. Chem. Phys.* **120**, 7877 (2004).
26. A. Heyden, A. Bell, and F. J. Keil, *J. Chem. Phys.* **123**, 224101 (2005).
27. Y. Tateyama, T. Ogitsu, K. Kusakabe, and S. Tsuneyuki, *Phys. Rev. B* **54**, 14994 (1996).
28. G. T. Barkema and N. Mousseau, *Phys. Rev. Lett.* **77**, 4358 (1996).
29. J. Klimeš, D. R. Bowler, and A. Michaelides, *J. Phys.: Condens. Matter* **22**, 074203 (2010).
30. N. D. Mermin, *Phys. Rev.* **137**, A1441 (1965).
31. M. J. Gillan, *J. Phys.: Condens. Matter* **1**, 689 (1989).
32. R. M. Wentzcovitch, J. L. Martins, and P. B. Allen, *Phys. Rev. B* **45**, 11372 (1992).
33. C. Bartels and M. Karplus, *J. Phys. Chem. B* **102**, 865 (1998).
34. Y. Sugita and Y. Okamoto, *Chem. Phys. Lett.* **314**, 141 (1999).
35. Y. Sugita, A. Kitao, and Y. Okamoto, *J. Chem. Phys.* **113**, 6042 (2000).
36. P. Liu, B. Kim, R. A. Friesner, and B. J. Berne, *Proc. Natl. Acad. Sci. U.S.A.* **102**, 13749 (2005).
37. X. H. Huang, M. Hagen, B. Kim, R. A. Friesner, R. H. Zhou, and B. J. Berne, *J. Phys. Chem. B* **111**, 5405 (2007).
38. A. Laio and M. Parrinello, *Proc. Natl. Acad. Sci. U.S.A.* **99**, 12562 (2002).
39. A. Laio and F. L. Gervasio, *Rep. Prog. Phys.* **71**, 126601 (2008).
40. B. A. Berg and T. Neuhaus, *Phys. Lett. B* **267**, 249 (1991).
41. S. G. Itoh and Y. Okamoto, *J. Chem. Phys.* **124**, 104103 (2006).
42. H. Grubmuller, *Phys. Rev. E* **52**, 2893 (1995).
43. J. Lee, H. A. Scheraga, and S. Rackovsky, *J. Comput. Chem.* **18**, 1222 (1997).
44. Y. Q. Gao, L. J. Yang, Y. B. Fan, and Q. Shao, *Int. Rev. Phys. Chem.* **27**, 201 (2008).
45. Y. Q. Gao, *J. Chem. Phys.* **128**, 064105 (2008).
46. Y. Q. Gao, *J. Chem. Phys.* **128**, 134111 (2008).
47. L. J. Yang, Q. Shao, and Y. Q. Gao, *Prog. Chem.* **24**, 1199 (2012).
48. D. Frenkel and B. Smit, *Understanding Molecular Simulation: From Algorithms to Applications* (Academic Press; 2nd Ed., 2001).
49. D. Frenkel and B. M. Mulder, *Mol. Phys.* **55**, 1171 (1985).
50. A. Stroobants, H. N. W. Lekkerkerker, and D. Frenkel, *Phys. Rev. A* **36**, 2929 (1987).
51. D. Frenkel, H. N. W. Lekkerkerker, and A. Stroobants, *Nature* **332**, 822 (1988).
52. E. J. Meijer, D. Frenkel, R. A. LeSar, and A. J. C. Ladd, *J. Chem. Phys.* **92**, 7570 (1990).
53. E. J. Meijer and D. Frenkel, *J. Chem. Phys.* **94**, 2269 (1991).
54. M. D. Eldridge, P. A. Madden, and D. Frenkel, *Nature* **365**, 35 (1993).
55. D. Alfè, G. D. Price, and M. J. Gillan, *Phys. Rev. B* **64**, 045123 (2001).

56. O. Sugino and R. Car, *Phys. Rev. Lett.* **74**, 1823 (1995).
57. G. A. de Wijs, G. Kresse, and M. J. Gillan, *Phys. Rev. B* **57**, 8223 (1998).
58. D. Alfè, G. A. de Wijs, G. Kresse, and M. J. Gillan, *Int. J. Quantum Chem.* **77**, 871 (2000).
59. A. Kupperman, *Potential Energy Surfaces and Dynamical Calculations,* D. Truhlar, ed. (Plenum, New York, 1981), pp. 375–420.
60. K. Liu, *Ann. Rev. Phys. Chem.* **52**, 139 (2001).
61. M. H. Qiu and Z. F. Ren *et al.*, *Science* **311**, 1440 (2006).
62. Y. T. Lee, *Science* **236**, 793 (1987).
63. R. P. Feynman, *Phys. Rev.* **76**, 769 (1949).
64. R. P. Feynman, *Phys. Rev.* **90**, 1116 (1953).
65. R. P. Feynman, *Phys. Rev.* **91**, 1291 (1953).
66. R. P. Feynman, *Phys. Rev.* **91**, 1301 (1953).
67. R. P. Feynman and A. R. Hibbs, *Quantum Mechanics and Path Integrals* (McGraw-Hill Inc., 1965).
68. D. Chandler and P. G. Wolynes, *J. Chem. Phys.* **74**, 4078 (1981).
69. M. Parrinello and A. Rahman, *J. Chem. Phys.* **80**, 860 (1984).
70. D. M. Ceperley, *Rev. Mod. Phys.* **67**, 279 (1995).
71. E. L. Pollock and D. M. Ceperley, *Phys. Rev. B* **30**, 2555 (1984).
72. D. M. Ceperley and E. L. Pollock, *Phys. Rev. Lett.* **56**, 351 (1986).
73. E. L. Pollock and D. M. Ceperley, *Phys. Rev. B* **36**, 8343 (1987).
74. B. J. Berne and D. Thirumalai, *Annu. Rev. Phys. Chem.* **37**, 401 (1986).
75. D. Thirumalai and B. J. Berne, *J. Chem. Phys.* **81**, 2512 (1984).
76. D. Thirumalai and B. J. Berne, *Chem. Phys. Lett.* **116**, 471 (1985).
77. R. D. Coalson, *J. Chem. Phys.* **83**, 688 (1985).
78. J. D. Doll, D. L. Freeman, and T. L. Beck, *Adv. Chem. Phys.* **78**, 61 (1990).
79. N. Makri, *Comput. Phys. Commun.* **63**, 389 (1991).
80. J. S. Cao and G. A. Voth, *J. Chem. Phys.* **100**, 5093 (1994).
81. J. S. Cao and G. A. Voth, *J. Chem. Phys.* **100**, 5106 (1994).
82. J. S. Cao and G. A. Voth, *J. Chem. Phys.* **101**, 6157 (1994).
83. J. S. Cao and G. A. Voth, *J. Chem. Phys.* **101**, 6168 (1994).
84. G. A. Voth, *Adv. Chem. Phys.* **93**, 135 (1996).
85. I. R. Craig and D. E. Manolopoulos, *J. Chem. Phys.* **121**, 3368 (2004).
86. I. R. Craig and D. E. Manolopoulos, *J. Chem. Phys.* **122**, 084106 (2005).
87. T. F. Miller III and D. E. Manolopoulos, *J. Chem. Phys.* **122**, 184503 (2005).
88. I. R. Craig and D. E. Manolopoulos, *J. Chem. Phys.* **123**, 034102 (2005).
89. T. F. Miller III and D. E. Manolopoulos, *J. Chem. Phys.* **123**, 154504 (2005).
90. I. R. Craig and D. E. Manolopoulos, *Chem. Phys.* **322**, 236 (2006).
91. B. J. Braams and D. E. Manolopoulos, *J. Chem. Phys.* **125**, 124105 (2006).
92. S. Habershon, B. J. Braams, and D. E. Manolopoulos, *J. Chem. Phys.* **127**, 174108 (2007).
93. T. E. Markland and D. E. Manolopoulos, *J. Chem. Phys.* **129**, 024105 (2008).
94. G. M. Torrie and J. P. Valleau, *Chem. Phys. Lett.* **28**, 578 (1974).
95. G. M. Torrie and J. P. Valleau, *J. Comput. Phys.* **77**, 187 (1977).

96. B. Roux, *Comput. Phys. Commun.* **91**, 275 (1995).
97. M. P. Allen and D. J. Tildesley, *Computer Simulation of Liquids* (Oxford University Press, USA, 1989).
98. D. R. Hartree, *Proc. Cambridge Phil. Soc.* **24**, 89 (1928).
99. V. Fock, *Z. Physik* **61**, 126 (1930).
100. J. A. Pople and P. K. Nesbet, *J. Chem. Phys.* **22**, 571 (1954).
101. A. Szabo and N. S. Ostlund, *Modern Quantum Chemistry: Introduction to Advanced Electronic Structure Theory* (Dover Pub. Inc., New York, 1996).
102. C. Mller and M. S. Plesset, *Phys. Rev.* **46**, 618 (1934).
103. F. Coester, *Nucl. Phys.* **7**, 421 (1958).
104. F. Coester and H. Kummel, *Nucl. Phys.* **17**, 477 (1960).
105. P. Hohenberg and W. Kohn, *Phys. Rev.* **136**, B864 (1964).
106. W. Kohn and L. J. Sham, *Phys. Rev.* **140**, A1133 (1965).
107. L. H. Thomas, *Proc. Cambridge Phil. Soc.* **23**, 542 (1927).
108. E. Fermi, *Rend. Accad. Naz. Lincei* **6**, 602 (1927).
109. E. Fermi, *Z. Physik* **48**, 73 (1928).
110. P. A. M. Dirac, *Proc. Cambridge Phil. Soc.* **26**, 361 (1930).
111. U. von Barth and L. Hedin, *J. Phys.: Condens. Maths* (*c.f.* [13]) **5**, 1629 (1972).
112. O. Gunnarsson, B. I. Lundqvist, and J. W. Wilkins, *Phys. Rev. B* **10**, 1319 (1974).
113. R. O. Jones and O. Gunnarsson, *Rev. Mod. Phys.* **61**, 689 (1989).
114. M. Levy, *Proc. Natl. Acad. Sci. U.S.A.* **76**, 6062 (1979).
115. M. Levy, *Phys. Rev. A* **26**, 1200 (1982).
116. A. Shimony and H. Feshbach, *Physics as Natural Philosophy* (MIT Press, Cambridge, 1982), p. 111.
117. E. H. Lieb, *Int. J. Quant. Chem.* **24**, 243 (1983).
118. R. M. Dreizler and J. da Providencia, *Density Functional Methods in Physics* (Plenum, New York, 1985), p. 31.
119. R. M. Dreizler and E. K. U. Gross, *Density Functional Theory* (Springer-Verlag, Berlin, Heidelberg, 1990).
120. A. K. Rajagopal and J. Callaway, *Phys. Rev. B* **7**, 1912 (1973).
121. L. N. Oliveira, E. K. U. Gross, and W. Kohn, *Phys. Rev. Lett.* **60**, 2430 (1988).
122. S. Kurth, M. Marques, M. Lüders, and E. K. U. Gross, *Phys. Rev. Lett.* **83**, 2628 (1999).
123. J. P. Perdew and K. Schmidt, *Density Functional Theory and Its Applications to Materials*, V. Van Doren *et al.*, eds. (American Institute of Physics, New York, 2001).
124. G. D. Mahan, *Many-Particle Physics* (Plenum Press, New York, 1990).
125. D. M. Ceperley and B. J. Alder, *Phys. Rev. Lett.* **45**, 566 (1980).
126. J. P. Perdew and A. Zunger, *Phys. Rev. B* **23**, 5048 (1981).
127. J. P. Perdew and Y. Wang, *Phys. Rev. B* **33**, 8800 (1986).
128. J. P. Perdew and Y. Wang, *Phys. Rev. B* **40**, 3399 (1989).
129. J. P. Perdew, K. Burke, and M. Ernzerhof, *Phys. Rev. Lett.* **77**, 3865 (1996).

130. J. P. Perdew, K. Burke, and Y. Wang, *Phys. Rev. B* **54**, 16533 (1996).
131. W. Kohn, A. D. Becke, and R. G. Parr, *J. Chem. Phys.* **100**, 12974 (1996).
132. A. D. Becke, *J. Chem. Phys.* **102**, 8554 (1997).
133. J. P. Perdew, A. Ruzsinszky, J. M. Tao, V. N. Staroverov, G. E. Scuseria, and G. I. Csonka, *J. Chem. Phys.* **123**, 062201 (2005).
134. J. C. Slater and K. H. Johnson, *Phys. Rev. B* **5844** (1972).
135. J. F. Janak, *Phys. Rev. B* **18**, 7165 (1978).
136. C. O. Almbladh and U. von Barth, *Phys. Rev. B* **31**, 3231 (1985).
137. S. Lizzit, A. Baraldi, A. Groso, K. Reuter, M. V. Ganduglia-Pirovano, C. Stamp, M. Scheer, M. Stichler, C. Keller, W. Wurth *et al.*, *Phys. Rev. B* **63**, 205419 (2001).
138. J. P. Perdew, D. C. Langreth, and V. Sahni, *Phys. Rev. Lett.* **38**, 1030 (1977).
139. V. L. Moruzzi, J. F. Janak, and A. R. Williams, *Calculated Electronic Properties of Metals* (Pergamon Pr., Oxford, 1978).
140. A. Görling and M. Levy, *Phys. Rev. A* **50**, 196 (1994).
141. A. Görling, *Phys. Rev. B* **53**, 7024 (1996).
142. S. Sharma, J. K. Dewhurst, and C. Ambrosch-Draxl, *Phys. Rev. Lett.* **95**, 136402 (2005).
143. M. Städele, J. A. Majewski, P. Vogl, and A. Göorling, *Phys. Rev. Lett.* **79**, 2089 (1997).
144. T. Kotani, *Phys. Rev. Lett.* **74**, 2989 (1995).
145. J. P. Perdew and M. Levy, *Phys. Rev. Lett.* **51**, 1884 (1983).
146. L. J. Sham and M. Schlüter, *Phys. Rev. Lett.* **51**, 1888 (1983).
147. L. J. Sham and M. Schlüter, *Phys. Rev. B* **32**, 3883 (1985).
148. M. Grüning, A. Marini, and A. Rubio, *J. Chem. Phys.* **124**, 154108 (2006).
149. D. J. Singh, *Planewaves, Pseudopotential and the LAPW method* (Kluwer Academic Publisher, Norwell, Massachusetts, 1994).
150. D. R. Hamann, M. Schlüter, and C. Chiang, *Phys. Rev. Lett.* **43**, 1494 (1979).
151. G. B. Bachelet, D. R. Hamann, and M. Schlüter, *Phys. Rev. B* **26**, 4199 (1982).
152. D. Vanderbilt, *Phys. Rev. B* **41**, 7892 (1990).
153. K. Laasonen, A. Pasquarello, R. Car, C. Lee, and D. Vanderbilt, *Phys. Rev. B* **47**, 10142 (1993).
154. M. Fuchs and M. Scheer, *Comput. Phys. Commun.* **119**, 67 (1999).
155. D. R. Hamann, *Phys. Rev. B* **40**, 2980 (1989).
156. N. Troullier and J. L. Martins, *Phys. Rev. B* **43**, 1993 (1991).
157. L. Kleinman and D. M. Bylander, *Phys. Rev. Lett.* **48**, 1425 (1982).
158. U. von Barth and C. D. Gelatt, *Phys. Rev. B* **21**, 2222 (1980).
159. S. G. Louie, S. Froyen, and M. L. Cohen, *Phys. Rev. B* **26**, 1738 (1982).
160. W. Ku and A. G. Eguiluz, *Phys. Rev. Lett.* **89**, 126401 (2002).
161. P. Puschnig and C. Ambrosch-Draxl, *Phys. Rev. B* **66**, 165105 (2002).

162. R. Gomez-Abal, X. Z. Li, M. Scheer, and C. Ambrosch-Draxl, *Phys. Rev. Lett.* **101**, 106404 (2008).

163. X. Z. Li, R. Gomez-Abal, H. Jiang, C. Ambrosch-Draxl, and M. Scheer, *New J. Phys.* **14**, 023006 (2012).

164. J. C. Slater, *Phys. Rev.* **51**, 846 (1937).

165. P. de Ciccio, *Phys. Rev.* **153**, 931 (1967).

166. N. Elyashar and D. D. Koelling, *Phys. Rev. B* **13**, 5362 (1976).

167. O. K. Andersen, *Phys. Rev. B* **12**, 3060 (1975).

168. E. Sjöstedt, L. Nordström, and D. J. Singh, *Solid State Commun.* **114**, 15 (2000).

169. G. K. H. Madsen, P. Blaha, K. Schwarz, E. Sjöstedt, and L. Nordström, *Phys. Rev. B* **64**, 195134 (2001).

170. D. Singh, *Phys. Rev. B* **43**, 6388 (1991).

171. M. E. Rose, *Elementary Theory of Angular Momentum* (John Wiley and Sons, 1957).

172. M. Kara and K. Kurki-Suonio, *Acta Crystallograca A* **37**, 201 (1981).

173. J. Deslippe, G. Samsonidze, D. A. Strubbe, M. Jain, M. L. Cohen, and S. G. Louie, *Comput. Phys. Commun.* **183**, 1269 (2012).

174. M. R. A. Shegelski, *Am. J. Phys.* **72**, 676 (2004).

175. A. L. Fetter and J. D. Walecka, *Quantum Theory of Many-Particle Systems* (McGraw-Hill Inc., 1971).

176. L. Hedin, *Phys. Rev.* **139**, A796 (1965).

177. L. Hedin and S. Lundqvist, *Solid State Phys.: Advances in Research and Applications* **23**, 1 (1969).

178. F. Aryasetiawan and O. Gunnarsson, *Rep. Prog. Phys.* **61**, 237 (1998).

179. J. Schwinger, *Proc. Natl. Acad. Sci. U.S.A.* **37**, 452 (1951).

180. G. Pratt, *Phys. Rev.* **118**, 462 (1960).

181. G. Pratt, *Rev. Mod. Phys.* **35**, 502 (1963).

182. L. Hedin, *Bull. Am. Phys. Soc.* **8**, 535 (1963).

183. E. K. U. Gross, R. Runge, and O. Heinonen, *Many-Particle Theory* (Adam Hilger, 1991).

184. D. Straub, L. Ley, and F. J. Himpsel, *Phys. Rev. Lett.* **54**, 142 (1985).

185. M. S. Hybertsen and S. G. Louie, *Phys. Rev. Lett.* **55**, 1418 (1985).

186. J. E. Northrup, M. S. Hybertsen, and S. G. Louie, *Phys. Rev. Lett.* **59**, 819 (1987).

187. C. Petrillo and F. Sacchetti, *Phys. Rev. B* **38**, 3834 (1988).

188. E. L. Shirley, Z. J. Zhu, and S. G. Louie, *Phys. Rev. B* **56**, 6648 (1997).

189. O. Zakharov, A. Rubio, X. Blase, M. L. Cohen, and S. G. Louie, *Phys. Rev. B* **50**, 10780 (1994).

190. X. J. Zhu and S. G. Louie, *Phys. Rev. B* **43**, 14142 (1991).

191. T. Kotani and M. van Schilfgaarde, *Solid State Commun.* **121**, 461 (2002).

192. C. Friedrich, A. Schindlmayr, S. Blügel, and T. Kotani, *Phys. Rev. B* **74**, 045104 (2006).

193. H. Jiang, R. Gomez-Abal, X. Z. Li, C. Meisenbichler, C. Ambrosch-Draxl, and M. Scheer, *Comput. Phys. Commun.* **184**, 348 (2013).

194. F. Gygi and A. Baldereschi, *Phys. Rev. Lett.* **62**, 2160 (1989).
195. A. Fleszar and W. Hanke, *Phys. Rev. B* **56**, 10228 (1997).
196. R. W. Godby, M. Schlüter, and L. J. Sham, *Phys. Rev. B* **36**, 6497 (1987).
197. J. Q. Wang, Z. Q. Gu, and M. F. Li, *Phys. Rev. B* **44**, 8707 (1991).
198. K. Delaney, P. Garcia-Gonzalez, A. Rubio, P. Rinke, and R. W. Godby, *Phys. Rev. Lett.* **93**, 249701 (2004).
199. W. Ku and A. G. Eguiluz, *Phys. Rev. Lett.* **93**, 249702 (2004).
200. M. van Schilfgaarde, T. Kotani, and S. V. Faleev, *Phys. Rev. B* **74**, 245125 (2006).
201. M. L. Tiago, S. Ismail-Beigi, and S. G. Louie, *Phys. Rev. B* **69**, 125212 (2004).
202. M. van Schilfgaarde, T. Kotani, and S. Faleev, *Phys. Rev. Lett.* **96**, 226402 (2006).
203. S. Faleev, M. van Schilfgaarde, and T. Kotani, *Phys. Rev. Lett.* **93**, 126406 (2004).
204. M. Shishkin and G. Kresse, *Phys. Rev. B* **75**, 235102 (2007).
205. M. Shishkin, M. Marsman, and G. Kresse, *Phys. Rev. Lett.* **99**, 246403 (2007).
206. T. Kotani and M. van Schilfgaarde, *Phys. Rev. B* **76**, 165106 (2007).
207. K. S. Thygesen and A. Rubio, *Phys. Rev. B* **77**, 115333 (2008).
208. F. Bruneval, N. Vast, and L. Reining, *Phys. Rev. B* **74**, 045102 (2006).
209. W. D. Schöne and A. G. Eguiluz, *Phys. Rev. Lett.* **81**, 1662 (1998).
210. C. Rostgaard, K. W. Jacobsen, and K. S. Thygesen, *Phys. Rev. B* **81**, 085103 (2010).
211. A. Stan, N. E. Dahlen, and R. van Leeuwen, *Europhys. Lett.* **76**, 298 (2006).
212. X. Z. Li, *All-electron G0W0 code based on FP-(L)APW+lo and applications*, Ph.D. Thesis (Free University of Berlin, 2008).
213. P. Blaha, K. Schwarz, G. K. H. Madsen, D. Kvasnicka, and J. Luitz, *WIEN2k, An Augmented Plane Wave Plus Local Orbitals Program for Calculating Crystal Properties* (Tchn. Universität Wien, Austria, 2002), ISBN 3-9501031-1-2.
214. URL http://exciting-code.org.
215. S. Sagmeister and C. Ambrosch-Draxl, *Phys. Chem. Chem. Phys.* **11**, 4451 (2009).
216. S. Sharma, J. K. Dewhurst, and C. Ambrosch-Draxl, *Phys. Rev. Lett.* **95**, 136402 (2005).
217. A. Baldereschi, *Phys. Rev. B* **7**, 5212 (1973).
218. D. J. Chadi and M. L. Cohen, *Phys. Rev. B* **7**, 692 (1973).
219. D. J. Chadi and M. L. Cohen, *Phys. Rev. B* **8**, 5747 (1973).
220. H. J. Monkhorst and J. D. Pack, *Phys. Rev. B* **13**, 5188 (1976).
221. O. Jepsen and O. K. Andersen, *Solid State Commun.* **9**, 1763 (1971).
222. G. Lehmann, P. Rennert, M. Taut, and H. Wonn, *Phys. Status Solidi* **37**, K27 (1970).
223. J. Rath and A. J. Freeman, *Phys. Rev. B* **11**, 2109 (1975).
224. P. E. Blöchl, O. Jepsen, and O. K. Andersen, *Phys. Rev. B* **49**, 16223 (1994).

225. R. W. Godby, M. Schlüter, and L. J. Sham, *Phys. Rev. B* **37**, 10159 (1988).
226. T. Matsubara, *Prog. Theor. Phys.* **14**, 351 (1955).
227. M. M. Rieger, L. Steinbeck, I. D. White, H. N. Rojas, and R. W. Godby, *Comput. Phys. Commun.* **117**, 211 (1999).
228. R. Car and M. Parrinello, *Phys. Rev. Lett.* **55**, 2471 (1985).
229. L. Verlet, *Phys. Rev.* **159**, 98 (1967).
230. W. C. Swope, H. C. Andersen, P. H. Berens, and K. R. Wilson, *J. Chem. Phys.* **76**, 637 (1982).
231. R. W. Hockney and J. W. Eastwood, *Computer Simulations Using Particles* (McGraw-Hill, New York, 1981).
232. A. Rahman, *Phys. Rev.* **136**, A405 (1964).
233. W. F. Vangunsteren and H. J. C. Berendsen, *Mol. Phys.* **34**, 1311 (1977).
234. J. L. Lebowitz, J. K. Percus, and J. Verlet, *Phys. Rev.* **153**, 250 (1967).
235. P. S. Y. Cheung, *Mol. Phys.* **33**, 519 (1967).
236. J. R. Ray and H. W. Graben, *Mol. Phys.* **43**, 1293 (1981).
237. J. R. Ray and H. W. Graben, *Phys. Rev. A* **44**, 6905 (1991).
238. H. W. Graben and J. R. Ray, *Phys. Rev. A* **43**, 4100 (1991).
239. H. C. Andersen, *J. Chem. Phys.* **72**, 2384 (1980).
240. S. Nose, *J. Chem. Phys.* **81**, 511 (1984).
241. S. Nose, *Mol. Phys.* **52**, 255 (1984).
242. W. G. Hoover, *Phys. Rev. A* **31**, 1695 (1985).
243. W. G. Hoover, *Phys. Rev. A* **34**, 2499 (1986).
244. S. Nose, *Mol. Phys.* **57**, 187 (1986).
245. G. J. Martyna, M. L. Klein, and M. E. Tuckerman, *J. Chem. Phys.* **97**, 2635 (1992).
246. M. E. Tuckerman and G. J. Martyna, *J. Phys. Chem. B* **104**, 159 (2000).
247. M. Parrinello and A. Rahman, *Phys. Rev. Lett.* **45**, 1196 (1980).
248. M. Parrinello and A. Rahman, *J. Appl. Phys.* **52**, 7182 (1981).
249. G. J. Martyna, D. J. Tobias, and M. L. Klein, *J. Chem. Phys.* **101**, 4177 (1994).
250. J. M. Haile and H. W. Graben, *J. Chem. Phys.* **73**, 2412 (1980).
251. J. M. Haile and H. W. Graben, *Mol. Phys.* **40**, 1433 (1980).
252. J. R. Ray and H. W. Graben, *Phys. Rev. A* **34**, 2517 (1986).
253. J. R. Ray and H. W. Graben, *J. Chem. Phys.* **75**, 4077 (1981).
254. M. Ceriotti, G. Bussi, and M. Parrinello, *Phys. Rev. Lett.* **102**, 020601 (2009).
255. P. E. Blöchl and M. Parrinello, *Phys. Rev. B* **45**, 9413 (1992).
256. D. Quigley and M. I. J. Probert, *J. Chem. Phys.* **120**, 11432 (2004).
257. W. G. Hoover, K. Aoki, C. G. Hoover, and S. V. de Groot, *Physica D-Nonlinear Phenomena* **187**, 253 (2004).
258. W. G. Hoover, *Phys. Rev. A* **34**, 2499 (1986).
259. A. Kolb and B. Dunweg, *J. Chem. Phys.* **111**, 4453 (1999).
260. G. Kresse and J. Hafner, *Phys. Rev. B* **47**, 558 (1993).
261. G. Kresse and J. Hafner, *Phys. Rev. B* **49**, 14251 (1994).
262. G. Kresse and J. Furthmüller, *Comput. Mat. Sci.* **6**, 15 (1996).

263. G. Kresse and J. Furthmüller, *Phys. Rev. B* **54**, 11169 (1996).
264. J. Chen, X. Z. Li, Q. F. Zhang, A. Michaelides, and E. G. Wang, *Phys. Chem. Chem. Phys.* **15**, 6344 (2013).
265. M. Tuckerman, K. Laasonen, M. Sprik, and M. Parrinello, *J. Phys. Chem.* **99**, 5749 (1995).
266. M. Tuckerman, K. Laasonen, M. Sprik, and M. Parrinello, *J. Chem. Phys.* **103**, 150 (1995).
267. R. Vuilleumier and D. Borgis, *J. Phys. Chem. B* **102**, 4261 (1998).
268. R. Vuilleumier and D. Borgis, *J. Chem. Phys.* **111**, 4251 (1999).
269. U. W. Schmitt and G. A. Voth, *J. Phys. Chem. B* **102**, 5547 (1998).
270. U. W. Schmitt and G. A. Voth, *J. Chem. Phys.* **111**, 9361 (1999).
271. G. A. Voth, *Acc. Chem. Res.* **39**, 143 (2006).
272. O. Markovitch, H. N. Chen, S. Izvekov, F. Paesani, G. A. Voth, and N. Agmon, *J. Phys. Chem. B* **112**, 9456 (2008).
273. S. Woutersen and H. J. Bakker, *Phys. Rev. Lett.* **96**, 138305 (2006).
274. J. M. Headrick, E. G. Diken, R. S. Walters, N. I. Hammer, R. A. Christie, J. Cui, E. M. Myshakin, M. A. Duncan, M. A. Johnson, and K. D. Jordan, *Science* **308**, 1765 (2005).
275. C. J. T. de Grotthuss, *Ann. Chim.* **58**, 54 (1806).
276. N. Agmon, *Chem. Phys. Lett.* **244**, 456 (1995).
277. S. Scandolo, *Proc. Natl. Acad. Sci. U.S.A.* **10**, 3051 (2003).
278. S. A. Bonev, E. Schwegler, T. Ogitsu, and G. Galli, *Nature* **431**, 669 (2004).
279. H. Y. Liu and Y. M. Ma, *Phys. Rev. Lett.* **110**, 025903 (2013).
280. D. Alfé, M. J. Gillan, and G. D. Price, *Nature* **405**, 172 (2000).
281. X. Z. Li, B. Walker, M. I. J. Probert, C. J. Pickard, R. J. Needs, and A. Michaelides, *J. Phys.: Condens. Matter* **25**, 085402 (2013).
282. M. A. Morales, C. Pierleoni, E. Schwegler, and D. M. Ceperley, *Proc. Natl. Acad. Sci. U.S.A.* **107**, 12799 (2010).
283. M. A. Morales, J. M. McMahon, C. Pierleoni, and D. M. Ceperley, *Phys. Rev. B* **87**, 184107 (2013).
284. M. A. Morales, J. M. McMahon, C. Pierleoni, and D. M. Ceperley, *Phys. Rev. Lett.* **110**, 065702 (2013).
285. S. Azadi and W. M. C. Foulkes, *Phys. Rev. B* **88**, 014115 (2013).
286. X. Z. Li, M. I. J. Probert, A. Alavi, and A. Michaelides, *Phys. Rev. Lett.* **104**, 066102 (2010).
287. V. Tozzini, *Curr. Opin. Struct. Biol.* **15**, 144 (2005).
288. M. Christen *et al.*, *J. Comput. Chem.* **26**, 1719 (2005).
289. J. G. Kirkwood, *J. Chem. Phys.* **3**, 300 (1935).
290. J. Shen and J. A. McCammon, *Chem. Phys.* **158**, 191 (1991).
291. M. Mezei, P. K. Mehorotra, and D. L. Beveridge, *J. Am. Chem. Soc.* **107**, 2239 (1985).
292. C. Haydock, J. C. Sharp, and F. G. Prendergast, *Biophys. J.* **57**, 1269 (1990).
293. T. B. Woolf and B. Roux, *J. Am. Chem. Soc.* **116**, 5916 (1994).
294. T. B. Woolf, S. Crouzy, and B. Roux, *Biophys. J.* **67**, 1370 (1994).

295. S. Kumar, D. Bouzida, R. H. Swendsen, P. A. Kollman, and J. M. Rosenberg, *J. Comput. Chem.* **13**, 1011 (1992).

296. M. Mezei, *J. Comput. Phys.* **68**, 237 (1987).

297. G. H. Paine and H. A. Scheraga, *Biopolymers* **24**, 1391 (1985).

298. R. W. W. Hooft, B. P. van Eijck, and J. Kroon, *J. Chem. Phys.* **97**, 6690 (1992).

299. R. Rajamani, K. J. Naidoo, and J. L. Gao, *J. Comput. Chem.* **24**, 1175 (2003).

300. C. Bartels and M. Karplus, *J. Comput. Chem.* **18**, 1450 (1997).

301. M. Iannuzzi and M. Parrinello, *Phys. Rev. Lett.* **93**, 025901 (2004).

302. A. Barducci, R. Chelli, P. Procacci, V. Schettino, F. Gervasio, and M. Parrinello, *J. Am. Chem. Soc.* **128**, 2705 (2006).

303. A. Laio, A. Rodriguez-Fortea, and F. L. Gervasio, *J. Phys. Chem. B* **109**, 6714 (2005).

304. R. Martonak, A. Laio, and M. Parrinello, *Phys. Rev. Lett.* **90**, 075503 (2003).

305. B. Ensing, M. De Vivo, Z. W. Liu, P. Moore, and M. L. Klein, *Acc. Chem. Res.* **39**, 73 (2006).

306. F. L. Gervasio, A. Laio, and M. Parrinello, *J. Am. Chem. Soc.* **127**, 2600 (2005).

307. S. Piana and A. Laio, *J. Phys. Chem. B* **111**, 4553 (2007).

308. C. Tsallis, *J. Stat. Phys.* **52**, 479 (1988).

309. I. Fukuda and H. Nakamura, *Phys. Rev. E* **71**, 046708 (2005).

310. F. G. Wang and D. P. Landau, *Phys. Rev. E* **64**, 056101 (2001).

311. F. G. Wang and D. P. Landau, *Phys. Rev. Lett.* **86**, 2050 (2001).

312. L. J. Yang and Y. Q. Gao, *J. Chem. Phys.* **131**, 214109 (2009).

313. O. G. Mouritsen, *Computer Studies of Phase Transitions and Critical Phenomena* (Springer, Berlin, 1984).

314. L. J. Lauhon and W. Ho, *Phys. Rev. Lett.* **85**, 4566 (2000).

315. Y. V. Suleimanov, *J. Phys. Chem. C* **116**, 11141 (2012).

316. L. Masgrau *et al.*, *Science* **312**, 237 (2006).

317. C. R. Pudney *et al.*, *J. Am. Chem. Soc.* **135**, 2512 (2013).

318. J. R. Rommel *et al.*, *J. Phys. Chem. B* **116**, 13682 (2012).

319. M. E. Tuckerman, D. Marx, M. L. Klein, and M. Parrinello, *J. Chem. Phys.* **104**, 5579 (1996).

320. D. Marx and M. Parrinello, *J. Chem. Phys.* **104**, 4077 (1996).

321. D. Marx, M. E. Tuckerman, J. Hutter, and M. Parrinello, *Nature* **397**, 601 (1999).

322. X. Z. Li, B. Walker, and A. Michaelides, *Proc. Natl. Acad. Sci. U.S.A.* **108**, 6369 (2011).

323. J. Chen, X. Z. Li, and Q. F. Zhang *et al.*, *Nat. Commun.* **4**, 2064 (2013).

324. M. A. Morales, J. M. McMahon, C. Pierleoni, and D. M. Ceperley, *Phys. Rev. Lett.* **110**, 065702 (2013).

325. M. A. Morales, J. M. McMahon, C. Pierleoni, and D. M. Ceperley, *Phys. Rev. B* **87**, 184107 (2013).

326. M. Tachikawa and M. Shiga, *J. Am. Chem. Soc.* **127**, 11908 (2005).

327. A. Kaczmarek, M. Shiga, and D. Marx, *J. Phys. Chem.* A **113**, 1985 (2009).
328. P. Pechukas, *Ann. Rev. Phys. Chem.* **32**, 159 (1981).
329. W. H. Miller, *Acc. Chem. Res.* **9**, 306 (2005).
330. J. D. Doll and A. F. Voter, *Ann. Rev. Phys. Chem.* **38**, 413 (1987).
331. G. Mills, H. Jonsson, and G. K. Schenter, *Surf. Sci.* **324**, 305 (1995).
332. R. Ramrez, C. P. Herrero, A. Antonelli, and E. R. Hernandez, *J. Chem. Phys.* **129**, 064110 (2008).
333. S. Habershon and D. E. Manolopoulos, *J. Chem. Phys.* **135**, 224111 (2011).
334. G. F. Reiter, J. Mayers, and P. Platzman, *Phys. Rev. Lett.* **89**, 135505 (2002).
335. G. F. Reiter, J. C. Li, J. Mayers, T. Abdul-Redah, and P. Platzman, *Braz. J. Phys.* **34**, 142 (2004).
336. C. Andreani, D. Colognesi, J. Mayers, G. F. Reiter, and R. Senesi, *Adv. Phys.* **54**, 377 (2005).
337. J. A. Morrone, V. Srinivasan, D. Sebastiani, and R. Car, *J. Chem. Phys.* **126**, 234504 (2007).
338. J. A. Morrone and R. Car, *Phys. Rev. Lett.* **101**, 017801 (2008).
339. J. A. Morrone, L. Lin, and R. Car, *J. Chem. Phys.* **130**, 204511 (2009).
340. L. Lin, J. A. Morrone, R. Car, and M. Parrinello, *Phys. Rev. Lett.* **105**, 110602 (2010).
341. L. Lin, J. A. Morrone, and R. Car, *J. Stat. Phys.* **145**, 365 (2011).
342. M. Ceriotti and D. E. Manolopoulos, *Phys. Rev. Lett.* **109**, 100604 (2012).
343. R. W. Hall and B. J. Berne, *J. Chem. Phys.* **81**, 3641 (1984).
344. S. Habershon, G. S. Fanourgakis, and D. E. Manolopoulos, *J. Chem. Phys.* **129**, 074501 (2008).
345. B. Chen, I. Ivanov, M. Klein, and M. Parrinello, *Phys. Rev. Lett.* **91**, 215503 (2003).
346. V. Buch, *J. Chem. Phys.* **97**, 726 (1992).
347. P. Sandler, J. O. Jung, M. M. Szczesniak, and V. Buch, *J. Chem. Phys.* **101**, 1378 (1994).
348. J. K. Gregory and D. C. Clary, *Chem. Phys. Lett.* **228**, 547 (1994).
349. J. K. Gregory and D. C. Clary, *J. Phys. Chem.* **100**, 18014 (1996).
350. M. F. Herman, E. J. Bruskin, and B. J. Berne, *J. Chem. Phys.* **76**, 5150 (1982).
351. M. E. Tuckerman, *Statistical Mechanics: Theory and Molecular Simulations* (Oxford University Press, USA, 2010).
352. A. Nakayama and N. Makri, *J. Chem. Phys.* **125**, 024503 (2006).
353. J. S. Shao and N. Makri, *J. Phys. Chem.* A **103**, 7753 (1999).
354. J. S. Shao and N. Makri, *J. Phys. Chem.* A **103**, 9479 (1999).
355. Q. Shi and E. Geva, *J. Phys. Chem.* A **107**, 9059 (2003).
356. Q. Shi and E. Geva, *J. Phys. Chem.* A **107**, 9070 (2003).
357. J. Liu and N. Makri, *Chem. Phys.* **322**, 23 (2006).
358. M. H. Beck, A. Jackle, G. A. Worth, and H. D. Meyer, *Phys. Rep.* **324**, 1 (2000).
359. S. C. Althorpe and D. C. Clary, *Annu. Rev. Phys. Chem.* **54**, 493 (2003).

360. D. H. Zhang, M. A. Collins, and S. Y. Lee, *Science* **290**, 961 (2000).
361. M. A. Collins, *Theo. Chem. Acc.* **108**, 313 (2002).
362. O. Vendrell, F. Gatti, and H. D. Meyer, *Angew. Chem. Int. Ed.* **46**, 6918 (2007).
363. G. Baym and N. D. Mermin, *J. Math. Phys.* **2**, 232 (1961).
364. E. Rabani, G. Krilov, and B. J. Berne, *J. Chem. Phys.* **112**, 2605 (2000).
365. G. Krilov, E. Sim, and B. J. Berne, *Chem. Phys.* **268**, 21 (2001).
366. E. Sim, G. Krilov, and B. J. Berne, *J. Phys. Chem. A* **105**, 2824 (2001).
367. E. Rabani, D. R. Reichman, G. Krilov, and B. J. Berne, *Proc. Natl. Acad. Sci. U.S.A.* **99**, 1129 (2002).
368. D. R. Reichman and E. Rabani, *Phys. Rev. Lett.* **87**, 265702 (2001).
369. E. Rabani and D. R. Reichman, *Phys. Rev. E* **65**, 036111 (2002).
370. E. Rabani and D. R. Reichman, *J. Chem. Phys.* **116**, 6271 (2002).
371. D. R. Reichman and E. Rabani, *J. Chem. Phys.* **116**, 6279 (2002).
372. E. Rabani and D. R. Reichman, *J. Chem. Phys.* **120**, 1458 (2004).
373. E. Rabani and D. R. Reichman, *Annu. Rev. Phys. Chem.* **56**, 157 (2005).
374. D. Kim, J. D. Doll, and J. E. Gubernatis, *J. Chem. Phys.* **106**, 1641 (1997).
375. B. J. Berne, *J. Stat. Phys.* **43**, 911 (1986).
376. H. Wang, X. Sun, and W. H. Miller, *J. Chem. Phys.* **108**, 9726 (1998).
377. X. Sun, H. Wang, and W. H. Miller, *J. Chem. Phys.* **109**, 7064 (1998).
378. J. Liu and W. H. Miller, *J. Chem. Phys.* **125**, 224104 (2006).
379. J. Liu and W. H. Miller, *J. Chem. Phys.* **127**, 114506 (2007).
380. J. A. Poulsen, G. Nyman, and P. J. Rossky, *J. Chem. Phys.* **119**, 12179 (2003).
381. J. A. Poulsen, G. Nyman, and P. J. Rossky, *J. Phys. Chem. A* **108**, 8743 (2004).
382. J. A. Poulsen, G. Nyman, and P. J. Rossky, *J. Phys. Chem. B* **108**, 19799 (2004).
383. J. A. Poulsen, G. Nyman, and P. J. Rossky, *Proc. Natl. Acad. Sci. U.S.A.* **102**, 6709 (2005).
384. J. S. Cao and G. A. Voth, *J. Chem. Phys.* **99**, 10070 (1993).
385. R. P. Feynman and H. Kleinert, *Phys. Rev. A* **34**, 5080 (1986).
386. M. J. Gillan, *Phys. Rev. Lett.* **58**, 563 (1987).
387. M. J. Gillan, *J. Phys. C: Solid State Phys.* **20**, 3621 (1987).
388. B. C. Garrett and D. G. Truhlar, *J. Phys. Chem.* **83**, 1052 (1979).
389. B. C. Garrett and D. G. Truhlar, *J. Phys. Chem.* **87**, 4553 (1983).
390. W. H. Miller, *J. Chem. Phys.* **61**, 1823 (1974).
391. B. C. Garrett, D. G. Truhlar, R. S. Grev, and A. W. Magnuson, *J. Phys. Chem.* **84**, 1730 (1980).
392. J. W. Tromp and W. H. Miller, *J. Phys. Chem.* **90**, 3482 (1986).
393. G. A. Voth, D. Chandler, and W. H. Miller, *J. Chem. Phys.* **91**, 7749 (1989).
394. M. Messina, G. K. Schenter, and B. C. Garrett, *J. Chem. Phys.* **98**, 8525 (1993).
395. N. F. Hansen and H. C. Andersen, *J. Chem. Phys.* **101**, 6032 (1994).
396. J. Cao and G. J. Martyna, *J. Chem. Phys.* **104**, 2028 (1996).

397. T. D. Hone, P. J. Rossky, and G. A. Voth, *J. Chem. Phys.* **124**, 154103 (2006).

398. A. Pérez, M. E. Tuckerman, and M. H. Müser, *J. Chem. Phys.* **130**, 184105 (2009).

399. S. D. Ivanov, A. Witt, M. Shiga, and D. Marx, *J. Chem. Phys.* **132**, 031101 (2010).

400. J. Morales and K. Singer, *Mol. Phys.* **73**, 873 (1991).

401. A. Hodgson and S. Haq, *Surf. Sci. Rep.* **64**, 381 (2009).

402. G. Held and D. Menzel, *Surf. Sci.* **316**, 92 (1994).

403. A. Michaelides and P. Hu, *J. Am. Chem. Soc.* **123**, 4235 (2001).

404. P. J. Feibelman, *Science* **295**, 99 (2002).

405. S. Völkening, K. Bedürftig, K. Jacobi, J. Wintterlin, and G. Ertl, *Phys. Rev. Lett.* **83**, 2672 (1999).

406. C. Clay, S. Haq, and A. Hodgson, *Phys. Rev. Lett.* **92**, 046102 (2004).

407. T. Schiros, L. A. Näslund, K. Andersson, J. Gyllenpalm, G. S. Karlberg, M. Odelius, H. Ogasawara, L. G. M. Pettersson, and A. Nilsson, *J. Phys. Chem. C* **111**, 15003 (2007).

408. L. Giordano, J. Goniakowski, and J. Suzanne, *Phys. Rev. Lett.* **81**, 1271 (1998).

409. Y. D. Kim, R. M. Lynden-Bell, A. Alavi, J. Stultz, and D. W. Goodman, *Chem. Phys. Lett.* **352**, 318 (2002).

410. B. Meyer, D. Marx, O. Dulub, U. Diebold, M. Kunat, D. Langenberg, and C. Wöll, *Angew. Chem. Int. Ed.* **43**, 6641 (2004).

411. A. Michaelides, A. Alavi, and D. A. King, *Phys. Rev. B.* **69**, 113404 (2004).

412. M. Benoit, D. Marx, and M. Parrinello, *Nature* **392**, 258 (1998).

413. E. Schwegler, M. Sharma, F. Gygi, and G. Galli, *Proc. Natl. Acad. Sci. U.S.A.* **105**, 14779 (2008).

414. S. J. Clark, M. D. Segall, C. J. Pickard, P. J. Hasnip, M. J. Probert, K. Refson, and M. C. Payne, *Z. Kristallogr.* **220**, 567 (2005).

415. C. Sachs, S. Völkening, J. Wintterlin, and G. Ertl, *Science* **293**, 1635 (2001).

416. K. Bedürftig, S. Völkening, Y. Wang, J. Wintterlin, K. Jocobi, and G. Ertl, *J. Chem. Phys.* **123**, 064711 (2005).

417. K. Ando and J. T. Hynes, *J. Phys. Chem. B* **101**, 10464 (1997).

418. A. R. Ubbelohde and K. J. Gallagher, *Acta Crystallogr.* **8**, 71 (1955).

419. E. Matsushita and T. Matsubara, *Prog. Theor. Phys.* **67**, 1 (1982).

420. S. Raugei and M. L. Klein, *J. Am. Chem. Soc.* **125**, 8992 (2003).

421. D. C. Clary, D. M. Benoit, and T. van Mourik, *Acc. Chem. Res.* **33**, 441 (2000).

422. A. C. Legon and D. J. Millen, *Chem. Phys. Lett.* **147**, 484 (1988).

423. C. Swalina, Q. Wang, A. Chakraborty, and S. Hammes-Schier, *J. Phys. Chem. A* **111**, 2206 (2007).

424. K. Wendler, J. Thar, S. Zahn, and B. Kirchner, *J. Phys. Chem. A* **114**, 9529 (2010).

425. S. S. Xantheas and T. H. Dunning, *J. Chem. Phys.* **99**, 8774 (1993).

426. E. Cubero, M. Orozco, P. Hobza, and F. J. Luque, *J. Phys. Chem. A* **103**, 6394 (1999).

427. S. Habershon, T. E. Markland, and D. E. Manolopoulos, *J. Chem. Phys.* **131**, 024501 (2009).

428. J. Ireta, J. Neugebauer, M. Scheer, A. Rojo, and M. J. Galvan, *J. Phys. Chem. B* **107**, 1432 (2003).

429. V. Srinivasan and D. Sebastiani, *J. Phys. Chem. C* **115**, 12631 (2011).

430. M. E. Tuckerman, D. Marx, M. L. Klein, and M. Parrinello, *Science* **275**, 817 (1997).

431. H. Ishibashi, A. Hayashi, M. Shiga, and M. Tachikawa, *ChemPhysChem* **9**, 383 (2008).

432. E. Wigner and H. B. Huntington, *J. Chem. Phys.* **3**, 764 (1935).

433. H. K. Mao and R. J. Hemley, *Rev. Mod. Phys.* **66**, 671 (1994).

434. K. A. Johnson and N. W. Ashcroft, *Nature* **403**, 632 (2000).

435. E. Babaev, A. Sudbo, and N. W. Ashcroft, *Nature* **431**, 666 (2004).

436. S. Deemyad and I. F. Silvera, *Phys. Rev. Lett.* **100**, 155701 (2008).

437. P. Loubeyre, F. Occelli, and R. LeToullec, *Nature* **416**, 613 (2002).

438. M. I. Eremets and I. A. Troyan, *Nat. Mater.* **10**, 927 (2011).

439. C. S. Zha, Z. X. Liu, and R. J. Hemley, *Phys. Rev. Lett.* **108**, 146402 (2012).

440. L. Dubrovinsky, N. Dubrovinskaia, V. B. Prakapenka, and A. M. Abakumov, *Nat. Commun.* **3**, 1163 (2012).

441. A. F. Goncharov, R. J. Hemley, and H. K. Mao, *J. Chem. Phys.* **134**, 174501 (2011).

442. V. E. Fortov, R. I. Ilkaev, V. A. Arinin, V. V. Burtzev, V. A. Golubev, I. L. Iosilevskiy, V. V. Khrustalev, A. L. Mikhailov, M. A. Mochalov, V. Y. Ternovoi *et al.*, *Phys. Rev. Lett.* **99**, 185001 (2007).

443. W. J. Nellis, S. T. Weir, and A. C. Mitchell, *Science* **273**, 936 (1996).

444. N. W. Ashcroft, *J. Phys.: Condens. Matter* **12**, A129 (2000).

445. P. Cudazzo, G. Profeta, A. Sanna, A. Floris, A. Continenza, S. Massidda, and E. K. U. Gross, *Phys. Rev. Lett.* **100**, 257001 (2008).

446. J. M. McMahon and D. M. Ceperley, *Phys. Rev. B* **84**, 144515 (2011).

447. T. Ogitsu, E. Schwegler, F. Gygi, and G. Galli, *Phys. Rev. Lett.* **91**, 175502 (2003).

448. D. Alfè, *Phys. Rev. B* **68**, 064423 (2003).

449. C. J. Pickard, M. Martinez-Canales, and R. J. Needs, *Phys. Rev. B* **85**, 214114 (2012).

450. C. J. Pickard, M. Martinez-Canales, and R. J. Needs, *Phys. Rev. B* **86**, 059902 (2012).

451. J. M. McMahon and D. M. Ceperley, *Phys. Rev. Lett.* **106**, 165302 (2011).

452. H. Y. Liu, H. Wang, and Y. M. Ma, *J. Phys. Chem. C* **116**, 9221 (2012).

453. R. T. Howie, C. L. Guillaume, T. Scheler, A. F. Goncharov, and E. Gregoryanz, *Phys. Rev. Lett.* **108**, 125501 (2012).

454. E. R. Hernandez, A. Rodriguez-Prieto, A. Bergara, and D. Alfè, *Phys. Rev. Lett.* **104**, 185701 (2010).

455. V. Adamo and V. Barone, *J. Chem. Phys.* **110**, 6158 (1999).

456. J. Klimes, D. R. Bowler, and A. Michaelides, *J. Phys.: Condens. Matter* **22**, 022201 (2010).

457. J. Klimes, D. R. Bowler, and A. Michaelides, *Phys. Rev. B* **83**, 195131 (2011).

458. M. Dion, H. Rydberg, E. Schroder, D. C. Langreth, and B. I. Lundqvist, *Phys. Rev. Lett.* **92**, 246401 (2004).

459. M. Dion, H. Rydberg, E. Schroder, D. C. Langreth, and B. I. Lundqvist, *Phys. Rev. Lett.* **95**, 109902 (2005).

460. N. W. Ashcroft, *Phys. Rev. Lett.* **21**, 1748 (1968).

461. P. B. Allen and R. C. Dynes, *Phys. Rev. B* **12**, 905 (1975).

462. URL http://www.quantum-expresso.org.

463. J. Bardeen, L. N. Cooper, and J. R. Schrieer, *Phys. Rev.* **108**, 1175 (1957).

464. R. Szczesniak and M. W. Jarosik, *Solid State Commun.* **149**, 2053 (2009).

465. M. Abramowitz and I. A. Stegun, *Handbook of Mathematical Functions* (Dover Pub. Inc., New York, 1972).

Acknowledgements

The authors would like to acknowledge a number of people, who, from the time of their education, have been of great help in the preparation of this book. We are grateful to J. B. Xia for his patience in reading a large part of the book and giving comments/suggestions. We are also grateful to a number of people for their help during the education of each of the authors. Specifically, XZL is deeply indebted to J. B. Xia, M. Scheffler, A. Michaelides, R. Gomez-Abal, C. Draxl, H. Jiang, E. K. U. Gross, R. J. Needs, D. Manolopoulos, D. Alfè, M. J. Gillan, C. J. Pickard, M. I. J. Probert, A. Alavi, P. J. Hu, X. G. Ren, W. X. Li, P. Rinke, V. Blum, K. Reuter, Q. M. Hu, H. Wu, X. L. Hu, L. M. Liu, and all his colleagues in FHI (Berlin) and UCL (London). EGW wants to thank G. Allan (ISEN) and D. S. Wang (IOP) for their kind help. During the writing of this manuscript in Peking University, we are also grateful to Y. Q. Gao, L. J. Yang, W. J. Xie, G. Sun, J. Liu, J. S. Shao, Q. Shi, H. T. Quan, J. Chen, Y. X. Feng, W. Fang, W. An, L. M. Xu, J. Feng, and all our colleagues/friends in China for the illuminating discussions. Finally, XZL wants to thank his wife, his late parents, his parents-in-law, his daughter, his sister and her family for all the support in the past years. The book would not have been completed without the support from them.

Xin-Zheng Li & Enge Wang

Index